Cognitive
Radio
Networks

Edited by

Yang Xiao · Fei Hu

Cognitive
Radio
Networks

CRC Press
Taylor & Francis Group
Boca Raton London New York

CRC Press is an imprint of the
Taylor & Francis Group, an **informa** business

AN AUERBACH BOOK

CRC Press
Taylor & Francis Group
6000 Broken Sound Parkway NW, Suite 300
Boca Raton, FL 33487-2742

First issued in paperback 2019

© 2009 by Taylor & Francis Group, LLC
CRC Press is an imprint of Taylor & Francis Group, an Informa business

No claim to original U.S. Government works

ISBN-13: 978-1-4200-6420-9 (hbk)
ISBN-13: 978-0-367-38604-7 (pbk)

Library of Congress Cataloging-in-Publication Data

Cognitive radio networks / editors, Yang Xiao and Fei Hu.
    p. cm.
    Includes bibliographical references and index.
    ISBN 978-1-4200-6420-9 (alk. paper)
    1. Cognitive radio networks. 2. Wireless LANs. 3. Software radio. I. Xiao, Yang, 1966- II. Hu, Fei, 1972-

TK5103.4815.C64 2008
621.39'81--dc22

2008015660

Visit the Taylor & Francis Web site at
http://www.taylorandfrancis.com

and the CRC Press Web site at
http://www.crcpress.com

Contents

Preface

According to the statistics of the Federal Communications Commission (FCC), temporal and geographical variations in the utilization of the assigned spectrum range from 15% to 85%. The limited available radio spectrum and the inefficiency in spectrum usage necessitate a new communication paradigm to exploit the existing spectrum dynamically. One of the most efficient paradigms is the cognitive radio network (CRN) which allows intelligent radios to provide unlicensed users with opportunistic access to the licensed bands without interfering with the existing users. In a CRN, an unlicensed user can use spectrum sensing hardware/software to sense idle spectrum, select the best available channel, coordinate access to this channel with other unlicensed users, and vacate the channel when a licensed user needs that channel.

To exemplify the importance of CRNs, let's consider the case of natural disasters that may temporarily disable the existing communication infrastructure (either wired Internet or wireless networks). Emergency personnel working in the disaster areas need to establish emergency networks. The emergency communication requires a significant amount of the radio spectrum for transmitting a large volume of traffic including voice, video and data. CRNs can enable the efficient usage of the existing spectrum without the need for an expensive infrastructure. In addition, a CRN can enable military radios to choose arbitrary intermediate frequency (IF) bandwidth and modulation/coding schemes to adapt to the variable radio environment in the battlefield.

Although some initial research activities have been conducted in this field, there is a need for a systematic reference book that presents clear definitions, functions, and current research challenges of CRNs. This book covers CRN essentials including spectrum sensing, spectrum handoff, spectrum sharing, and CRN routing schemes.

The purpose of this book is to provide state-of-the-art approaches and novel technologies for cognitive wireless radio networks, covering a range of topics in the various areas so that it will be an excellent reference book for researchers in these areas. It comprises contributions of many prominent researchers working in the cognitive radio field around the world. The book is divided into five parts: physical layer, medium access control, routing layer, cross-layer considerations, and advanced topics in cognitive radio networks, and includes 17 chapters on a wide range of topics. The advanced topics include ultra-wideband (UWB) systems, models, new applications, and system control in cognitive radio networks. The book will shed light on future research in these areas.

This book is made possible by the great efforts of our publishers and contributors. First, we are indebted to the contributors, who have sacrificed many days and nights to put together these excellent chapters for our readers. Second, we owe our special thanks to our publishers and their staff members. Without their encouragement and quality work, this book would not have been possible. Finally, we would like to thank our families for their support.

Yang Xiao
Department of Computer Science
The University of Alabama

Fei Hu
Department of Electrical
and Computer Engineering
The University of Alabama

About the Editors

Dr. Yang Xiao worked in industry as a MAC (Medium Access Control) architect involved in IEEE 802.11 standard enhancement work before he joined the Department of Computer Science at the University of Memphis in 2002. He is currently with the Department of Computer Science at the University of Alabama. Dr. Xiao was a voting member of the IEEE 802.11 Working Group from 2001 to 2004, is an IEEE Senior Member, and a member of the American Telemedicine Association. He currently serves as Editor-in-Chief for the *International Journal of Security and Networks (IJSN)*, the *International Journal of Sensor Networks (IJSNet)*, and the *International Journal of Telemedicine and Applications (IJTA)*. He serves as a referee/reviewer for many funding agencies, as well as a panelist for the NSF and a member of the Canada Foundation for Innovation (CFI) Telecommunications expert committee. He has served on technical program committees for more than 100 conferences, including INFOCOM, ICDCS, MOBIHOC, ICC, GLOBECOM, WCNC. He also serves as an associate editor for several journals, e.g., *IEEE Transactions on Vehicular Technology*. His research areas are security, telemedicine, sensor networks, and wireless networks. He has published more than 200 papers in major journals (more than 50 in various IEEE journals/magazines), refereed conference proceedings, and authored book chapters related to these research areas. Dr. Xiao's research has been supported by the U.S. National Science Foundation (NSF) and U.S. Army Research.

Dr. Fei Hu is an associate professor in the Electrical and Computer Engineering Department at the University of Alabama. His research interests are wireless networks, wireless security, and their applications in biomedicine. His research has been supported by the NSF, Cisco, Sprint, and other sources. He obtained his first PhD at Shanghai Tongji University, China, in 1999, and a second PhD at Clarkson University, Potsdam, New York in 2002, both in the field of computer engineering. He has published over 100 journal/conference papers and book chapters. He is also an editor for five international journals.

List of Contributors

Arvin Agah
Department of Electrical Engineering
and Computer Science
The University of Kansas
Lawrence, Kansas, U.S.A.

Lutfa Akter
WiCom Group
Department of Electrical
and Computer Engineering
The Kansas State University
Manhattan, Kansas, U.S.A.

S. Anand
Department of Electrical
and Computer Engineering
Stevens Institute of Technology
Castle Point at Hudson
Hoboken, New Jersey, U.S.A.

Chee Wei Ang
Institute for Infocomm Research
Singapore

Brett Barker
Department of Electrical Engineering
and Computer Science
The University of Kansas
Lawrence, Kansas, U.S.A.

R. Chandramouli
Department of Electrical
and Computer Engineering
Stevens Institute of Technology
Castle Point at Hudson
Hoboken, New Jersey, U.S.A.

Tao Chen
VTT Technical Research
Centre of Finland
Otaniemi, Finland

Chunxiao Chigan
Department of Electrical
and Computer Engineering
Michigan Technological University
Houghton, Michigan, U.S.A.

Kaushik R. Chowdhury
Broadband Wireless Networking
Laboratory
Georgia Institute of Technology
Atlanta, Georgia, U.S.A.

Cristina Comaniciu
Stevens Institute of Technology
Castle Point on Hudson
Hoboken, New Jersey, U.S.A.

Luca De Nardis
INFO-COM Department
University of Rome La Sapienza
Rome, Italy

Maria-Gabriella Di Benedetto
INFO-COM Department
University of Rome La Sapienza
Rome, Italy

De-cun Dong
Transportation Engineering Institute
Tongji University
Shanghai, China

Linda E. Doyle
Centre for Telecommunications
  Value-Chain Research
Lloyd Institute
University of Dublin, Trinity College
Dublin, Ireland

Joseph B. Evans
Information and Telecommunication
  Technology Center (ITTC)
The University of Kansas
Lawrence, Kansas, U.S.A.

Benjamin J. Ewy
Information and Telecommunication
  Technology Center (ITTC)
The University of Kansas
Lawrence, Kansas, U.S.A.

Anh Tuan Hoang
Institute for Infocomm Research
Singapore

Ekram Hossain
Department of Electrical
  and Computer Engineering
University of Manitoba
Winnipeg, Canada

Y. Thomas Hou
Bradley Department of Electrical
  and Computer Engineering
Virginia Polytechnic Institute
  and State University
Blacksburg, Virginia, U.S.A.

Fei Hu
Department of Electrical and
  Computer Engineering
The University of Alabama
Tuscaloosa, Alabama, U.S.A.

Katia Jaffrès-Runser
INSA Lyon & Stevens Institute
  of Technology
Castle Point on Hudson Hoboken
New Jersey, U.S.A.

Vikram Krishnamurthy
Department of Electrical
  and Computer Engineering
University of British Columbia
Vancouver, British Columbia, Canada

Won-Yeol Lee
Broadband Wireless Networking
  Laboratory
Georgia Institute of Technology
Atlanta, Georgia, U.S.A.

Victor C.M. Leung
Department of Electrical
  and Computer Engineering
University of British Columbia
Vancouver, British Columbia, Canada

Ying-Chang Liang
Institute for Infocomm Research
Singapore

Mark Lifson
Department of Computer
 Engineering
Rochester Institute of Technology
Rochester, New York, U.S.A.

Xin Liu
Department of Computer Science
University of California
Davis, California, U.S.A.

Balasubramaniam Natarajan
WiCom Group
Department of Electrical
 and Computer Engineering
Kansas State University
Manhattan, Kansas, U.S.A.

Timothy Newman
Information and Telecommunication
 Technology Center (ITTC)
The University of Kansas
Lawrence, Kansas, U.S.A.

Nie Nie
Stevens Institute of Technology
Castle Point at Hudson
Hoboken, New Jersey, U.S.A.

Dusit Niyato
Department of Electrical
 and Computer Engineering
University of Manitoba
Winnipeg, Canada

Qixiang Pang
Department of Electrical
 and Computer Engineering
University of British Columbia
Vancouver, British Columbia, Canada

Rahul Patibandla
Department of Computer
 Engineering
Rochester Institute of Technology
Rochester, New York, U.S.A.

Edward Peh
Institute for Infocomm Research
Singapore

Laxminarayana S. Pillutla
Department of Electrical
 and Computer Engineering
University of British Columbia
Vancouver, British Columbia, Canada

Muthukumaran Pitchaimani
Information and Telecommunication
 Technology Center (ITTC)
The University of Kansas
Lawrence, Kansas, U.S.A.

Caterina Scoglio
WiCom Group
Department of Electrical
 and Computer Engineering
Kansas State University
Manhattan, Kansas, U.S.A.

Yi Shi
Bradley Department of Electrical and
 Computer Engineering
Virginia Polytechnic Institute
 and State University
Blacksburg, Virginia, U.S.A.

Paul D. Sutton
Centre for Telecommunications
 Value-Chain Research
Lloyd Institute
University of Dublin, Trinity College
Dublin, Ireland

Zhi Tian
Department of Electrical
  and Computer Engineering
Michigan Technological University
Houghton, Michigan, U.S.A.

Mehmet C. Vuran
Department of Computer Science and
  Engineering
University of Nebraska–Lincoln
Lincoln, Nebraska, U.S.A.

Wei Wang
Department of Computer Science
University of California
Davis, California, U.S.A

Alexander M. Wyglinski
Department of Electrical
  and Computer Engineering
Worcester Polytechnic Institute
Worcester, Massachusetts, U.S.A.

Hong Xiao
Department of Mathematics
University of California
Davis, California, U.S.A.

Yang Xiao
Department of Computer Science
University of Alabama
Tuscaloosa, Alabama, U.S.A.

Yonghong Zeng
Institute for Infocomm Research
Singapore

Honggang Zhang
Zhejiang University
Hangzhou, China

Xiaofei Zhou
China Netcom Group
Hong Kong, China

Chao Zou
Department of Electrical
  and Computer Engineering
Michigan Technological
  University
Houghton, Michigan, U.S.A.

PHYSICAL LAYER OF COGNITIVE RADIO NETWORKS

I

Chapter 1

Spectrum Sensing Algorithms for Cognitive Radio Networks

Won-Yeol Lee, Kaushik R. Chowdhury,
and Mehmet C. Vuran

Contents

1.1 Introduction

The design principle for cognitive radio (CR) networks regards the cognitive radio users as *visitors* in the spectrum they occupy. This necessitates efficient spectrum management functionalities to occupy idle channels without causing interference with the primary users, and leave these channels when primary user activity is detected. The successful operation of these principles relies on the CR users' ability to be aware of their surroundings, which is accomplished through *spectrum sensing* solutions.

The main objective of spectrum sensing is to provide more spectrum access opportunities to CR users without interference to the primary networks.[1] Cognitive radio hardware should be able to opportunistically identify portions of the spectrum with reduced licensed user activity and use them for communication. However, these licensed channels, also defined as *primary bands*, should be immediately vacated if the legitimate or *primary users* (PR) are detected. Thus, accurate sensing of the wireless spectrum is a key challenge in realizing CR technology. Since CR networks are responsible for detecting the transmission of primary networks and avoiding interference with them, they should intelligently sense the primary band to avoid interference with the transmission of primary users. This necessitates support from the physical layer of the cognitive radio architecture, as well as intelligent algorithms that are implemented in software.

Depending upon the deployed architecture, the spectrum sensing methodology may differ. As an example, in infrastructure-based networks, where base stations control the communication in their surroundings, the individual cognitive radio devices can sense the environment and report back to the base station. However, as the base station footprint may span several kilometers, the network environment may be different for each of the sensing devices. The base station may have to adopt a common set of channels that do not conflict with PR users throughout its coverage area. On the other hand, in distributed networks, which are usually characterized by a mesh setting, the mesh router (MR) has a comparatively smaller coverage area, and channel measurements by associated mesh clients (MCs) may not significantly differ along the diameter of its footprint. However, it may not directly receive information from neighboring MRs, leading to a suboptimal channel occupancy knowledge.

In this chapter, the design choices and existing solutions in spectrum sensing are discussed in detail. More specifically, various network architectures that affect the design of spectrum sensing are discussed. Moreover, existing solutions in this domain are discussed with open research issues that remain with regard to effective spectrum sensing in CR networks. The rest of this chapter is organized as follows. The architectural aspects of spectrum sensing in terms of the infrastructure- and mesh-based networks are discussed in Section 1.2. The various sensing techniques existing in the literature are described in Section 1.3. The optimal sensing framework for infrastructure-based networks is then described in Section 1.4. Section 1.5 details the algorithms used for sensing in a wireless mesh network environment. The open research issues for spectrum sensing in CR networks are discussed in Section 1.6. Finally, we conclude the paper in Section 1.7.

1.2 Infrastructure and Wireless Mesh-Based Cognitive Radio Network Architectures

In this section, we shall describe the key challenges and highlight the research issues for both infrastructure-based and wireless mesh-based architectures.

1.2.1 Infrastructure-Based Cognitive Radio Network Architecture

The components of the infrastructure-based (or centralized) CR network architecture, as shown in Figure 1.1, can be classified in two groups as the *primary network* and the *CR network*.

The *primary network* is referred to as the legacy network that has an exclusive right to a certain spectrum band. Examples include the common cellular and TV broadcast networks. In contrast the *CR network* does not have a license to operate in the desired band. Hence, the spectrum access is allowed only in an opportunistic manner. The following are the basic components of primary networks:[1]

- *Primary user:* A primary user has a license to operate in a certain spectrum band. This access can only be controlled by the primary base station and should not be affected by the operations of any other CR users. Primary users do not need any modification or additional functions for coexistence with CR base stations and CR users.

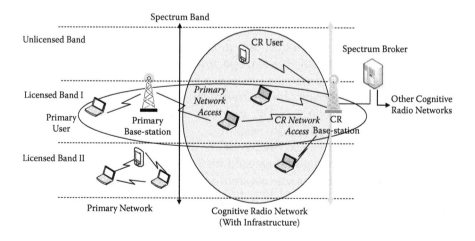

Figure 1.1 Infrastructure-based CR network architecture.

■ *Primary base station:* A primary base station is a fixed infrastructure network component that has a spectrum license, such as a base station transceiver system (BTS) in a cellular system. In principle, the primary base station does not have any CR capability for sharing spectrum with CR users.

The basic elements of the CR network are defined as follows:[1]

■ *CR user:* A CR user has no spectrum license. Hence, additional functionalities are required to share the licensed spectrum band. In infrastructure-based networks, the CR users may be able to only sense a certain portion of the spectrum band through local observations. They do not make a decision on spectrum availability and just report their sensing results to the CR base station.
■ *CR base station:* A CR base station is a fixed infrastructure component with CR capabilities. It provides single-hop connection without spectrum access licenses to CR users within its transmission range and exerts control over them. Through this connection, a CR user can access other networks. It also helps in synchronizing the sensing operations performed by the different CR users. The observations and analysis performed by the latter are fed to the central CR base station so that the decision on the spectrum availability can be made.
■ *Spectrum broker:* A spectrum broker (or scheduling server) is a central network entity that plays a role in sharing the spectrum resources among different CR networks. It is not directly engaged in spectrum sensing. It just manages the spectrum allocation among different networks according to the sensing information collected by each network.

1.2.2 *Wireless Mesh-Based Cognitive Radio Network Architecture*

While wireless mesh networks (WMNs) can be broadly categorized as *distributed* networks, they exhibit various levels of node-level independence. There are three major classifications of WMNs: (i) backbone, (ii) mesh-client, and (iii) hybrid,[2] based on how the MCs are linked to the MRs. These architectures are shown in a combined manner in Figure 1.2. We next discuss each of these from the spectrum sensing perspective.

1.2.2.1 *Backbone Mesh*

These are the most commonly used type of mesh networks, with the MRs serving as the *access point* supporting several users in a residential setting or

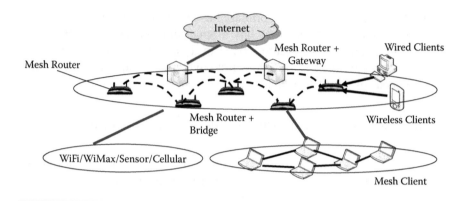

Figure 1.2   Backbone, mesh-client, and hybrid WMN architectures.[2]

along a road. The MCs can only connect to the MRs, and thus each MR along with its associated nodes forms a *clusterlike* arrangement. The MRs serve as the routing backbone, forwarding packets till the *gateway* that provides the final link to the Internet is reached. These MRs may serve other networks with independent communication standards, such as a WiMax network in Figure 1.2.

Some research challenges associated with backbone networks are

- The MCs in a cluster may be directed by the MR to sense the environment independently of the others. In a dense deployment environment, several MCs may monitor the same channel conditions. Hence, dividing the sensing task is more efficient, particularly in those cases where a large number of channels need to be sensed for PR user activity. Metrics that assign this sensing task on the node density, location, and PR statistics need to be devised.

- Data from several different MCs can be used to improve the sensing accuracy. The wireless channel is inherently unreliable and may be affected by external ambient noise. Thus, by basing the sensing decision on multiple dissimilar sets of nodes, the effect of localized noise can be eliminated. The measure of redundancy in the readings and choosing the appropriate sets of MCs merit further thought.

- Different techniques exist in PR user detection. As an example, transmitter detection, receiver detection, and more advanced schemes like cyclostationary detection may be used. Each of these methods works best under its own set of optimal environmental conditions. By assigning a different detection technique to the MCs and collecting the results at the MR, the benefits of these diverse schemes can be seen in the same CR network.

1.2.2.2   Client Mesh

The client mesh consists of several MCs connected to each other without MR support. They form ad hoc networks for peer-to-peer connection and are not connected to the Internet. In the absence of total central control, the sensing task is particularly complicated. Network-wide messaging and pairwise network sensing during transmission are the best suited mechanisms to sense the channel and then disseminate this information in the network. This architecture is seldom explored in the literature and constitutes the hardest scenario for implementing a CR network.

1.2.2.3   Hybrid Mesh

This architecture strikes a balance between the clusterlike and totally centralized approaches. Here, MCs are free to either associate themselves with an MR in a cluster, or form their own ad hoc network. The ad hoc network is connected to the backbone via any closest peer, and this poses some interesting research issues:

- A symbiotic relationship can be envisaged between the backbone and the ad hoc components of the mesh network. As the individual peer nodes can be spread widely around the network, they can perform channel sensing on the network edge. This information is then conveyed back to the backbone MR, and in return, the node is allowed to forward packets towards the gateway through that MR. This raises issues of devising feasible economic models, and also this supply-reward structure allows a check on nodes submitting malicious or intentionally misleading sensed information.
- Assigning of the sensing task by the MR to these fringe ad hoc networks needs a suitable handshaking mechanism calling for an interaction between application and lower-level sensing layers.

1.3   Overview of Spectrum Sensing

A cognitive radio is designed to be aware of and sensitive to the changes in its surroundings, which makes spectrum sensing an important requirement for the realization of CR networks. Spectrum sensing enables CR users to adapt to the radio environment by determining currently unused spectrum portions, so-called *spectrum holes*, without causing interference to the primary network.

Generally, spectrum sensing techniques can be classified into four groups: (1) primary transmitter detection, (2) cooperative detection, (3) primary receiver detection, and (4) interference temperature management, as described in the following.

1.3.1 Primary Transmitter Detection

Since CR users are usually assumed not to have any real-time interaction with the primary transmitters and receivers, they cannot know the exact information on current transmissions within the primary networks. Thus, in transmitter detection, in order to distinguish between used and unused spectrum bands, CR users detect the signal from a primary transmitter through only the local observations of CR users, as shown in Figure 1.3. Thus, CR users should have the capability to determine if a signal from the primary transmitter is locally present in a certain spectrum. A basic hypothesis model for transmitter detection can be defined as follows:

$$x(t) = \begin{cases} n(t) & H_0 \\ hs(t) + n(t) & H_1 \end{cases} \tag{1.1}$$

where $x(t)$ is the signal received by the CR user, $s(t)$ is the transmitted signal of the primary user, $n(t)$ is a zero-mean additive white Gaussian noise (AWGN), and h is the amplitude gain of the channel. H_0 is a null hypothesis, which states that there is no licensed user signal in a certain spectrum band. On the other hand, H_1 is an alternative hypothesis, which indicates that there exists some primary user signal.

Three schemes are generally used for the transmitter detection: *matched filter detection, energy detection,* and *feature detection.*[3]

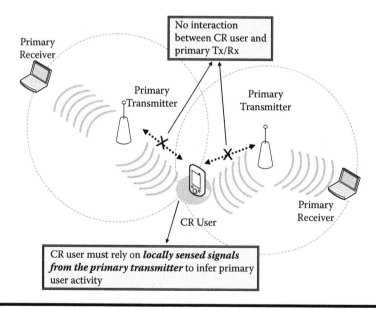

No interaction between CR user and primary Tx/Rx

Primary Receiver

Primary Transmitter

Primary Transmitter

Primary Receiver

CR User

CR user must rely on *locally sensed signals from the primary transmitter* to infer primary user activity

Figure 1.3   Primary transmitter detection.

- *Matched filter detection:* When the information of the primary user signal is known to the CR user, the optimal detector in stationary Gaussian noise is the matched filter for maximizing the signal-to-noise ratio (SNR) in the presence of additive stochastic noise. This detection method requires only $O(1/\text{SNR})$ samples to achieve a detection error probability constraint.[3] However, the matched filter requires not only *a priori* knowledge of the characteristics of the primary user signal, but also synchronization between the primary transmitter and the CR user. If this information is not accurate, then the matched filter performs poorly.
- *Energy detection:* If the receiver cannot gather sufficient information about the primary user signal, for example, if only the power of the random Gaussian noise is known to the receiver, the optimal detector is an energy detector. In the energy detection scheme, CR users sense the presence/absence of the primary users through the energy of the received primary signal. In order to measure the energy of the received primary signal, the received signal is squared and integrated over the observation interval. Finally, the output of the integrator is compared with a threshold to decide if a primary user is present.

  The energy detector requires $O(1/\text{SNR}^2)$ samples for a given detection error probability. Thus, if CR users need to detect weak signals (SNR: -10dB to -40 dB), energy detection suffers from longer detection time compared to matched filter detection.[3]

  While the energy detector is easy to implement, it can only determine the presence of the signal but cannot differentiate signal types. Thus, the energy detector often generates false detection triggered by unintended signals. Another shortcoming is that since energy detection depends only on the SNR of the received signal, its performance is susceptible to uncertainty in noise power. If the noise power is uncertain and can take any value in the range of x dB, the energy detector will not be able to detect the signal reliably when the SNR is less than the threshold $10 \log_{10} 10^{x/10} - 1$, called an SNR wall.[4]
- *Feature detection:* Modulated signals are, in general, coupled with sine wave carriers, pulse trains, repeating spreading, hopping sequences, or cyclic prefixes, which result in built-in periodicity. Thus, these modulated signals are characterized by *cyclostationarity*, since their mean and autocorrelation exhibit periodicity. The feature detector exploits this inherent periodicity in the primary user's signal by analyzing a spectral correlation function.[5] The main advantage of the feature detection scheme is its robustness to the uncertainty in noise power. The feature detector distinguishes between the noise energy and the modulated signal energy, which is the result of the

fact that the noise is a wide-sense stationary signal with no correlation, while the modulated signals are cyclostationary with spectral correlation due to the built-in periodicity. Furthermore, since the feature detector is also capable of differentiating different types of signals, it can tolerate false alarms caused by external signals, such as those from other CR users or interference. Therefore, a cyclostationary feature detector can perform better than an energy detector in differentiating different signal types. However, it is computationally complex and requires significantly longer observation time.

1.3.2 Cooperative Transmitter Detection

Because of the lack of interaction between the primary users and the CR users, transmitter detection techniques rely on the weak signals from only the primary transmitters. Hence, transmitter detection techniques alone cannot avoid causing interference to primary receivers because of the lack of primary receiver information as depicted in Figure 1.4(a). Moreover, transmitter detection models cannot prevent the hidden terminal problem. A CR user (transmitter) can have a good line-of-sight to a CR receiver, but may not be able to detect the primary transmitter due to shadowing as shown in Figure 1.4(b). Therefore, sensing information from other users is required for more accurate primary transmitter detection; this is referred to as *cooperative detection*. Cooperative detection is theoretically more accurate, since the uncertainty in a single user's detection can be minimized through collaboration.[6] Moreover, multipath fading and shadowing effects can be mitigated so that the detection probability is improved in a heavily shadowed environment.

Assume there are three CR users as illustrated in Figure 1.5. Since CR user 2 receives a weak signal (with a low SNR) due to multipath fading, it cannot detect the signal of the primary transmitter. CR user 1 is in the shadowing area so it cannot detect the primary user, either. Only CR user 3 detects the signal of the primary user correctly. In this case, CR users 1 and 2 will cause interference if they base their decision to transmit on their local observations. However, by exchanging sensing information with CR user 3, CR users 1 and 2 can detect the existence of the primary user even though they are under fading and shadowing environments.

As explained above, in traditional cooperative detection, the spectrum band is decided to be available only if no primary user activity is detected. Even if only one primary user's activity is detected, CR users cannot use this spectrum band.[6] From this detection criterion, the cooperative detection probability $\overline{P}_d^c$ of N CR users is obtained by $\overline{P}_d^c = 1 - (1 - \overline{P}_d)^N$, where $\overline{P}_d$ is the detection probability of the individual CR user.

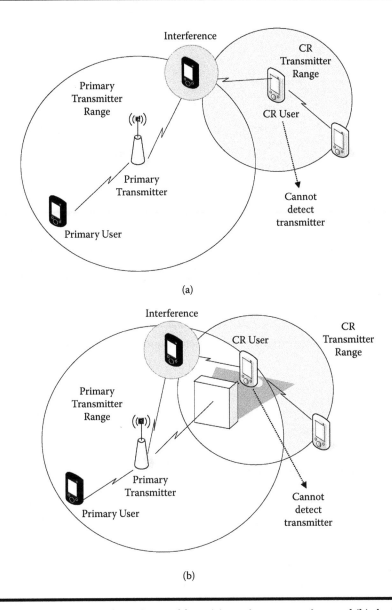

(a)

(b)

Figure 1.4   Transmitter detection problem: (a) receiver uncertainty and (b) shadowing uncertainty.[1]

While this decision strategy surely increases the detection probability, it also increases the cooperative false alarm probability, $\overline{P}_f^c = 1 - (1 - \overline{P}_f)^N$, where $\overline{P}_f$ is the false alarm probability of the individual CR user, which leads to losing more spectrum opportunities. Furthermore, cooperative

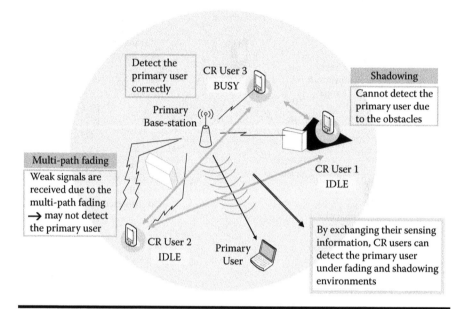

Figure 1.5 Cooperative transmitter detection under highly faded and shadowed environment.

approaches cause adverse effects on resource-constrained networks due to the overhead traffic.

1.3.3 Primary Receiver Detection

Although cooperative detection reduces the probability of interference, the most efficient way to detect spectrum holes is to detect the primary users that are receiving data within the communication range of a CR user. As depicted in Figure 1.6, the primary receiver usually emits local oscillator (LO) leakage power from its RF front-end when it receives signals from the primary transmitter. In order to determine the spectrum availability, a primary receiver detection method exploits this LO leakage power instead of the signal from the primary transmitter, and detects the presence of the primary receiver directly.[7] Such an approach may be feasible for TV receivers only, or need further hardware such as a supporting sensor network in the area with the primary receivers.

1.3.4 Interference Temperature Management

Traditionally, interference can be controlled at the transmitter through the radiated power, out-of-band emissions, and location of individual

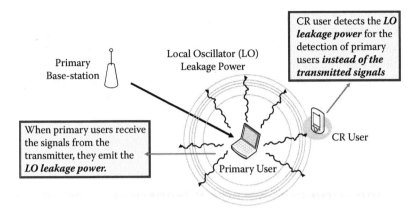

Figure 1.6   Primary receiver detection.

transmitters. However, interference actually takes place at the receivers, as shown in Figure 1.4(a). Recently, a new model for measuring interference, referred to as *interference temperature*, has been introduced by the FCC.[8]

Figure 1.7 shows the signal of the primary transmitter designed to operate out to the distance at which the received power approaches the level of the noise floor. The noise floor is location-specific depending on the additional interfering signals at that point. As shown in Figure 1.7, this model suggests an interference temperature limit, which is the amount of new interference that the primary receiver could tolerate. As long as CR users do not exceed this limit, they can use the spectrum band.

Although this model is best fitted for the objective of spectrum sensing, the difficulty lies in accurately determining the interference temperature

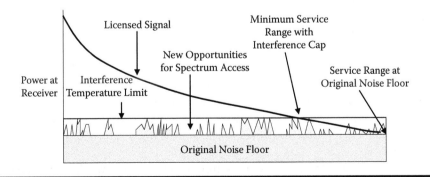

Figure 1.7   Interference temperature management.[8]

limit for each location-specific case. There is no practical way for a CR user to measure or estimate the interference temperature, since CR users have difficulty in distinguishing between actual signals from the primary user and noise/interference. Also, with the increase in the interference temperature limit, the SNR at the primary receiver decreases, resulting in a decrease in the primary network's capacity and coverage, as depicted in Figure 1.7.

1.4 Optimal Sensing Framework for Infrastructure-Based CR Networks

In this section, we assume CR networks have a centralized network entity such as a base station in infrastructure-based networks. This centralized network entity can communicate with all CR users within its coverage area and decide the spectrum availability of its coverage. By considering these architectural characteristics, we will introduce the optimal spectrum sensing framework for infrastructure-based CR networks.

The main objective of spectrum sensing is to provide more spectrum access opportunities to CR users without interference to the primary networks.[1] Since CR networks are responsible for detecting the transmissions of primary networks and avoiding interference with them, they should intelligently sense the primary band to avoid interference with the transmissions of primary users. Thus, *sensing accuracy* has been considered as the most important factor to determine the performance of CR networks. Hence, recent research has been focused on improving sensing accuracy for interference avoidance.

Ideally, to avoid interference with the primary users, CR users should monitor the spectrum continuously through the radio frequency (RF) front-end. However, in reality, the RF front-end cannot differentiate between primary user signals and CR user signals.[9] While feature detection is known to be capable of identifying the modulation types of the primary signal, it requires longer processing time and higher computational complexity[10] with energy detection. CR users are not able to perform the transmission and sensing tasks at the same time. CR users need a *periodic sensing* structure where sensing and transmission operations are performed in a periodic manner with separate observation and transmission periods as shown in Figure 1.8. In this structure, a CR base station needs to coordinate the transmission of all CR users in such a way that CR users stop their transmissions during the sensing time to prevent false alarms triggered by unintended CR signals.

This periodic sensing structure introduces the following design issues:[11]

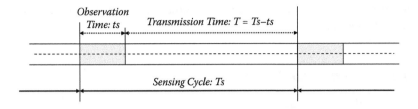

Figure 1.8   Periodic sensing structure.

■ *Interference avoidance:* Interference avoidance in CR networks depends on the sensing accuracy, which is determined by the observation time. However, in periodic sensing, CR users cannot sense the spectrum bands during the transmission time, which leads to an increase in interference.

■ *Transmission efficiency:* During the observation period, the transmission of CR users is not allowed, which inevitably decreases the transmission opportunities of CR users, leading to the so-called *transmission efficiency* issue.

As explained above, there is a tradeoff between interference and transmission efficiency. For interference avoidance, the observation time needs to be long enough to achieve sufficient detection accuracy, i.e., longer observation time leads to higher sensing accuracy, and hence to less interference. But as the observation time becomes longer, the transmission time of CR users will be decreased. Conversely, while a longer transmission time enhances transmission efficiency, it causes higher interference due to the lack of sensing information. Hence, *observation time*, t_s, and *transmission time*, $T_s - t_s$, are the main sensing parameters that influence both transmission efficiency and interference avoidance. Thus, the proper selection of these sensing parameters is the most critical factor influencing the performance of CR networks.

In order to solve both the interference avoidance and transmission efficiency problems in infrastructure-based CR networks, an optimal sensing framework is introduced to maximize spectrum access opportunities under the constraint of interference and sensing resource limitations.[12] This work is extended to multi-spectrum environments by a novel sensing resource allocation method developed to maximize the spectrum access opportunities of CR users. Finally, in order to exploit the sensing accuracy gain obtained by multi-user cooperation, an adaptive and cooperative decision method is used for the sensing parameters, where the transmission time can be optimized adaptively to the number of users. In the following subsections, we will explain how the optimal sensing framework works in infrastructure-based CR networks.

1.4.1 System Model

1.4.1.1 Network Model

The infrastructure-based architecture has several advantages in developing more accurate spectrum sensing methods, as we now enumerate. First, as explained in Section 1.3, with the transmitter detection scheme CR networks cannot avoid interference at the nearby primary receivers; the so-called receiver uncertainty problem.[1] Hence, in order to reduce the receiver uncertainty, the base station can collect and exploit sensing information of CR users inside its coverage. In the presence of a centralized base station, even if only one of the CR radio users detects the presence of the primary user, the message can be broadcast to all other cognitive radio users. Thus, the time needed to vacate an occupied band is expected to be very short. Second, due to the limitations in spectrum sensing, all CR users should concurrently undertake the sensing task so that they themselves do not introduce interference. The need for this synchronization can be easily filled through base-station support.

1.4.1.2 Primary User Activity

Since primary user activity is closely related to the performance of CR networks, the estimation of this activity is a very crucial issue in spectrum sensing. We assume that primary user activity can be modeled as exponentially distributed inter-arrivals.[13] In this model, the primary user traffic can be modeled as a two-state birth–death process, with death rate α and birth rate β. An ON (busy) state represents the period used by primary users, and an OFF (idle) state represents the unused period. Since each user arrival is independent, each transition follows the Poisson arrival process. Thus, the lengths of ON and OFF periods are exponentially distributed. From the primary user activity model, we can estimate the probabilities of ON and OFF states as follows:

$$P_{on} = \frac{\beta}{\alpha + \beta}$$

$$P_{off} = \frac{\alpha}{\alpha + \beta}$$

(1.2)

where P_{on} is the probability of the period used by primary users and P_{off} is the probability of the idle period.

1.4.2 Spectrum Sensing Framework

Generally, CR networks have multiple cooperating CR users. This multi-user environment helps CR users to enhance the sensing accuracy, which is time-varying and is a function of the number of users. Furthermore, CR users are allowed to exploit multiple spectrum bands. Practically, CR users do not have enough sensing transceivers to sense all the available spectrum bands. In order to adapt these various sensing environments efficiently, we introduce an optimal spectrum sensing framework, which is illustrated in Figure 1.9. This framework consists of the *optimization of sensing parameters* in a single spectrum band, *spectrum selection and scheduling*, and an *adaptive and cooperative sensing* method.[12]

The detailed scenario for the optimal sensing framework is as follows. According to the radio characteristics, base stations initially determine the optimal sensing parameters of each spectrum band through the *sensing parameter optimization phase*. When CR users join the CR networks, through *spectrum selection and scheduling* methods, the base stations select the best spectrum bands for sensing and configure sensing schedules according to the number of transceivers and the optimized sensing parameters. Then, CR users begin to monitor spectrum bands continuously with the optimized sensing schedule and report sensing results to the base station. Using these sensing results, the base station determines the spectrum availability. If the base station detects any changes that affect the sensing performance, sensing parameters need to be reoptimized and announced to the CR users through the *adaptive and cooperative sensing phase*.

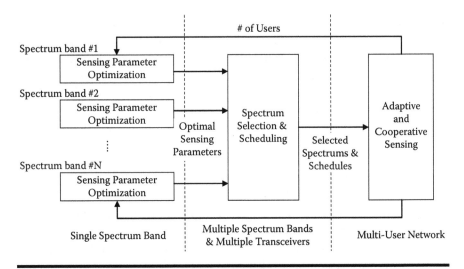

Figure 1.9 Overall architecture of the optimal sensing framework.[12]

1.4.3 Sensing Parameter Optimization in a Single Spectrum Band

In this section, we first introduce a sensing parameter optimization method to maximize the transmission efficiency subject to the interference constraint.

Assume that a single CR user monitors a single spectrum band. It alternately senses the spectrum and transmits data with the observation time t_s and transmission time T. To determine these sensing parameters accurately, we need to consider the interference constraint and the transmission efficiency at the same time. The *transmission efficiency*, η, is the ratio of the transmission time over the entire sensing cycle, defined as follows:

$$\eta = \frac{T}{T + t_s} \tag{1.3}$$

The objective of spectrum sensing is to achieve accurate detection probability, as well as high transmission efficiency. Since both metrics are related to the sensing parameters T and t_s, the sensing parameter decision can be expressed as the optimization problem to maximize the transmission efficiency satisfying interference constraint T_P as follows:

Find: T, t_s

$$\text{Maximize: } \eta = \frac{T}{T + t_s} \tag{1.4}$$

Subject to: $T_I \leq T_P$

where the *interference ratio* T_I is the expected fraction of interference caused by the transmissions of CR users, and the *maximum outage ratio* T_P is the maximum fraction of interference that primary networks can tolerate.

In order to solve this optimization problem, we investigated the relationship between sensing parameters and the expected interference. In the following subsections, the maximum *a posteriori* (MAP) energy detection model is explained first, and then an analytical interference model is introduced.

1.4.3.1 Maximum a Posteriori (MAP) Energy Detection Model for Spectrum Sensing

Although the MAP detector is known to be optimal,[14] a maximum likelihood (ML) detection has been widely used for energy detection without considering the probabilities of ON and OFF states.[6,15,16] In this subsection, we explain MAP-based energy detection and its decision criterion based on primary user activities.

From the definition of MAP detection and the hypothesis model explained in Section 1.3, the detection probability P_d and false alarm probability P_f can be expressed as follows:

$$P_d(\lambda) = \Pr[Y > \lambda | H_1] P_{on} = \overline{P_d} \cdot P_{on}$$

$$(1.5)$$

$$P_f(\lambda) = \Pr[Y > \lambda | H_0] P_{off} = \overline{P_f} \cdot P_{off}$$

where λ is a decision threshold of MAP detection, P_{on} is the probability of the period used by primary users and P_{off} is the probability of the idle period.

Generally, the decision threshold λ can be determined by the minimum probability of error decision rule as $f(\lambda | H_1) P_{on} = f(\lambda | H_0) P_{off}$, where $f(y | H_1)$ and $f(y | H_0)$ are probability density functions of the received signal through the occupied spectrum and the idle spectrum, respectively. This method minimizes the total error probabilities including false alarms and missed detection. Though sometimes one of the error probabilities may be greater than the other, in reality, the false alarm probability also affects the interference, explained in Section 1.4.3.2. Furthermore, in spectrum sensing, the detection of opportunities is as important as that of the primary signals. Hence, instead of minimizing the total error probability, this method emphasizes the balance of both the probabilities as follows:

$$P_{on} - P_d(\lambda) = P_f(\lambda)$$

$$(1.6)$$

This method enables balancing between the interference T_I and the lost spectrum opportunity T_L, which is the expected fraction of undetected idle spectrum portions caused by the detection errors and the false alarms.

1.4.3.2 Analytical Model for Interference

In order to optimize sensing parameters satisfying the interference constraint, we need to specify the relation between the interference ratio T_I and sensing parameters, as explained in Equation 1.4. Hence, we investigated an analytical interference model as a function of primary user activities and detection statistics, as explained in Section 1.4.3.1.

In the periodic sensing structure, interference can be expected to occur in the following cases:

■ *Interference on busy state sensing, I_{on}:* When the spectrum band is busy but CR users do not detect the primary user signal, CR users begin to transmit and interference can occur during the transmission period.

■ *Interference on idle state sensing, I_{off}:* Even though the spectrum band is idle and CR users detect it correctly, there still exists the possibility that primary user activity appears during the transmission period.

Here I_{on} and I_{off} can be expressed as functions of the primary user activity parameters α and β and the transmission time T.

Therefore, the total expected interference T_I can be obtained by adding the expected interference on both busy state and idle state as follows:

$$T_I = \frac{E[I_{on}] + E[I_{off}]}{T \cdot P_{on}}$$

$$= \frac{(P_{on} - P_d) \cdot I_{on} + (P_{off} - P_f) \cdot I_{off}}{T \cdot P_{on}} \tag{1.7}$$

Since P_d and P_f are functions of the observation time, t_s, and primary user activities, T_I can be rewritten as a function of sensing parameters and primary user activities. By combining Equations (1.4) and (1.7), we can solve the optimization problem and finally obtain the optimal sensing parameters.

In this section, we investigated how to derive optimal observation time and transmission time that maximize the transmission efficiency under the interference constraint. In order to extend this optimization method to a multi-spectrum/multi-user network environment, additional functionalities need to be developed, which will be explained in the following subsections.

1.4.4 Spectrum Selection and Scheduling for Spectrum Sensing on Multiple Spectrum Bands

In the following subsections, we explain how to extend the sensing parameter optimization method to multiple spectrum bands, especially focusing on spectrum selection and sensing scheduling issues.

1.4.4.1 Spectrum Selection

In the previous section, we explained how to find the optimized parameters for single-band/single-user sensing. However, in reality, in order to mitigate the fluctuating nature of opportunistic spectrum access, CR users are supposed to exploit multiple available spectrum bands showing different characteristics. As explained in Section 1.4.3, multiple spectrum bands have different optimal observation and transmission times according to their characteristics. If CR users are required to exploit all available spectrum bands, the number of sensing transceivers can be expressed as

$$\sum_{i \in A} \frac{t_{s,i}^{\mathrm{op}}}{T_i^{\mathrm{ou}} + t_{s,i}^{\mathrm{op}}}$$

where A is a set of all available spectrum bands and $t_{s,i}^{\mathrm{op}}$ and T_i^{op} represent optimal observation and sensing times of spectrum band i. However, since CR users generally have a finite number of transceivers, it is not practical to monitor all available spectrum bands. Instead of exhaustive sensing, selective sensing is more feasible in CR networks. To select the spectrum bands properly under the sensing resource constraint, we define a new notion, *opportunistic sensing capacity*, which represents the expected transmission capacity of spectrum band i that CR users can achieve, and which can be defined as follows:

$$C_{o,i} = \rho_i \eta_i W_i P_{\mathrm{off},i} \tag{1.8}$$

where η_i, W_i, and $P_{\mathrm{off},i}$ represent the transmission efficiency, the bandwidth, and the idle state probability of the spectrum band i. The coefficient ρ_i is the normalized capacity of the spectrum band i (bit/sec/Hz) depending on the modulation and channel coding schemes. In order to reflect the dynamic nature of spectrum bands in CR networks, $C_{o,i}$ considers the transmission efficiency and the probability of the idle state. Then a spectrum selection method can be expressed as the optimization problem to maximize the opportunistic sensing capacity of CR networks subject to the number of sensing transceivers.[12] Once spectrum bands are selected, the transceiver is required to be scheduled for spectrum sensing, which is explained in the following subsection.

1.4.4.2 Sensing Scheduling for Multiple Spectrum Bands

Another practical sensing problem in multi-spectrum networks is that each spectrum band has different optimized sensing cycles $T_i^{\mathrm{op}} + t_{s,i}^{\mathrm{op}}$. It is possible that the heterogeneous sensing cycles of each transceiver conflict during this operation, thus degrading the transmission capacity in CR networks. Hence, a novel sensing scheduling method needs to be developed satisfying optimal sensing cycles of each spectrum.

If there are multiple spectrum bands for the same sensing slot, CR users choose one of the spectrum bands based on the *opportunity cost*. The *opportunity cost* is defined as the sum of the expected opportunistic sensing capacities of the spectrum bands to be blocked if one of the competing bands is selected. In the *least cost first serve (LCFS) scheduling* algorithm,[12] the current sensing task is assigned to one of the competing spectrum bands to minimize the opportunity cost. For fair scheduling, the LCFS method considers not only the opportunity cost for the future sensing time but also blocked capacity in the past. Through these procedures, the LCFS algorithm assigns the current time slot to the spectrum band in such a way that the sum of the opportunity cost and the blocked opportunistic capacity of other spectrum bands is minimized.

1.4.5 Adaptive and Cooperative Sensing/Optimal Sensing in Multi-User Networks

In infrastructure-based CR networks, CR users sense spectrum bands at each location and report the sensing results to the base station periodically. Then, the base station decides the availability of the spectrum bands inside its coverage and allocates the available spectrum bands to users. These sensing data show a high spatial correlation, which can be used to enhance the spectrum sensing accuracy through cooperation.

However, in order to exploit this cooperative gain, the base station should consider the following issues. First, since the cooperative scheme can enhance the detection probability, the expected interference ratio is less than that originally estimated in the sensing parameter optimization. Thus, the optimal parameters are no longer valid. Second, the cooperation gain has a time-varying characteristic (derived in Section 1.4.3) according to the number of users involved in the cooperation. Furthermore, the number of primary user activity regions will affect the cooperative gain. In this section, we extend our proposed optimal sensing method to the multi-user network by considering all above issues and propose an adaptive and cooperative sensing scheme.

As explained in Section 1.3.2, the traditional cooperation scheme increases both detection and false alarm probabilities. Thus, to solve this problem, a new cooperative gain is defined for the decision on spectrum availability.[12] Assume that the base-station gathers sensing information from N CR users. Then, the number of detections follows the binomial distribution $B(N, \overline{P}_d)$. Similarly, the number of false alarms also shows the binomial distribution $B(N, \overline{P}_f)$. Thus, in order to determine the detection threshold N_{th}, it is necessary to balance between the detection error probability and the false alarm probability as follows:

$$P_{on}(1 - P_{bd}(N_{th})) = P_{off} \cdot P_{bf}(N_{th}) \tag{1.9}$$

where P_{bd} is the binomial cumulative distribution function (CDF) of the number of detections, and P_{bf} is the binomial CDF of the number of false alarms.

In order to use this cooperative scheme, all CR users should be located in the same primary user activity region. If there are multiple primary user activities, the base station should calculate the cooperative detection probability of each region separately. Then the cooperation gain is obtained as follows:

$$\overline{P}_d^c = 1 - \prod_{i=1}^{N_{\text{corr}}} \left(1 - \overline{P}_{d,i}^c\right)$$

$$\overline{P}_f^c = 1 - \prod_{i=1}^{N_{\text{corr}}} \left(1 - \overline{P}_{f,i}^c\right)$$
(1.10)

where N_{corr} is the number of primary user activity regions in the CR network coverage, and $\overline{P}_{d,i}^c$ and $\overline{P}_{f,i}^c$ represent the cooperative detection and false alarm probabilities of the primary user activity region i, respectively. In this case, it is only if none of the regions detects the primary signals that the spectrum is determined to be available, which shows the same pattern as the traditional cooperation approach.

Through the cooperation gain explained above, both detection and false alarm probabilities can be improved.

$$P_d^c = \overline{P}_{on}\overline{P}_d^c = P_{on} \sum_{i=N_{tb}}^{N} \binom{N}{i} \overline{P}_d^i (1 - \overline{P}_d)^{N-i}$$

$$P_f^c = P_{off}\overline{P}_f^c = P_{off} \sum_{i=N_{tb}}^{N} \binom{N}{i} \overline{P}_f^i (1 - \overline{P}_f)^{N-i}$$
(1.11)

However, since both detection and false alarm probabilities change, the optimal sensing parameters need to be reoptimized according to the cooperation gain. Since the optimal observation time t_s is already considered for the calculation of the cooperation gain, only transmission time T is affected and needs to be reoptimized. Usually the number of sensing data varies over time due to user mobility and user transmission. Whenever it changes, the base station reoptimizes the transmission time, which improves the transceiver utilization while maintaining the same interference level as the noncooperative sensing approach.

1.5 Spectrum Sensing Framework for Wireless Mesh Networks

In this section, we describe a distributed sensing approach for wireless mesh networks (WMNs).[17] Our design aims to embed intelligent spectrum sensing capability in a standard mesh scenario without advanced hardware/software techniques such as (i) wideband filters that can sense bandwidth of several hundreds of MHz, and (ii) the use of *a priori*

Table 1.1   Spectral Overlap as a Function of Channel Separation

Ω	0	1	2	3	4	5	6
Spectral Overlap	1	0.8	0.5	0.2	0.1	0.001	0

knowledge of the sender's transmission content.[1] Mesh networks are already being planned to provide Internet connectivity on a city-wide basis. As an example, a startup firm, Meraki, is in the process of setting up a WMN in the city of San Francisco by installing mesh routers (MRs) on the rooftops of volunteering individuals.[18] Such efforts will entail the use of off-the-shelf components to provide a low-cost solution, possibly based on the widely used 802.11 standard. We base the subsequent discussion and our solution for spectrum sensing on these hardware constraints in order to realize practical mesh networks.

1.5.1   Primary User Channels

Our architecture assumes fixed primary users whose locations are known beforehand. As an example, television stations or amateur stationary radio operators could be considered in this model. The channels currently used by the primary users are unknown and the mesh network components coordinate to identify the occupied channels. We consider a channel structure and separation similar to the one existing in the 2.4 GHz ISM band. Here, a transmission on any one given channel introduces some leakage power in the adjacent channels. The extent or the ratio of this power overlap (Ω) depends upon the separation between channels and the spectral mask specified by the standard. As an example, sample values for Ω for IEEE 802.11b with $\Delta f = 5$ MHz are listed in Table 1.1. These values are obtained through empirical measurements using two Linux laptops equipped with NETGEAR MA401 802.11 b wireless cards in an indoor setting. Note that the television stations have a large coverage radius and different portions of the mesh network may observe different channel usages.

1.5.2   Mesh Network Architecture

As seen in Figure 1.10, the standard mesh network architecture consists of several clusters of nodes. All uplink/downlink flows are directed from the individual mesh clients (MCs) towards their own mesh routers (MRs) and forwarded in a multi-hop fashion towards the gateway connected to the Internet. The MCs may be mobile and the cluster membership may change over time, though the MRs must have a fixed location. Through localization schemes or with the help of a GPS device, the MCs of a cluster should be

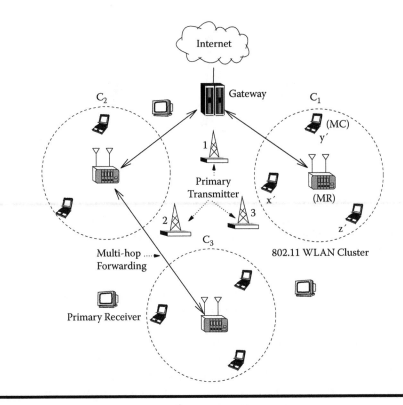

Figure 1.10 Example of mesh architecture with mesh routers (MRs) and mesh clients (MCs) under them. The mesh components are located in a spatially over-lapped region with primary transmitter stations.

able to determine the distance from the different primary stations. Stressing the need to use standard hardware components, we assume that each MR and MC is equipped with a single IEEE 802.11b-based transceiver. Using typical assumptions of a frequency-agile radio,[1] this is tunable to any one of the allowed frequencies (channels) in the primary or the secondary band at a given time. In addition, the MRs are considered to be resource-rich and can undertake computational overheads. MR-MR multi-hop links are achieved through out-of-band communication on a different channel, and are not subject to contention from the remaining nodes.[2]

1.5.3 Spectrum Sensing by Mesh Clients

A typical mesh network is constrained in terms of its sensing time. The MRs may depute the task of sensing to the MCs under them, but switching to individual channels and measuring the received power in each of them introduces significant time delay. Also, because of the leakage power, it

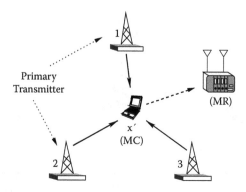

Figure 1.11 MC x' senses the channel when three primary stations are in the neighborhood. The received power is the sum of the individual transmitted powers scaled by the spectral overlap factor.

becomes difficult to ascertain whether the received power in a channel is due to several transmitters on adjacent channels.

In our scheme, the MC tunes to a single pre-decided primary channel and senses the received power for the entire duration available. This is essentially a superposition of the received power due to several transmitters. Thus, from Figure 1.11, let P_1, P_2, and P_3 be the individual received powers from the three primary transmitters. The MC x' measures the summation of the three powers and reports this to its MR. Such a measurement is also carried out at the other MCs of the cluster. This received power in the measurement channel is actually the transmitted power scaled by the spectral overlap factor I and adjusted for path loss by the propagation constants α and β. For this particular example, three MCs independently measure the total received power, and this information is used to formulate linear equations at the MR. These equations are solved to calculate the unknowns I and α, which are functions of the primary channels in use. By placing the known center frequencies of the primary channels in these functions and observing how closely the values match with the observed results, the channels in use can be identified with reasonable accuracy. This process is easily scaled for n different primary transmitters. As there are n unknowns (specifically n unknown channels), we need to solve n linear equations simultaneously. Thus, at least this many MCs must be present in the cluster for the success of this approach. Typically, each MR may support nearly a hundred clients and there are only a limited number of television stations, making this assumption reasonable from the practical standpoint.

We can improve the performance of our approach by taking multiple sets of readings, each set comprising n equations that are solved simultaneously. The channels identified are averaged over all the sets for accuracy. Also, the wireless channel introduces a finite amount of noise

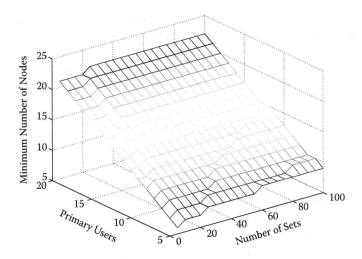

Figure 1.12 Minimum number of sensing nodes required for a given number of required sets and primary transmitters.

and the received power fluctuates around a mean level. If this noise can be estimated or is known for a given environment, then the receiver could be modified to cancel out its effects in the final sensed power value used for calculations. Both of these effects are explored in Ref. [17].

The relationship between the number of primary users (M), the user-defined metric of number of sets that determines accuracy (S), and the minimum required number of sensing MCs (n) is shown in Figure 1.12. This graph is computed as follows: We need a minimum of n sensing MCs such that we can generate a new set of M linear equations by choosing M nodes out of n each time. As S sets are required, $^{n}C_{M} = S$. We observe that the minimum MCs needed for a given number of sets scales nearly linearly with the increasing number of primary users. However, there is no significant overhead introduced in terms of the number of nodes communicating their sensed values even for a very large number of sets. From the observed values, we see that even for 100 sets, with 20 independently transmitting primary transmitters, 22 sensing MCs are sufficient.

While centralized base station-oriented schemes and distributed mesh network-based CR technology look promising, there are still several challenges that need to be addressed, which we outline in the next section.

1.6 Spectrum Sensing Challenges

Spectrum sensing constitutes one the most important components of the cognitive radio operation as highlighted in this chapter. The accuracy and the overhead of the spectrum sensing are two main issues in this area. The solutions discussed so far in this chapter provide valuable insight into the challenges and potential solutions in spectrum sensing. Nevertheless, there still exist several open research challenges that need to be investigated for the development of accurate and efficient spectrum sensing solutions. We discuss these challenges in detail in this section.

1.6.1 Interference Temperature Measurement

Interference temperature measurement techniques, as discussed in Section 1.3, provide a way of interference avoidance without the need for accurate primary transmitter detection techniques. Instead, the interference at the primary receivers is limited through an interference temperature limit. The difficulty of this receiver detection model lies in effectively measuring the interference temperature and determining the limit.

A CR user is naturally aware of its transmitted power level and can determine its precise location with the help of a positioning system. With this ability, however, its transmission could cause significant interference at a neighboring receiver on the same frequency. Currently, there exists no practical way for cognitive radios to measure or estimate the interference temperature at nearby primary receivers. Since primary receivers are usually passive devices, a CR user cannot be aware of the precise locations of primary receivers. Furthermore, if CR users cannot measure the effects of their transmissions on all possible receivers, a useful interference temperature measurement may not be feasible.

Another important challenge is the determination of the interference temperature limit. Since CR users try to control their transmissions according to this limit, an accurate model for the determination of the interference temperature limit is necessary. However, the optimal value of this limit may depend on the density and the traffic characteristics of the primary users. Furthermore, the physical layer characteristics, such as the modulation and transmitted power, as well as the operating frequency of the primary users, also affect the limit. Consequently, adaptive techniques for the determination of the interference temperature limit are necessary.

1.6.2 Multi-User CR Networks

CR networks usually reside in a multi-user environment, which consists of multiple CR users and primary users. Furthermore, CR networks can also be co-located with other CR networks competing for the same spectrum band. However, current interference models[8,19] do not consider the effect of multiple CR users. A multi-user environment makes it more difficult to sense the primary users and to estimate the actual interference. First, the effects of the transmissions of other CR users are unknown to a specific CR user. Consequently, it is hard to estimate the total interference that would be caused at a primary receiver. Second, the transmissions of other CR users may prevent a specific CR user from detecting the activity of a primary transmitter and lead the CR user to regard the primary user transmission as noise. This leads to degradation in sensing accuracy. Spectrum sensing functions should be developed considering the possibility of a multi-user/network environment. In order to solve the multi-user problem, the cooperative detection schemes can be considered, which exploit the spatial diversity inherent in a multi-user network.

1.6.3 Physical Layer Constraints

Spectrum sensing techniques require efficient physical layer capabilities in terms of wideband sensing and rapid spectrum switching. However, the constraints of the physical layer need to be known to design practical sensing algorithms. As discussed in Section 1.4, the fact that the cognitive radio cannot sense and transmit simultaneously is one of the factors in the design of spectrum sensing algorithms. This fact has been considered in Ref. [12] to optimally schedule the transmission and sensing without degrading the sensing accuracy. As an alternative, the effect of using multiple radios has been investigated in Ref. [20], where a two-transceiver operation is considered such that a transceiver always listens to the control channel for sensing. This operation improves the system performance; however, the complexity and device costs are high.

Another constraint is the limited spectrum sensing capabilities of cognitive radios. In other words, scanning the whole spectrum takes time. Since sensing consumes energy, this process has to be carefully scheduled. One of the main requirements of CR networks is the detection of the primary users in a very short time.[4,21] Since sensing time is important, OFDM-based CR networks are known to be an excellent fit for the physical architecture of CR networks.[1,15,22] Since multi-carrier sensing can be exploited in OFDM-based CR networks, the overall sensing time can be reduced. Once a primary user is detected in a single carrier, sensing in other carriers is not necessary. In Ref. [15] a power-based sensing algorithm in OFDM networks is proposed for detecting the presence of a primary user. It is shown that the overall detection time is reduced by collecting information from each carrier. However, this necessitates the use of a large number of carriers, which increases the design complexity. Hence, novel spectrum sensing algorithms need to be developed such that the number of samples needed to detect the primary user is minimized within a given detection error probability. In this sense, cooperative spectrum sensing mechanisms can be exploited to overcome the constraints of each cognitive radio. The superiority of cooperative techniques in terms of system performance has already been demonstrated in many studies.[23–26] On the other hand, such a collaboration increases the communication overhead and may lead to overall system performance degradation when channel capacity or energy consumption are considered. Consequently, effective spectrum sharing techniques that enable efficient collaboration between different CR nodes in terms of spectrum sensing information sharing are required.

1.6.4 Cooperative Sensing

Cooperative sensing constitutes one of the potential solutions for spectrum sensing in CR networks. Spectrum sensing accuracy for a single user increases with the sensing time. Considering the sensing capabilities of CR radios, however, an acceptable accuracy may be reached only after very long sensing times. The uncertainty in noise, however, prevents even infinite sensing times from being accurate in some cases.[27] This theoretical finding motivates cooperative sensing schemes.

Cooperative sensing, although more efficient, creates additional challenges for accurate spectrum sensing in CR networks. The communication requirement of the cooperating nodes necessitates cross-layer techniques that support joint design of spectrum sensing with spectrum sharing. Efficient communication and sharing techniques are necessary to alleviate the effects of communication overhead in cooperative sensing. To this end, dynamic common control channel techniques, which provide a common control channel for the CR users to exchange spectrum sensing information, may be required.[26] Moreover, efficient and distributed coordination solutions that partition the spectrum sensing tasks to various co-located CR users are required.

1.6.5 Mobility

Spectrum sensing techniques aim to provide a map of the spectrum in a CR user's vicinity. Consequently, efficient spectrum decision techniques can be used. However, if a CR user moves, the spectrum allocation map may change rapidly. Therefore, the spectrum allocation map constructed by the sensing algorithm may become obsolete with high mobility. Consequently, the CR user may need to perform spectrum sensing as they change location. This necessitates an adaptive spectrum sensing technology that is responsive to the mobility of the CR user.

1.6.6 Adaptive Spectrum Sensing

As explained in Section 1.4 and Section 1.5, the requirements of spectrum sensing solutions may depend on the network architecture. While centralized solutions focus on efficient information collection from multiple sensing devices and optimally allocating spectrum for users, distributed architectures lead to frequent information exchange between CR users. Consequently, the nature of the spectrum sensing solution may differ depending on the architecture. However, considering that CR user devices will need to adapt to any network setting, whether it is centralized or distributed, adaptive spectrum sensing solutions are crucial for rapid proliferation of the CR technology. As a result, a single CR device can be used in different network settings with a single, adaptive spectrum sensing solution.

Adaptive techniques are also necessary for different underlying physical layer functionalities. As explained above, physical layer constraints significantly affect the performance of spectrum sensing solutions. Moreover, it is clear that the realization of cognitive radio networks will lead to the implementation of different CR devices by different companies, similar to the current case with WLANs. To provide seamless spectrum sensing for higher networking layers, spectrum sensing solutions need to be adaptive to the physical layer capabilities.

1.6.7   Security

From the primary user point of view, CR users can be regarded as *malicious* devices that *eavesdrop* on the channel that the primary user is transmitting. In a sense, spectrum sensing techniques resemble eavesdropping attacks. In order to preserve the privacy of the users, spectrum sensing techniques need to carefully designed. This is particularly important considering the economics that lie behind the primary networks. Since each primary user owns the particular spectrum, the traffic flowing through this spectrum needs to be protected. Spectrum sensing techniques, however, necessitate the knowledge of the existence of primary users for efficient operation. Consequently, spectrum sensing techniques should be designed in such a way that they are aware of the *existence* of the ongoing traffic but cannot determine the *content* of the traffic. Moreover, these techniques need to be implemented so that any CR user that performs spectrum sensing will not be regarded as malicious by the already existing security protocols in primary networks.

1.7   Conclusions

In this chapter, we classify centralized and wireless mesh architectures while pointing out how their salient features could be leveraged from the spectrum sensing viewpoint. As part of an optimal centralized sensing framework, the need to balance the transmission and observation time is emphasized for maximizing the transmission efficiency under interference constraints. To address this concern, our solution is also extended for multi-spectrum environments. For distributed wireless mesh-based systems, we describe a sensing algorithm that is executed at the MRs and relies on localized information collected by the MCs under them. Furthermore, we list the research challenges related to spectrum sensing for both these architectures, which must be considered for practical deployment scenarios. We envisage that both centralized and distributed sensing solutions in the future will necessarily be developed in a cross-layer fashion, use learning algorithms to improve sensing accuracy, and have advanced hardware that will complement the algorithmic approach followed by the higher-layer networking protocols.

References

[1] I.F. Akyildiz, W.Y. Lee, M.C. Vuran, and S. Mohanty, Next generation/ dynamic spectrum access/cognitive radio wireless networks: A survey, *Computer Networks Journal "Elsevier Computer Networks Journal."* 50(13), pp. 2127–2159 (Sept. 2006).

[2] I. F. Akyildiz, X. Wang, and W. Wang, Wireless mesh networks: a survey, *Elsevier Computer Networks Journal.* 47(4), pp. 445–487 (Nov. 2005).

[3] D. Cabric, S. M. Mishra, and R. W. Brodersen. Implementation issues in spectrum sensing for cognitive radios. In *Proc. 38th Asilomar Conference on Signals, Systems and Computers 2004*, pp. 772–776 (Nov. 2004).

[4] A. Sahai, N. Hoven, and R. Tandra. Some fundamental limits on cognitive radio. In *Proc. Allerton Conference, Monticello, 2004* (Oct. 2004).

[5] M. Oner and F. Jondral, On the extraction of the channel allocation information in spectrum pooling systems, *IEEE Journal on Selected Areas in Communications.* 25(3), pp. 558–565 (Apr. 2007).

[6] S. M. Mishra, A. Sahai, and R. W. Brodersen. Cooperative sensing among cognitive radios. In *Proc. IEEE ICC 2006*, vol. 4, pp. 1658–1663 (June 2006).

[7] B. Wild and K. Ramchandran. Detecting primary receivers for cognitive radio applications. In *Proc. IEEE DySPAN 2005*, pp. 124–130 (Nov. 2005).

[8] FCC. ET Docket No 03-289 Notice of Proposed Rule Making and Order (Nov. 2003).

[9] S. Shankar. Spectrum agile radios: Utilization and sensing architecture. In *Proc. IEEE DySPAN 2005*, pp. 160–169 (Nov. 2005).

[10] Y. Hur, J. Park, W. Woo, J. S. Lee, K. Lim, C.-H. Lee, H. S. Kim, and J. Laskar. A cognitive radio (CR) system employing a dual-stage spectrum sensing technique: A multi-resolution spectrum sensing (MRSS) and a temporal signature detection (TSD) technique. In *Proc. IEEE Globecom 2006* (Nov. 2006).

[11] I. F. Akyildiz, W. Y. Lee, M. C. Vuran, and S. Mohanty. A survey on spectrum management in cognitive radio networks. *IEEE Communications Magazine.* 46(4), pp. 40–48 (Apr. 2008).

[12] W. Y. Lee and I. F. Akyildiz. Optimal spectrum sensing framework for cognitive radio networks. To appear in *IEEE Transactions an Wireless Communications* (2008).

[13] K. Sriram and W. Whitt, Characterizing superposition arrival processes in packet multiplexers for voice and data, *IEEE Journal on Selected Areas in Communications*. SAC-4, pp. 833–846 (Sept. 1986).

[14] J. G. Proakis, *Digital Communications*. (New York: McGraw-Hill, 2001), 4th edition.

[15] H. Tang. Some physical layer issues of wide-band cognitive radio system. In *Proc. IEEE DySPAN 2005*, pp. 151–159 (Nov. 2005).

[16] Y. Chen, Q. Zhao, and A. Swami. Joint design and separation principle for opportunistic spectrum access. In *Proc. IEEE Asilomar Conference on Signals, Systems and Computers 2006* (Oct. 2006).

[17] K. R. Chowdhury and I. F. Akyildiz, Cognitive wireless mesh networks with dynamic spectrum access, *IEEE Journal on Selected Areas in Communications*. 26(1), pp. 161–181 (Jan. 2008).

[18] A free mesh network for San Francisco. [Online]. MIT Technology Review. URL http://www.technologyreview.com/Infotech/19260/page1/.

[19] T. X. Brown. An analysis of unlicensed device operation in licensed broadcast service bands. In *Proc. IEEE DySPAN 2005*, pp. 11–29 (Nov. 2005).

[20] S. Sankaranarayanan, P. Papadimitratos, A. Mishra, and S. Hershey. A bandwidth sharing approach to improve licensed spectrum utilization. In *"Proc. IEEE DySPAN 2005,"* pp. 279–288 (Nov. 2005).

[21] G. Ganesan and Y. G. Li. Agility improvement through cooperative diversity in cognitive radio networks. In *Proc. IEEE GLOBECOM 2005*, pp. 2505–2509 (Nov. 2005).

[22] T. A. Weiss, J. Hillenbrand, A. Krohn, and F. K. Jondral. Efficient signaling of spectral resources in spectrum pooling systems. In *Proc. 10th Symposium on Communications and Vehicular Technology (SCVT) 2003* (Nov. 2003).

[23] L. Cao and H. Zheng. Distributed spectrum allocation via local bargaining. In *Proc. of IEEE SECON 2005*, Santa Clara, CA, USA (Sept. 2005).

[24] J. Huang, R. A. Berry, and M. L. Honig. Spectrum sharing with distributed interference compensation. In *Proc. IEEE DySPAN 2005*, pp. 88–93 (Nov. 2005).

[25] L. Ma, X. Han, and C. C. Shen. Dynamic open spectrum sharing MAC protocol for wireless ad hoc network. In *Proc. IEEE DySPAN 2005*, pp. 203–213 (Nov. 2005).

[26] J. Zhao, H. Zheng, and G.-H. Yang. Spectrum agile radios: utilization and sensing architecture. In *"Proc. IEEE DySPAN 2005,"* pp. 259–268 (Nov. 2005).

[27] S. Srinivasa and S. A. Jafar, The throughput potential of cognitive radio: A theoretical perspective, *IEEE Communications Magazine*. pp. 73–79 (May, 2007).

Chapter 2

Spectrum Usage Modeling and Forecasting in Cognitive Radio Networks

Lutfa Akter, Balasubramaniam Natarajan,
and Caterina Scoglio

Contents

2.1 Introduction

We are currently experiencing a rapid growth in wireless network technologies along with a plethora of wireless devices competing for a limited amount of unlicensed spectrum. The FCC has realized that overcrowding of unlicensed bands is only going to worsen and is considering opening up licensed bands for opportunistic use by secondary radios/users [1]. Traditionally, licensed bands were exclusively reserved for use by the primary license holders (primary users). However, several recent measurements of spectrum usage indicate that many licensed spectrum bands remain relatively unused for most of the time [2,3]. Therefore, there exists an opportunity for radios with cognitive capabilities to use these licensed bands when primary users are absent, and hence improve overall spectrum usage efficiency.

The concept of cognitive radios was first introduced by Joseph Mitola [4]. A cognitive radio network (CRN) is built on the following principle: a network of secondary users (users without license) continuously sense the use of a spectrum band by primary users and opportunistically utilize the band when primary users are absent. Any secondary user (SU) in a CRN performs two main functions: (1) sensing spectrum usage to identify absence or presence of a primary user (PU), and (2) transmitting at appropriate power if the PU is absent.

Over the past five years, research efforts have extensively focused on the first task of sensing primary users. Different sensing techniques and the impact of cooperative sensing have been described in Ref. [5]. An experimental study on the performances of different sensing techniques has been presented in Ref. [6]. In Ref. [7], the optimum sensing time for the secondary users' throughput efficiency (in terms of airtime) has been developed. An appropriate model for both primary and secondary users is needed to accomplish the second task. In Ref. [7], the authors have assumed that primary users follow a continuous-time Markov chain. In Ref. [8], the authors have assumed that both primary and secondary users follow a continuous-time Markov chain. The authors in Refs. [7,8] have considered a spectrum band used by either a PU or an SU. Additionally, researchers have also investigated how spectral sharing among PUs and SUs on spectrum agility and fairness in cooperation results in improved performance [9,10]. Game-theoretic approaches for maintaining QoS in cognitive radio network architectures have also been proposed [11,12]. However, much of the prior work on QoS issues in CRNs is based on assumptions regarding availability of spectrum. Additionally, work in CRNs has primarily been focused on sensing primary users with very little emphasis on how multiple secondary users may compete for available spectrum. Therefore, forecasting the behavior of secondary users is equally critical to the successful operation of a CRN. For example, if a spectrum band is determined to be free and a large number of secondary users decide to use this spectrum band simultaneously, the QoS or bit-error-rate (BER) performance of the secondary users will degrade due to the high level of interference. Therefore, it is important to investigate strategies for enabling SUs to sense and predict the behavior of both primary and competing secondary users in a frequency band of interest.

In this chapter, we introduce an integrated modeling and forecasting approach that SUs in a CRN can use to predict spectrum usage/availability primarily based on power level measurements. Our modeling and forecasting setup incorporates traffic behavior of both primary and competing secondary users in a spectrum band of interest. Here, every SU that is interested in opportunistically using a channel first enters a learning and modeling phase in which its power level measurements are used to estimate traffic model parameters for both primary and other secondary users. In the next phase, the SU begins to actively participate in using the channel by exploiting the traffic model to predict the number of users in future instants of time. Specifically, by considering a continuous-time Markov chain traffic model for the PU and SUs similar to that used in Refs. [8,10], we propose a Kalman filter approach to estimate the number of primary and secondary users at a given time instant. Using these estimates, we determine robust upper bound forecasts of the number of primary and secondary users for a future time instant. Knowledge of the upper bound provides valuable information to an SU interested in using a spectrum band that is already being used by other secondary users. This approach is analogous to the method proposed for forecasting Internet traffic flows [13]. Using both simulated data and measured power levels in the 2.4 GHz unlicensed band, we demonstrate the implementation of both the modeling and forecasting aspects of the proposed approach. Furthermore, practical implementation also shows that the proposed upper bound forecast is robust to errors in model parameter estimates. It is important to remember that this chapter offers a modeling and forecasting tool and not algorithms for spectrum sharing among multiple secondary users. Additionally, note that even though we use the words spectrum band, frequency band, and channel interchangeably throughout the chapter, they convey the same meaning.

The rest of the chapter is organized as follows. In Section 2.2, we describe the proposed spectrum usage model in detail. Section 2.3 presents the Kalman filtering techniques to estimate the number of primary and secondary users, and Section 2.4 describes the proposed forecasting techniques for both primary and secondary users. Experimental results illustrating the application of the modeling and forecasting methods for two different sets of data—(1) simulated data from a CRN and (2) real-time measurement data from the 2.4 GHz industrial, scientific, and medical (ISM) band—are provided in Section 2.5. Finally, conclusions are presented in Section 2.6.

2.2 System Model

We consider L channels within a CRN. Each of the L channels can be used by either a PU or one or more secondary users as shown in Figure 2.1. Each SU uses the spectrum opportunistically. That is, every SU scans the spectrum at regular intervals and starts transmitting on a particular channel once it determines that the channel will not be used by a PU.

Typically, researchers focus on the sensing aspect of secondary users in order to determine whether a channel is available for transmission. In this chapter, in addition to determining the presence or absence of the primary user in a given channel, we also consider the impact of multiple secondary users utilizing a single channel. For example, if a channel is determined to be free and a large number of secondary users decide to use this channel simultaneously, the QoS or BER performance of the secondary users will be poor due to the high level of interference from other secondary users. Consequently, it is important for every SU to monitor the spectrum usage by other secondary users, and this aspect is captured in our proposed modeling and forecasting setup. Additionally, we assume that the spectrum usage levels of various channels are independent. Therefore, in the rest of the chapter, we restrict ourselves to the modeling and forecasting of spectrum use in one channel. It is easy to extend the idea presented in the following sections to the case of correlated channels.

In this work, we assume that both the PU and secondary users follow a Poisson arrival process with rates λ_p and λ_s, respectively. Their negative

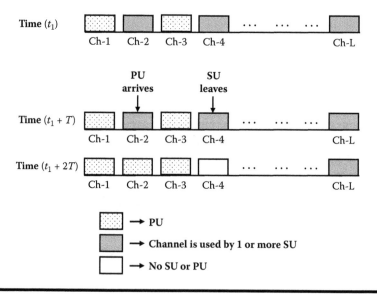

Figure 2.1 CRN operation.

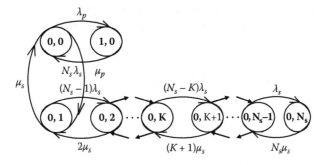

Figure 2.2   Two-dimensional state-transition-rate diagram of PU and SU.

exponential service time distributions have rates μ_p and μ_s, respectively. A similar model for arrival and departure processes of PU and secondary users has been assumed in Refs. [8,10]. The maximum numbers of PUs and SUs are N_p and N_s, respectively. For ease in presentation, N_p is assumed to be equal to 1, i.e., we assume that the PU is either present or absent. In other words, the PU follows a two-state ON-OFF Markov process. The two-dimensional state-transition-rate diagram for the CRN is shown in Figure 2.2. Each state in the model is denoted by (n_p, n_s), where n_p and n_s represent the numbers of primary users and secondary users, respectively. From this state-transition-rate diagram and concepts from queueing theory [14], the differential equations for the state probabilities $p_{i,k}(t)$ can be evaluated for both primary (with $i = p$) and secondary (with $i = s$) users. The state probability is defined as

$$p_{i,k}(t) \triangleq \text{prob}\{x_i(t) = k\}. \tag{2.1}$$

Since we have an ON-OFF traffic model for the PU, $k = 0$ or 1 when $i = p$. On the other hand, for secondary users, i.e., when $i = s$, k can vary from 0 to N_s provided that there is no primary user in the channel. In general, the differential equations for the state probabilities correspond to

$$\frac{dp_{i,0}(t)}{dt} = \mu_i p_{i,1}(t) - N_i \lambda_i p_{i,0}(t), \tag{2.2}$$

$$\frac{dp_{i,k}(t)}{dt} = (N_i - k + 1)\lambda_i p_{i,(k-1)}(t) + (k+1)\mu_i p_{i,(k+1)}(t)$$

$$-(k\mu_i + (N_i - k)\lambda_i)p_{i,k}(t), \quad 1 \le k < N_i, \tag{2.3}$$

$$\frac{dp_{i,N_i}(t)}{dt} = \lambda p_{i,(N_i-1)}(t) - N_i \mu_i p_{i,N_i}(t), \quad \text{for } i = p, \ s. \tag{2.4}$$

Typically, we are interested in determining the expected number of primary and secondary users in a given time instant. Therefore, $E\{x_i(t)\}$ can be written as

$$E\{x_i(t)\} = \sum_{k=0}^{N_i} k p_{i,k}(t), \quad \text{for } i = p, \ s. \tag{2.5}$$

Hence,

$$\frac{dE\{x_i(t)\}}{dt} = \sum_{k=0}^{N_i} k \frac{dp_{i,k}(t)}{dt}, \quad \text{for } i = p, \ s. \tag{2.6}$$

Let $\mathbf{L} = (0, 1, 2, \ldots, N_i)'$. From Equations (2.2)–(2.4), (2.6), we can write

$$\frac{dE\{x_i(t)\}}{dt} = \mathbf{L}' \dot{\mathbf{P}}_i = \mathbf{L}' \mathbf{Q}_i \mathbf{P}_i, \tag{2.7}$$

where

$$\mathbf{Q}_i = \begin{pmatrix} -N_i\lambda_i & \mu_i & 0 & . & 0 \\ N_i\lambda_i & -[(N_i-1)\lambda_i + \mu_i] & 2\mu_i & . & 0 \\ 0 & (N_i-1)\lambda_i & -[(N_i-2)\lambda_i + 2\mu_i] & . & 0 \\ 0 & . & . & . & 0 \\ 0 & 0 & & . & \lambda_i - N_i\mu_i \end{pmatrix}$$

and

$$\mathbf{P}_i = \begin{pmatrix} p_{i,0}(t) \\ p_{i,1}(t) \\ . \\ . \\ . \\ p_{i,N_i}(t) \end{pmatrix}, \quad \text{for } i = p, \ s.$$

In Equation (2.7), $(\cdot)'$ indicates matrix or vector transpose operator. $\mathbf{L}' \mathbf{Q}_i \mathbf{P}_i$ is obtained as $N_i\lambda_i - (\lambda_i + \mu_i)E\{x_i(t)\}$. Therefore,

$$\frac{dE\{x_i(t)\}}{dt} = N_i\lambda_i - (\lambda_i + \mu_i)E\{x_i(t)\}. \tag{2.8}$$

We assume that measurements are performed at discrete time instants mT, $m = 1, 2, 3, \ldots$, for a given value T. Using the initial condition that the number of users at time $t = (m-1)T$ is $x_i(m-1)$, the solution of Equation (2.8) is obtained as

$$E[x_i(m)|x_i(m-1)] = x_i(m-1)e^{-T(\lambda_i+\mu_i)} + \frac{N_i\lambda_i}{(\lambda_i+\mu_i)}[1 - e^{-T(\lambda_i+\mu_i)}]. \quad (2.9)$$

Therefore, it is possible to express the number of users (primary or secondary) at time mT in terms of the number of users (primary or secondary) at time $(m-1)T$ as

$$x_i(m) = A_i x_i(m-1) + B_i, \quad (2.10)$$

where

$$A_i = e^{-T(\lambda_i+\mu_i)} \quad (2.11)$$

and

$$B_i = \frac{N_i\lambda_i}{(\lambda_i+\mu_i)}[1 - e^{-T(\lambda_i+\mu_i)}]. \quad (2.12)$$

Equation (2.10) shows us the relationship between the number of users at two successive measurement instants, and in the most general case, Equation (2.10) corresponds to

$$x_i(m) = A_i x_i(m-1) + B_i u_i(m) + w_i(m), \quad \text{for } i = p, s. \quad (2.13)$$

Specifically, we can write down the state equations for primary and secondary users as

$$x_p(m) = A_p x_p(m-1) + B_p u_p(m) + w_p(m) \quad (2.14)$$

and

$$x_s(m) = A_s x_s(m-1) + B_s u_s(m) + w_s(m), \quad (2.15)$$

where $x_p(m)$ and $x_s(m)$ represent the numbers of primary and secondary users using the spectrum, respectively, at the measurement instant m. The parameters B_p and B_s relate the optional control inputs $u_p(m)$ and $u_s(m)$, respectively, to states. Equation (2.10) suggests that $u_p(m)$ and $u_s(m)$ are equal to 1 for our system model. The terms $w_p(m)$ and $w_s(m)$ are the process noise and are assumed to be zero-mean Gaussian noise with variances σ_p^2 and σ_s^2, respectively. The parameters A_p and A_s relate the states at previous and current measurement instants, in the absence of either a driving function or process noise. A_p and A_s are assumed to be constant over the analysis or vary very slowly.

The received power at a secondary user terminal during the measurement instant m consists of relative power level increments caused by both primary and secondary users, and in the most general case corresponds to

$$y(m) = C_p x_p(m) + C_s x_s(m) + D + v(m), \qquad (2.16)$$

where $y(m)$ is received power in dBm; C_p and C_s represent the relative increase in power level (in dB) due to the presence of primary and secondary users, respectively; D represents the background thermal noise; and $v(m)$ denotes the measurement noise, which may arise from miscalculation or misalignment of timings and is assumed to be zero-mean Gaussian noise with variance σ_v^2. The variable $y(m)$ is the only measurable variable in the system.

In a CRN, the primary user has the right of use. Therefore, modeling and forecasting tasks are more critical for secondary users. From a secondary

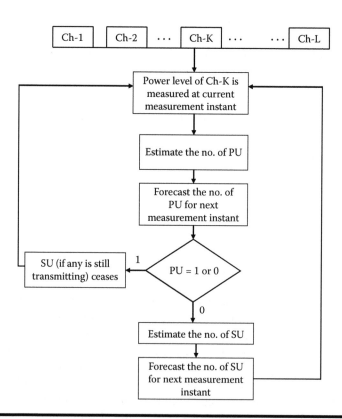

Figure 2.3 **Operation of the system model on a single channel's spectrum usage in active phase.**

user perspective, we assume two modes of operation. Once an SU decides to use a channel opportunistically, it will enter a "learning and modeling" phase. In this phase, the SU measures the power levels in the channel and estimates traffic parameters of primary and other secondary users. An example of this estimation process is discussed in Section 2.5, where a time-series-based approach is used for modeling. In the next phase, the SU becomes an "active" participant in the opportunistic spectrum sharing process. In the "active" phase, each secondary user continues to sense the received power level in order to forecast the use of spectrum bands by other secondary users. This process is illustrated in Figure 2.3.

2.3   Estimation of Spectrum Usage

In Section 2.2, we introduced the traffic models for the number of primary and secondary users using a given channel. In this section, we propose a Kalman-filter-based state estimation technique (i.e., estimating the number of primary and secondary users) based on the model from Section 2.2. The state estimation based on the Kalman filter is summarized below.

State Equation: The state of the CRN can be represented by a vector $\mathbf{x} = (x_p \ x_s)'$. The corresponding state equation is

$$\mathbf{x}(m) = \mathbf{A}\mathbf{x}(m-1) + \mathbf{B} + \mathbf{w}(m), \tag{2.17}$$

where

$$\mathbf{A} = \begin{pmatrix} A_p & 0 \\ 0 & A_s \end{pmatrix}, \quad \mathbf{B} = \begin{pmatrix} B_p \\ B_s \end{pmatrix},$$

and $\mathbf{w}(m)$ is a vector WGN with mean $\mathbf{0}$ and covariance matrix

$$\Sigma_w = \begin{pmatrix} \sigma_p^2 & 0 \\ 0 & \sigma_s^2 \end{pmatrix}.$$

Measurement Equation:

$$y(m) = \mathbf{C}'\mathbf{x}(m) + D + v(m), \tag{2.18}$$

where $\mathbf{C} = (C_p \ C_s)'$. Based on the state and measurement equations Equations (2.17) and (2.18), respectively, the Kalman filtering steps are given below:

Step 1. Initialization:

$$\widehat{\mathbf{x}}(0|0) = \begin{pmatrix} E(x_p(0)) \\ E(x_s(0)) \end{pmatrix} \quad \text{and} \quad \mathbf{M}(0|0) = \begin{pmatrix} \sigma_p^2(0) & 0 \\ 0 & \sigma_s^2(0) \end{pmatrix}. \quad (2.19)$$

Step 2. Prediction:

$$\widehat{\mathbf{x}}(m|m-1) = \mathbf{A}\widehat{\mathbf{x}}(m-1|m-1) + \mathbf{B}, \quad \forall\ m, \quad (2.20)$$

$$\mathbf{M}(m|m-1) = \mathbf{A}\mathbf{M}(m-1|m-1)\mathbf{A}' + \Sigma_w, \quad \forall\ m. \quad (2.21)$$

Step 3. Kalman gain vector calculation:

$$\mathbf{K}(m) = \mathbf{M}(m|m-1)\mathbf{C}(\mathbf{C}'\mathbf{M}(m|m-1)\mathbf{C} + \sigma_v^2)^{-1}, \quad \forall\ m. \quad (2.22)$$

Step 4. Correction:

$$\widehat{\mathbf{x}}(m|m) = \widehat{\mathbf{x}}(m|m-1) + \mathbf{K}(m)(y(m) - \mathbf{C}'\widehat{\mathbf{x}}(m|m-1) - D), \quad \forall\ m, \quad (2.23)$$

$$\mathbf{M}(m|m) = \{\mathbf{I} - \mathbf{K}(m)\mathbf{C}'\}\mathbf{M}(m|m-1), \quad \forall\ m. \quad (2.24)$$

From Equations (2.19)–(2.24), we estimate the number of primary and secondary users. The estimated value for the number of secondary users, denoted by $\widehat{x}_s(m|m)$, is reset to 0 if the predicted value for the number of primary users for the $(m+1)$st instant is 1. This is also depicted in the block diagram in Figure 2.3. The prediction methods for both primary and secondary users are described in the following section.

2.4 Forecasting Spectrum Usage

In this section, we describe the forecasting techniques used to predict the activity of primary and secondary users in the channel of interest.

One approach for forecasting is to determine the likely state at the next time instant given that we have the state estimate for the current instant. For this, we need to calculate the probability of transitioning to another state at time $(m+1)T$. The transitioning probabilities can be determined starting from the differential equation

$$\dot{\mathbf{P}}_i = \mathbf{Q}_i\mathbf{P}_i, \quad \text{for } i = p,\ s, \quad (2.25)$$

within the time interval $mT < t < (m+1)T$. The vector $\mathbf{P}_i$ and matrix $\mathbf{Q}_i$ are as defined in Section 2.2. Equation (2.25) governs the evolution of state transition probabilities. The solution of Equation (2.25) gives the state transition probabilities from the mth measurement instant. The existence of the solution of Equation (2.25) depends on two conditions. The first condition is the diagonalizability property of matrix $\mathbf{Q}_i$. The second one is nonpositive definiteness of matrix $\mathbf{Q}_i$. The matrix $\mathbf{Q}_i$ satisfies both conditions. It is reducible to a diagonal form $\mathbf{Q}_i = \mathbf{E}_i \Gamma_i \mathbf{E}_i^{-1}$, where Γ_i is a diagonal matrix with eigenvalues of $\mathbf{Q}_i$, and $\mathbf{E}_i$ is the matrix of corresponding right eigenvectors. The eigenvalues can be found to be $\gamma_l = -l(\lambda_i + \mu_i)$ for $l = 0, \ldots, N_i$. Hence, the solution, for $t \in (mT, (m+1)T]$, is given by

$$\mathbf{P}_i = \mathbf{E}_i e^{\Gamma_i t} \, \mathbf{F}_i, \quad \text{for } i = p, \, s, \tag{2.26}$$

where $\mathbf{F}_i$ is a constant vector determined from the initial condition (i.e, $\widehat{x}_i(m|m)$) as

$$\mathbf{F}_i = (e^{\Gamma_i mT})^{-1} \, \mathbf{E}_i^{-1} \, \mathbf{P}_{mT(i)}, \tag{2.27}$$

where $\mathbf{P}_{mT(i)}$ is a vector with all 0s except the $\widehat{x}_i(m|m)$)th element which is 1. Now, we compute state transition probability values for the instant $(m+1)T$ by integrating the time-varying state transition probability expressions (i.e., Equation (2.26)) [13]

$$\widetilde{\mathbf{P}}_i = \frac{1}{T} \int_{mT}^{(m+1)T} \mathbf{P}_i dt$$

$$= \frac{1}{T} \, \mathbf{E}_i \left(\int_{mT}^{(m+1)T} e^{\Gamma_i t} dt \right) \mathbf{F}_i$$

$$= \frac{1}{T} [\widetilde{p}_{i,0} \; \widetilde{p}_{i,1} \; \cdots \; \widetilde{p}_{i,N_i}]', \quad \text{for } i = p, \, s. \tag{2.28}$$

In the above integration notation, we have used the fact that the integral of a matrix is the integral of each element of the matrix. The elements $\widetilde{p}_{i,k}$ of the vector $\widetilde{\mathbf{P}}_i$ denote the probabilities of transitioning to state k at instant $(m+1)T$, for $i = p, \, s$. Based on the state transitioning probabilities, we can now forecast the number of primary and secondary users. For ease in presentation, the forecasting method for spectrum usage of SU is described first.

2.4.1 Forecasting Spectrum Usage of SU

Forecasting spectrum usage by other secondary users is critical for the following reasons. Each SU can now determine if a particular channel is overcrowded with secondary users. If it is, the channel may be avoided as its use may degrade the QoS. Additionally, by forecasting the number of secondary users in a channel, each SU can also determine how much power it needs to transmit without violating spectral emission limits (while maintaining its QoS).

In this work, we propose an upper bound forecasting technique based on the state transition probability matrix of SUs similar to the approach taken for forecasting the number of flows in Internet traffic [13]. We discuss the case when the number of PUs is 0 and we are interested in forecasting the number of secondary users at the next instant. The optimal estimate of the number of secondary users, i.e., $\hat{x}_s(m|m)$, at time instant m is used to forecast the number of secondary users at the $(m+1)$ instant. From the estimated number of secondary users $\hat{x}_s(m|m)$ at time mT, state transition probability values are computed from Equation (2.29) and then prediction for the $(m+1)$ instant is done. The predicted state of SU for the $(m+1)$ instant at time mT corresponds to the result in Ref. [13],

$$\tilde{x}_s(m) = \min_{x_s \in [\hat{x}_s(m|m), \ N_s]} x_s \ \text{s.t.} \ \tilde{p}_{s,k} < \beta. \tag{2.29}$$

Here, β is a preset probability value. Equation (2.29) can be understood with the help of an example. Suppose $\hat{x}_s(m|m)$ is obtained as 3, and N_s is 8. Based on the value of $\hat{x}_s(m|m)$, a state transition probability vector, i.e., $\mathbf{P}_s = \frac{1}{T}[\tilde{p}_{s,0} \ \tilde{p}_{s,1} \ \cdots \ \tilde{p}_{s,8}]'$, is obtained, which corresponds to the possible states of SU with number of users $[0 \ 1 \ \cdots \ 8]'$, respectively. By observing this state transition probability vector, one can determine multiple states for which $\tilde{p}_{s,k} < \beta$. All states with number of users greater than 5 might satisfy this condition. This effectively suggests that the probability of $x_s(m+1)$ being 5 or more is going to be negligible. Therefore, the upper bound for $x_s(m+1)$ should be 5. In general, the chosen state is the state with minimum number of users satisfying Equation (2.29). As a result, Equation (2.29) serves as a good upper bound for the number of secondary users at time $(m+1)$ based on measurements up to time m.

2.4.2 Forecasting Spectrum Usage of PU

We propose two ways to forecast the presence or absence of a primary user. The first method is Kalman-filter-based (KF-based) prediction. In this method, the number of primary users obtained from the prediction stage (Equation (2.20)) is used as the forecast number of primary users. For example, from the estimated value at the mth instant, $\hat{x}_p(m|m)$, the predicted value for the $(m+1)$ instant at the mth instant is taken as

$$\tilde{x}_p(m) = \hat{x}_p(m+1|m)$$

$$= A_p \hat{x}_p(m|m) + B_p. \tag{2.30}$$

The second method is based on state transition probabilities as given in Equation (2.28). In this method, state transition probabilities from the present estimated state $\hat{x}_p(m|m)$ are computed in the same way as described for SU above, and a state with higher state transition probability is taken as the predicted state $\tilde{x}_p(m)$. For the proposed system model in this chapter, the Markov chain of the PU shows that it two states—state 0 (number of PUs is 0) and state 1 (number of PUs is 1). Suppose $\hat{x}_p(m|m)$ is obtained as 1, and on this a state transition probability vector, i.e., $\tilde{\mathbf{P}}_p = \frac{1}{T}[\tilde{p}_{p,0} \ \tilde{p}_{p,1}]'$ is computed. The predicted state of the PU for the $(m+1)$ instant at time mT is proposed as

$$\tilde{x}_p(m) = \begin{cases} 1, & \text{if } \tilde{p}_{p,1} > \tilde{p}_{p,0}, \\ 0, & \text{otherwise.} \end{cases} \tag{2.31}$$

2.5   Experimental Results

In this section, we illustrate the potency of the proposed techniques for estimating and forecasting spectrum usage in a cognitive radio network through simulation. We show the performance of proposed methods on two different sets of data: (1) simulated data from a CRN, and (2) real time measurement data from the 2.4 GHz ISM band.

2.5.1   Simulated CRN

We consider a CRN where the PU follows an ON–OFF Markov process with arrival rate (λ_p) and departure rate (μ_p), 0.000625 sec^{-1} and 0.00125 sec^{-1}, respectively. For the primary user, λ_p is set smaller compared to μ_p to reflect the assumption that the primary user arrives less frequently but once it comes, it stays longer. The above choices for λ_p and μ_p reflect a 32% use of the channel by the PU on average. We assume that an SU is in its "active" phase, i.e., the SU knows all required traffic parameters. In our setup, secondary users follow a Markov process with 11 states. The arrival rate (λ_s) and departure rate (μ_s) are taken as 0.005 sec^{-1} and 0.005 sec^{-1}, respectively. The state noise variances σ_p^2 and σ_s^2 are considered to be 0.2 and 1, respectively. These parameters are summarized in Table 2.1.

The plots of evolution of primary and secondary users, $x_p(m)$ and $x_s(m)$, respectively, with time are shown in Figures 2.4(a) and 2.4(b). The number of measurement instants is 1001. The measurement interval T is 10 sec. At the terminal of an SU attempting to use this channel, the received power $y(m)$ is shown in Figure 2.4(c). The background noise level D and measurement noise variance σ_v^2 are assumed to be -135 dBm and 3, respectively. From Figure 2.4(c), it is evident that the arrival of a primary user results in a sudden increase in received power level as expected.

Table 2.1   Statistics of Primary and Secondary Users

	Primary User	Secondary User
Maximum number	$N_p = 1$	$N_s = 10$
Power level increase (dB)	$C_p = 30$	$C_s = 2$
State noise Variance	$\sigma_p^2 = 0.2$	$\sigma_s^2 = 1$
Arrival rate (sec^{-1})	$\lambda_p = 0.000625$	$\lambda_s = 0.005$
Departure rate (sec^{-1})	$\mu_p = 0.00125$	$\mu_s = 0.005$

From $y(m)$, we first estimate the number of primary and secondary users, $\hat{x}_p(m|m)$ and $\hat{x}_s(m|m)$, respectively, from Equations (2.19–2.24), and then use these estimates for forecasting. For estimation, the Kalman filter initialization parameters are set as

$$\hat{\mathbf{x}}(0|0) = \begin{pmatrix} \frac{B_p}{1-A_p} \\ \frac{B_s}{1-A_s} \end{pmatrix}$$

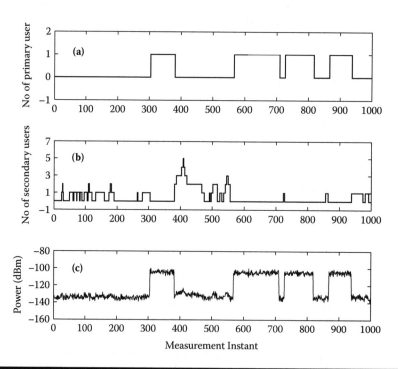

Figure 2.4   Evolution of primary and secondary users, $x_p(m)$, and $x_s(m)$, along with power level variation, $y(m)$, with time.

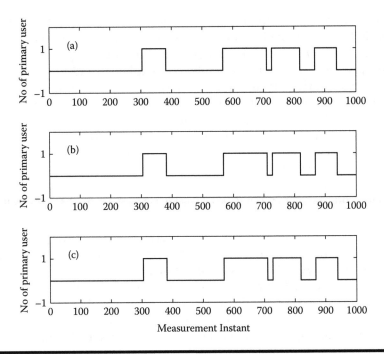

Figure 2.5 **Performance of prediction methods for the PU: (a) true PU activity (absence or presence), (b) prediction of activity by method 1, and (c) prediction of activity by method 2.**

and

$$\mathbf{M}(0|0) = \begin{pmatrix} \frac{\sigma_p^2}{1-A_p^2} & 0 \\ 0 & \frac{\sigma_s^2}{1-A_s^2} \end{pmatrix}.$$

After estimation, predictions for both the numbers of primary and secondary users are done.

The prediction performances for the activity of the PU are shown in Figure 2.5. Figure 2.5(a) shows the true activity of the PU. In Figure 2.5(b), the performance of the KF-based prediction (using (2.30)) for the activity of the PU is shown. This predictor solely depends on the estimated value at the present instant and the choices of A_p and B_p. In this setup, since A_p is 0.9814 and B_p is small, the predicted value for the next instant is equal to the current estimate. Therefore, with a single time instant lag, the KF predictor follows the presence or absence of the PU correctly. The state transition probability prediction method performs identically to the KF predictor as shown in Figure 2.5(c). This method computes the probabilities

$p_{p,k}(t)$ of transitioning to all possible states from the current state (Equations (2.26)–(2.27)). But it is very easy to prove that the probabilities $p_{p,k}(t)$, $t \in (mT, (m+1)T]$, as defined in Equation (2.1), are independent of m because arrival and departure times follow an exponential distribution (which is memoryless). This simplification reduces the computation effort and time. The calculation of $\mathbf{F}_p$ and $\tilde{\mathbf{P}}_p$ (in (2.26) and (2.27), respectively) can be performed off-line for all possible states. Then the forecast process only involves this table-lookup to determine the next instant's activity at each instant from a current instant's activity based on (2.31).

As shown in the block diagram in Figure 2.3, if the predicted value for the PU, $\tilde{x}_p(m)$, is 1, then any SU, if present, ceases to transmit. In order to reflect this in our simulation, $\hat{x}_s(m|m)$ is reset to 0 and no prediction is done when $\tilde{x}_p(m) = 1$. But if $\tilde{x}_p(m)$ is computed as 0, prediction for the SUs is done of the current-instant state estimate, $\hat{x}_s(m|m)$, using the proposed method in Section 2.4. To predict the number of secondary users, the probabilities $p_{s,k}(t)$ of transitioning to all possible states from the current state (see (2.26)–(2.27)) can be computed off-line for all possible states. Once again, these probabilities are independent of m. The forecast process only involves this table-lookup to determine the next state at each instant from the current state estimate using (2.29). The probability threshold β is fixed at 0.1 for this simulation. This value of β indicates that the system has less than 10% chance to exceed the predicted state. Figure 2.6 shows the predicted number of secondary users, $\tilde{x}_s(m)$, with the true number of secondary users, $x_s(m)$. For clarity, only 300 to 600 measurement instants are shown in this figure. As expected, $\tilde{x}_s(m)$ serves as a good upper bound predictor for the number of secondary users. For the same measurement window, the variation of predicted power level acting as an upper bound to true power level is shown in Figure 2.7.

The performance of the upper bound predictor for the SUs is affected by the preset value of β. For small values of β, the upper bound predictor performs satisfactorily. As β increases beyond a certain value, the upper bound begins to fail for a few time instants. As an example, Figure 2.8 shows the performance of the predictor for $\beta = 0.20$. From this figure, it is evident that as β increases, the quality of the upper bound predictor degrades.

The proposed estimation and forecasting process depends on the traffic parameters λ_p, μ_p, λ_s, and μ_s. The traffic parameters for the PU are easier to find than those for the SUs, as the PU follows ON–OFF traffic characteristics. Advances in sensing techniques [5,6] enable effective PU detection, which in turn can be used to estimate λ_p and μ_p.

Estimating traffic parameters for the SUs is more involved. One approach is to use time-series-based Yule–Walker estimation as illustrated in the next subsection. Therefore, it is important to evaluate the sensitivity of our proposed method to error in the λ_s and μ_s estimates. In our simulation setup,

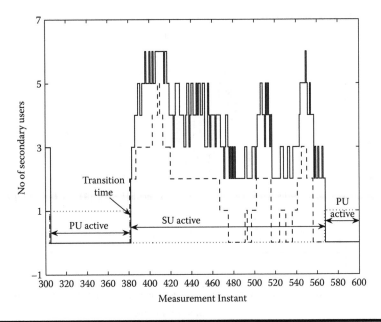

Figure 2.6 Performance of the predictor for SU; $(\cdots)$, $(--)$, and $(-)$ indicate the true activity of primary user, the true number of secondary users, and the predicted upper bound number of secondary users, respectively; $\beta = 0.1$.

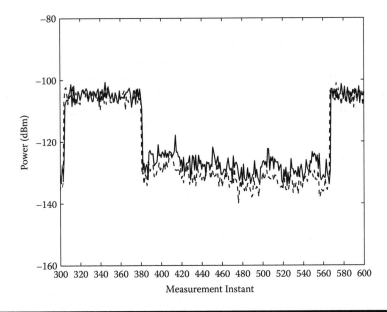

Figure 2.7 Variation of the predicted power level with the true power level: $(--)$ and $(-)$ indicate the true and predicted power levels, respectively.

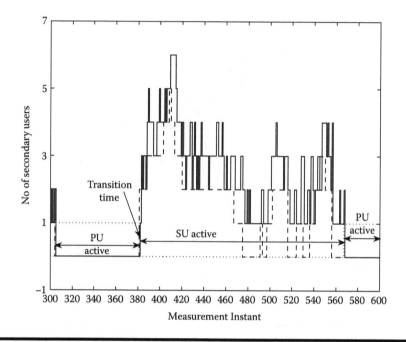

Figure 2.8 Performance of the predictor for the SUs: $(\cdots)$, $(--)$, and $(-)$ indicate the true activity of the primary user, the true number of secondary users, and the predicted upper bound number of secondary users, respectively; $\beta = 0.20$.

the sensitivity performance of the predictor for the SUs is evaluated by using erroneous values of λ_s and μ_s in Kalman filter estimation, instead of the values used to generate traffic. Figure 2.9 shows the predictor performance for the SUs with the true number of SUs where the value of λ_s is estimated as 25% less than its true value. From this figure, we observe that our proposed predictor continues to serve as a good upper bound for the number of SUs. Figure 2.10 shows the predictor performance where values of λ_s and μ_s are overestimated by the same amount; the performance of the predictor is still satisfactory. This shows that the predictor is relatively insensitive to inaccurate estimation of the values of λ_s and μ_s up to a certain limit.

2.5.2 Measured Data Analysis

In this setup, we show the effectiveness of the proposed techniques on power level measurements from the ISM band. Although this band is not allocated for CR operations, we use the measurements to illustrate how the proposed methods can be practically implemented. The ISM band power level measurements were taken in Chicago by Shared Spectrum Company [15].

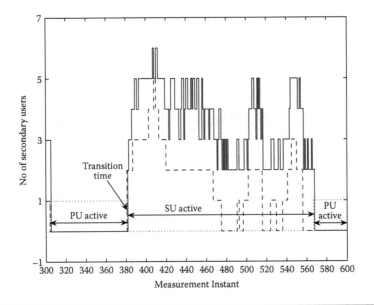

Figure 2.9 Sensitivity of the predictor for the SUs: λ_s is underestimated by 25%; $(\cdots)$, $(--)$, and $(-)$ indicate the true PU activity, the true number of secondary users, and the predicted upper bound number of secondary users, respectively.

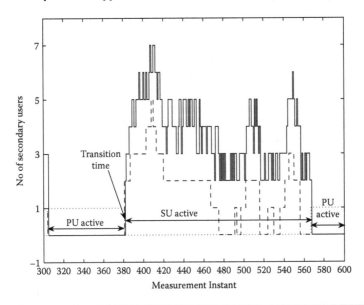

Figure 2.10 Sensitivity of the predictor for the SUs: both λ_s and μ_s are overestimated by 25%; $(\cdots)$, $(--)$, and $(-)$ indicate the true PU activity, the true number of secondary users, and the predicted upper bound number of secondary users, respectively.

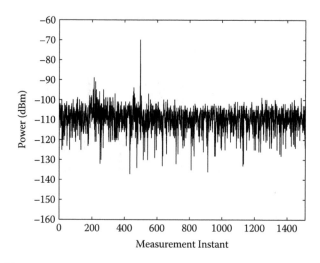

Figure 2.11 Received power level for carrier frequency 2412 MHz.

The ISM band ranges from 2.4 to 2.4835 GHz and is divided into 11 channels with carrier frequencies 2412, 2417, 2422, 2427, 2432, 2437, 2442, 2447, 2452, 2457, and 2462 MHz. This band is an unlicensed band. Since no primary user has exclusive access to these channels, we assume that the measured power is due to secondary users only. The received power measured for the 2412 MHz frequency is shown in Figure 2.11. The measurement interval, T, is 2 s. The data length is 1512. Any SU interested in using one of the channels needs to go through the "learning and modeling" phase first to find traffic parameters λ_s and μ_s for the corresponding channel. In this phase, the SU can use the time-series approach to get these parameters. Assuming that the models described in (2.15)–(2.16) (with no PU) hold, the received signal at the SU terminal can be written as

$$y(m) = A_s y(m-1) + C_s B_s + (1 - A_s)D + [C_s w_s(m) + v(m) - A_s v(m-1)].$$

(2.32)

Equation (2.32) indicates that the spectrum usage process is an autoregressive (AR) process of order 1. Any standard technique can be used to find the parameters of this AR process. In our case, we employ Yule–Walker estimation to find the model parameters. Employing some intuititive assumptions (e.g., $\lambda_s \leq \mu_s$, $\sigma_s^2 \leq \sigma_v^2$) along with Yule–Walker estimates—mean of the process, knowledge of D, N_s, C_s (considering N_s and C_s as regulatory constraints)—we can solve for all traffic parameters A_s, B_s, λ_s, μ_s, σ_s^2, and σ_v^2. Specifically, in our simulation, we solve for all parameters assuming $\lambda_s = \mu_s$ and $D = -130$ dB, and setting N_s and C_s to 20 and

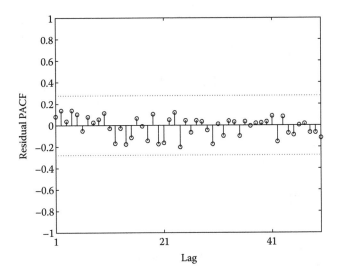

Figure 2.12   Plot of residual PACF values for carrier frequency 2412 MHz; the model parameters corresponding to this plot are $\lambda_s = \mu_s = 0.4544$ sec^{-1}.

2 dB, respectively. After finding model parameters, the fitness of the model is checked by finding residual partial autocorrelation function (PACF) values. In general, if 95% of the residual PACF values are within the acceptable bound, the AR model parameters are valid. An example plot of the PACF of the residuals (the differences between the true and the model-generated power level data) is shown in Figure 2.12, where ($\cdots$) and ($-$) indicate the acceptable PACF bound and residual PACF values, respectively. If the PACF values are satisfactory, the model parameters obtained in the "learning and modeling phase" are accepted and used in the "active phase."

In the "active phase," the SU attempts to use the channel opportunistically using the traffic parameters obtained in "learning and modeling phase." It is assumed that the accepted model parameters from "learning and modeling" phase do not change during our simulation time. Once again, (2.19)–(2.24) are used to estimate the number of secondary users, and on the basis of this estimation, (2.26)–(2.29) are used to predict the number of secondary users. Figures 2.13 and 2.14 show the comparison of the predicted number of users with the true number for two carrier frequencies, 2412 and 2437 MHz, respectively. These results show that the proposed forecasting technique also provides a good upper bound number of users for actually measured received power at a user terminal.

In summary, our simulation results illustrate that it is possible to effectively predict spectrum usage/availability in a CRN using a combination of modeling tools and filtering/estimation techniques.

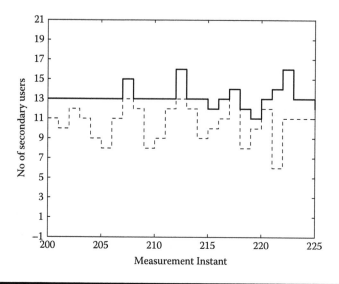

Figure 2.13   Performance of the upper bound predictor for carrier frequency 2412 MHz: (−−) **and** (−) **indicate the true and predicted upper bound number of users, respectively.**

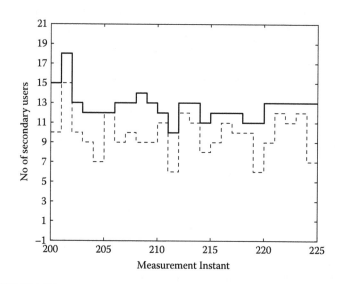

Figure 2.14   Performance of the upper bound predictor for carrier frequency 2437 MHz: (−−) **and** (−) **indicate the true and predicted upper bound number of users, respectively.**

2.6   Conclusion

In this chapter, we presented an integrated spectrum usage model and forecasting strategy for both primary and secondary users in CRNs. Assuming that primary and secondary users follow a continuous-time Markov chain model, we developed a Kalman filter approach to estimate the number of users based solely on power level measurements at the radio terminal of an opportunistic user. These estimates in turn were used to determine upper bound predictors for a future time instant. In a sense, this forecasting strategy provides secondary users not only the ability to predict whether a spectrum band will be used by a primary user, but also a means to determine whether a band will be overcrowded with secondary users. The effectiveness of the proposed architecture is demonstrated using experiments on both actually measured as well as simulated data.

References

[1]   Notice of proposed rule making and order (ET docket no. 03-322): Facilitating opportunities for flexible and reliable spectrum use employing cognitive radio technologies. Technical report, Federal Communications Commission.

[2]   Spectrum policy task force report (ET docket no. 02-135). Technical report, Federal Communications Commission.

[3]   Spectrum occupancy report for New York City during the Republican convention August 30-September 1, 2004. Technical report, Shared Spectrum Company.

[4]   J. Mitola and G. Q. Maguire, Cognitive radio: Making software radios more personal, *IEEE Personal Communications.* **6**(4), 13–18 (Aug. 1999).

[5]   D. Cabric, S. M. Mishra, and R. W. Brodersen. Implementation issues in spectrum sensing for cognitive radios. In *IEEE Asilomar Conference on Signals, Systems and Computers*, vol. 1, pp. 772–776 (Nov. 2004).

[6]   D. Cabric, A. Tkachenko, and R. W. Brodersen. Spectrum sensing measurements of pilot, energy and collaborative detection. In *IEEE Military Communications Conference*, pp. 1–7 (Oct. 2006).

[7]   A. Ghasemi and E. S. Sousa. Optimization of spectrum sensing for opportunistic spectrum access in cognitive radio network. In *IEEE Consumer Communications and Networking Conference*, pp. 1022–1026 (Jan. 2007).

[8]   P. K. Tang, Y. H. Chew, L. C. Ong, and M. K. Halder. Performance of secondary radios in spectrum sharing with prioritized primary access. In *IEEE Military Communications Conference*, pp. 1–7 (Oct. 2006).

[9]   S. S. Nandagopalan, C. Chou, K. Challapali, and S. Mangold. Spectrum agile radio: Capacity and QoS implications of dynamic spectrum assignment. In *IEEE Global Telecommunications Conference*, vol. 5, pp. 2510–2516 (Nov. 2005).

[10] Y. Xing, R. Chandramouli, S. Mangold, and S. S. Nandagopalan, Dynamic spectrum access in open spectrum in wireless networks, *IEEE Journal on Selected Areas in Communications*. **24**(3), 627–637 (Mar. 2006).

[11] Y. Xing and R. Chandramouli, QoS constrained secondary spectrum sharing, *First IEEE International Symposium on New Frontiers in Dynamic Spectrum Access Networks*. pp. 658–661 (Nov. 2005).

[12] Y. Xing, C. N. Mathur, M. A. Haleem, R. Chandramouli, and K. P. Subbalakshmi, Priority based dynamic spectrum access with QoS and interference temperature constraints, *IEEE International Conference on Communications*. **10**, 4420–4425 (June 2006).

[13] T. Anjali, C. Scoglio, and G. Uhl, A new scheme for traffic estimation and resource allocation for bandwidth brokers, *Computer Networks*. **41**(6), 761–777 (Apr. 2003).

[14] L. Kleinrock, *Queueing Systems*. John Wiley and Sons, 1975.

[15] M. McHenry, D. McCloskey, D. Roberson, and J. MacDonald. Chicago spectrum occupancy measurements. Technical report (Jan. 2006).

Chapter 3

Exploring Opportunistic Spectrum Availability in Wireless Communication Networks

Wei Wang, Xin Liu, and Hong Xiao

Contents

3.1   Introduction

The radio spectrum is among the most heavily regulated and expensive natural resources around the world. In Europe, the 3G spectrum auction yielded 35 billion dollars in England and 46 billion in Germany. The question is whether spectrum is really this scarce. Although almost all spectrum suitable for wireless communications has been allocated, preliminary studies and general observations indicate that much of the radio spectrum is not in use for a significant amount of time, and at a large number of locations. For instance, experiments conducted by Shared Spectrum Company indicate 62% percent "white space" (unused space) below the 3GHz band, even in the most crowded area near downtown Washington, DC, where both governmental and commercial spectrum usage are intensive.[1] In the experiment, a band is counted as white space if it is wider than 1MHz and remains unoccupied for 10 minutes or longer. Furthermore, spectrum usage levels vary dramatically in time, geographic locations, and frequency. A lot of the precious spectrum (below 5GHz), that is worth billions of dollars, and is perfect for wireless communications sits there silently. The large proportion of white space indicates that opportunistic or dynamic spectrum usage may significantly mitigate the spectrum scarcity.

In this paper, we focus on the *opportunistic exploration of the white space* by users other than the primary licensed ones on a noninterfering or leasing basis. Such usage is being enabled by regulatory policy initiatives and radio technology advances. First, both the Federal Communications Commission (FCC) and the federal government have made important initiatives towards more flexible and dynamic spectrum usage, e.g., Refs. [2–6]. Furthermore, opportunistic spectrum sharing is enabled by software-defined-radio or cognitive-radio technologies, where these technology advances provide the capability for a radio device to sense and operate on a wide range of frequencies using appropriate communication mechanisms, and thus enable dynamic and more intense spectrum reuse in space, time, and frequency dimensions.

We focus on the study of the secondary users who observe the channel availability dynamically and explore it opportunistically. Here, secondary users refer to spectrum users who are not the owners of the spectrum and operate on the basis of agreements/etiquette imposed by the primary users/owners of the spectrum. We study the impact of the opportunistic spectrum availability on the secondary users who explore the spectrum when allowed by the primary users of the spectrum. (Note that the secondary users may have their own licensed/allocated bandwidth where they are primary users, which is not the concern of this chapter.) Because of the traffic load and the distribution of the primary users, the available channels observed by the secondary users are time-varying and location-dependent. We study the impact of the characteristics of primary users on the spectrum's opportunistic availability. We present a general framework to model the correlation between primary and secondary users, and introduce a new metric to capture the impact of potential opportunistic spectrum sharing. We also propose several distributed spectrum access algorithms and study their performance under the above-mentioned time-varying and location-dependent channel availabilities.

3.1.1 Related Work

Channel allocation schemes have been studied in the literature for over three decades. Various static, dynamic, and hybrid channel allocation schemes have been proposed. In Ref. [7], the authors present a comprehensive survey of the channel assignment strategies. In addition, channel allocation has been modeled as a graph coloring, list coloring, or maximum packing problem (e.g., Refs. [7–14] and references therein).

The main differences between our work and the traditional approaches are multifold: (1) the allocation requirement is different, (2) channel availability is location-dependent and time-varying in this work, and (3) our algorithms may need to operate under the assumption of limited information exchange.

First, our objective is different. The traditional channel allocation problem is closely coupled with call admission control/handoff in circuit-switched cellular networks. When a call is generated (or a handoff occurs), a channel allocation algorithm is triggered to find a channel for the new call. The objective is to minimize the channel usage, while each call is assigned to one channel. Similarly, in most literature in list coloring, the objective is to find a coloring scheme such that each node can be assigned one channel from its channel list. On the other hand, the objective here is to fully utilize the available spectrum that is unused by the primary users while avoiding possible interference between neighboring nodes. The underlying assumption is that secondary users have elastic data traffic and can fully utilize the available bandwidth.

Second, the spectrum availability is different. In cellular networks, the available channels observed by the users are usually uniform and static because these channels are statically allocated to them. In our work, the available channels are location-dependent and time-varying because they are determined by the activities of the primary users. This makes our problem different from the network planning and graph coloring problems that are used to model the channel allocation in cellular networks.

Third, in our model, the channel availability may change before a channel allocation algorithm converges. Therefore, the channel allocation algorithms may have to work under scenarios where users have limited feedback information from their neighbors. The algorithm design needs to take this into account.

The multichannel wireless network is another emerging research area related to our work[15-18]. The major difference between our work and research in multichannel wireless networks is that the available channels in multichannel networks are also statically allocated for all the nodes thus they are not location-dependent or time-varying. Another difference is that we assume in our work that neighboring nodes cannot use the same channel simultaneously, while in multichannel networks, neighboring nodes can and need to share the same channel (the average throughput is lower if the channel is shared). Last, our focus is on the handling of the dynamics in channel availability.

3.1.2 Contributions

This chapter studies wireless networks with opportunistic spectrum availability and access. The contributions of our work are as follows:

■ We present a framework to model the *location-dependent* and *time-varying* channel availability observed by secondary users that is incurred by the activities of primary users. To the best of our knowledge, this is the first chapter to identify and model these two unique characteristics in such wireless networks. We also propose a new metric, the *effective nonopportunistic bandwidth*, to capture the inherent properties of white space.

■ Building on this framework, we formulate the channel allocation problem as a list coloring problem and develop several distributed algorithms for channel sharing. We validate their performance under scenarios with location-dependent and time-varying channel availabilities. It is worth noting that our algorithms work well under scenarios where secondary users have limited capability of information exchange.

3.1.3 Organization

The chapter is organized as follows: We first introduce the framework and a new performance metric in Section 3.2. Using the framework, we formulate the channel allocation problem as a graph coloring problem in Section 3.3. Then we propose several distributed approaches for opportunistic spectrum sharing with various degrees of complexity in Section 3.4. A numerical study is presented in Section 3.5, followed by conclusions in Section 3.6.

3.2 A Framework for Opportunistic Spectrum Sharing

We first introduce a model for the channel availability observed by the secondary users. Note that such availabilities are location-dependent and time-varying, which is incurred by the activities of the primary users. We abstract each network topology into a graph, where vertexes represent wireless users such as wireless lines, WLANs, or cells, and edges represent interference between vertexes. In particular, if two vertexes are connected by an edge in the graph, we assume that these two nodes cannot use the same piece of spectrum simultaneously. In addition, we associate with each vertex a set, which represents the available spectra at this location. Because of the differences in the geographical locations of the vertex, the sets of spectra of different nodes may be different. Furthermore, a node may observe time-varying channel availability due to the traffic load variation of the primary users.

In Figure 3.1, we show a model of such a network. The five circles labeled 1–5 represent five different secondary or opportunistic users. There are three frequency bands, namely A, B, and C, which are communication channels that are opportunistically available to the secondary users (1–5 in this figure). We assume that all channels have the same bandwidth, which can be generalized easily. In addition, four primary users I–IV are present, using bands B, A, B, and C, respectively. Due to the sharing agreement, channels used by primary users cannot be utilized by secondary users in the vicinity. Therefore, we assume that nodes within a certain range of each primary user I–IV cannot reuse the same frequency. In other words, if a vertex is within the dashed circle of a specific primary user, it cannot access that band used by the primary user. For instance, node 2 is within the interference range of primary user I, who uses channel B. Therefore, channel B is not available for node 2. As a consequence, each node has access to

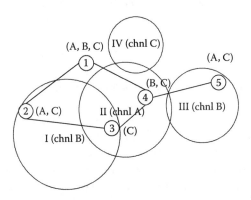

Figure 3.1   Color graph of network topology.

a different set of spectrum bands. In our figure, the available channels are (A,B,C) at vertex 1, (A,C) at vertex 2, etc. The resource allocation problem is how they should share these bands.

Note that Figure 3.1 shows a snapshot of the network. Time-varying channel availability is introduced by the mobility of users (both primary and secondary) and the traffic load variation of primary users. For instance, in the numerical results, we introduce the time-varying channel availability for secondary users by varying the usage of primary users. We consider a time-slotted system. In a generic time slot, if a primary user occupies one channel, it will keep the same channel in the next time slot with probability p_{11}; if a primary user is idle on one channel, it will occupy a channel in the next time slot with probability p_{01}. Then the channel availability of a secondary user varies at each time slot depending on all the primary users. Such a policy can be considered as an approximate exponential ON/OFF traffic model for the primary users. Other traffic models can be introduced similarly.

When channel availabilities change, secondary users need to adjust their channel allocation accordingly. They may also need to exchange information with neighboring nodes. However, secondary users may have limited capability of information exchange and experience delay during information exchange because secondary users coexist in an *ad hoc* manner. This is different from cellular systems where dedicated (and private) signaling channels exist between cells. Limited information exchange imposes an additional challenge on opportunistic channel sharing algorithms.

Our framework can be generalized so that more sophisticated and realistic scenarios are taken into account. For instance, a more sophisticated interference model can be introduced, such that two connected secondary users may be able to use the same channel but with lower data rates because of the co-channel interference. In the model described above, the underlying assumption is that the size of the node, which could be a WLAN or a cell, is small compared to the distance between neighboring nodes. On the other hand, the location and transmission power of each user in a WLAN can be taken into account at the cost of complexity. In addition, we can generalize the communication range of a user from a unit disk to a more general shape due to the randomness in the propagation environment.

We make the following assumptions in the paper. We assume that channel availabilities are given to secondary users, and focus on how to share the opportunistic spectrum among these users. We acknowledge that it is a challenging problem in itself to decide whether a spectrum can be used by the secondary users. The proposals for solving this problem include carrier sensing, signal-strength-based methods, beacon-based methods (a beacon indicates the availability or unavailability of a spectrum), location- and database-based mechanisms (a node has location information and can check a database about the availability of a spectrum), etc. Furthermore, we assume that secondary users have elastic data traffic and can fully utilize the spectrum being allocated. This is the benchmark case in terms of spectrum requirements. On the other hand, our proposed algorithms can easily adapt to the cases where nodes have (different) limited requirements. In particular, a node can stop requiring more channels after all its needs have been satisfied.

3.2.1   Effective Nonopportunistic Bandwidth

Consider the following question: suppose that there is 62% white space under 3G. If we can fully utilize such white space, is it equivalent to gaining an additional spectrum band of 0.62×3GHz? The answer is that it depends. To address this question, we introduce the notion of *effective nonopportunistic bandwidth*. This is defined as the equivalent nonopportunistic bandwidth required to achieve the same performance as in the case of opportunistic spectrum availability. A nonopportunistic channel is referred to as a channel that is always available to all (secondary) users in consideration, which is how spectrum is allocated in the traditional command-and-control manner. This metric is designed to study the impact of the opportunistic availability of the spectrum.

We elaborate the idea with a naïve example next. Consider a simple network with only two nodes, which cannot use the same channel simultaneously due to interference. Consider a channel with bandwidth W that is opportunistically available to these two nodes. Assume the channel is available at each node with probability p independently. Suppose that a user obtains one unit of throughput per unit of spectrum. Then the total throughput gained by the two users is:

$$W(p^2 \times 1 + 2p(1 - p) \times 1 + (1 - p)^2 \times 0) = Wp(2 - p).$$

The first term is the case where the channel is available to both nodes and only one of them can use it due to interference. The second term is the case where only one of the users has the spectrum availability and uses it. The last term is the case where the spectrum is available to neither of the two users. To achieve the same total throughput, $B_e = Wp(2 - p)$ units of nonopportunistic bandwidth are required. To elaborate, when a spectrum of B_e is available to both users in the traditional way (i.e., the spectrum is always available to both users), the throughput is $Wp(2 - p)$ because only one of the users can use it at any given time. Thus, we claim that $B_e = Wp(2 - p)$ is the effective nonopportunistic bandwidth in this simple example. Suppose $W = 3$GHz and $p = 62\%$. In this example, we see that the impact of 62% white space under 3G is equivalent to $Wp(2 - p) = 2.76$GHz of spectrum in the traditional way.

Note that spectrum is not being "created" by secondary users. Instead, they simply explore the spectrum holes generated by the given usage pattern of primary users. Primary users are legacy users, whose behavior we do not change. The inherent characteristics of the primary users, such as communication range, transmission power, traffic pattern, node density, and topology, determine the spectrum opportunities of secondary users. The notion of effective nonopportunistic bandwidth is a metric to quantify the potential of such spectrum opportunities for a given topology of secondary users.

The intuition is similar to that of the effective bandwidth used to capture the statistical multiplexing gain. However, what is being captured here is the degree of spatial reuse given the correlation of spectrum availabilities at secondary users. For instance, in the above example, if two secondary users are very close and also observe the same spectrum opportunity, then $B_e = Wp$ instead of $Wp(2-p)$ in the independent case. In general, because of the characteristics of primaries mentioned above, users observe different channel availabilities, which in general yield higher gain, which is quantified by the effective nonopportunistic bandwidth.

The above example is oversimplified, but serves to illustrate the idea. In general, the effective nonopportunistic bandwidth is difficult to calculate because the network topologies are more complicated and the availabilities of channels at different nodes are correlated. Thus, we quantify such effects using numerical results in Section 3.5. In summary, this metric is introduced to capture the inherent impact of opportunistic spectrum availability.

3.3 Problem Formulation

Using the model described earlier, we formulate the channel allocation problem as a graph coloring problem. We abstract the network as an undirected graph $G = (V, E, L)$, where vertexes represent users and edges represent interference, so that no channels (frequency bands) can be assigned simultaneously to any adjacent nodes. For simplicity, we assume that the interference graph is the same for all frequency bands. This can be generalized to the case where each frequency has its own interference graph, a possible scenario due to the different propagation properties in the environment associated with individual bands. Furthermore, let K be the number of available channels in G. Although it is possible that different channels have different bandwidths, we treat all channels the same for simplicity. We also refer to the graph G as the interference graph. In the paper, we use "channel" and "color" interchangeably.

Let $N = |V|$ denote the total number of users. Let edges be represented by the $N \times N$ matrix $E = \{e_{ij}\}$, where $e_{ij} = 1$ if there is an edge between vertexes i and j, and $e_{ij} = 0$ implies that i and j may use the same frequencies. Note that since G is an undirected graph, E is symmetric. In a similar notation, we represent the availability of frequencies at vertexes of G by an $N \times K$ matrix $L = \{l_{ik}\}$, which we refer to as the coloring matrix. In particular, $l_{ik} = 1$ means that color (channel) k is available at vertex i, and $l_{ik} = 0$ otherwise. For instance, Figure 3.1 is represented by the matrices

$$E = \begin{bmatrix} 1 & 1 & 0 & 1 & 0 \\ 1 & 1 & 1 & 0 & 0 \\ 0 & 1 & 1 & 1 & 0 \\ 1 & 0 & 1 & 1 & 1 \\ 0 & 0 & 0 & 1 & 1 \end{bmatrix}, \qquad L = \begin{bmatrix} 1 & 1 & 1 \\ 1 & 0 & 1 \\ 0 & 0 & 1 \\ 0 & 1 & 1 \\ 1 & 0 & 1 \end{bmatrix}.$$

Let us denote a color/channel assignment policy by an $N \times K$ matrix $S = \{s_{ik}\}$, where $s_{ik} = 0$ or 1, and $s_{ik} = 1$ if color k is assigned to node i, and 0 otherwise. We call S a feasible assignment if the assignments satisfy the interference graph constraint and the color availability constraint. More specifically, for any node i, we have $s_{ik} = 0$ if $l_{ik} = 0$ (i.e., a color can be assigned only if it is available at the node). Furthermore,

$$s_{ik} s_{jk} e_{ij} = 0, \quad \forall i, \quad j = 1, \ldots, N, \quad k = 1, \ldots, K.$$

In other words, two connected nodes cannot be assigned the same colors.

The objective of the resource allocation is to maximize the spectrum utilization. This problem can be formally represented as the following nonlinear integer programming problem.

$$\underset{S}{\text{maximize}} \quad \sum_{i=1}^{N} \sum_{k=1}^{K} s_{ik} \qquad (3.1)$$

$$\text{subject to} \quad s_{ik} \leq l_{ik},$$

$$s_{ik} s_{jk} e_{ij} = 0,$$

$$s_{ik} = 0, 1,$$

for all $i, j = 1, \ldots, N, k = 1, \ldots, K$. The above problem is sometimes referred to as a list multicoloring problem. When time is taken into account, a time index can be introduced into the equation where the objective is to maximize the utilization averaged over time and the three constraints are satisfied at each time instance.

The corresponding decision list coloring problem is formulated below.

Problem 3.1 *(DListColor Problem) Given a graph $G = (V, E, L)$ and a positive integer B, is there a solution such that*

$$\sum_{i=1}^{N}\sum_{k=1}^{K} s_{ik} > B, \qquad (3.2)$$

with the same set of constraints as in Equation (3.1)?

Proposition 3.1
The DListColor problem is NP-complete.

PROOF This problem is clearly in NP since once a valid coloring assignment S is obtained, condition (3.2) may be verified in $O(|V| \cdot K)$ time.

We now show that the *maximum clique* problem* can be reduced to the *DListColor* problem in polynomial time, and that the maximum clique problem has a solution if and only if *DListColor* has a solution.

Let $G = (V, E)$ be the undirected graph of the maximum clique problem. We construct the graph $G' = (V', E', L)$ for our *DListColor* problem, such that $V' = V$, and E' is the complementary set of E. Furthermore, the color matrix L is of dimension $|V| \times 1$, where $L = [1, 1, \dots, 1]^T$. Since any pair of nodes connected in G are not connected in G' and vice versa, we cannot simultaneously assign nodes in G' the same color if these nodes form a clique in G. Therefore, there exists a clique in G of size at least m if and only if there is no solution for *DListColor* for $B = |V| - m$. This reduction is obviously polynomial-time. ☐

3.3.1 Color Decoupling

The list coloring problem may be reduced to a set of maximum-size clique problems when fairness is not a consideration. In other words, in the process of finding the maximum in Equation (3.1), nodes are allowed to be assigned zero channels. The problem of assigning each node a set of colors may be solved by coloring the graph in sequence with individual colors:

$$\underset{S}{\text{maximize}} \sum_{i=1}^{N}\sum_{k=1}^{K} s_{ik} \Leftrightarrow \sum_{k=1}^{K} \underset{S_k}{\text{maximize}} \sum_{i=1}^{N} s_{ik}, \qquad (3.3)$$

where S_k denotes the channel allocation with respect to channel (color) k. More specifically, S_k is the kth column in the assignment matrix S. Note that the equality in (3.3) does not hold in general situations, e.g., a graph

---

* A clique is a fully connected subgraph; i.e., a clique consists of a set of nodes any pair of which has an edge in between.

coloring problem that requires each node to be colored with nonempty colors. Note that when fairness is taken into account, e.g., each node has to be assigned at least one color, then the decoupling property does not apply.

3.4 Proposed Algorithms

In this section, we discuss several approaches to the resource allocation problem formulated above. We prefer distributed algorithms because of their robustness and scalability. We use a brute force search algorithm with which we find optimal solutions to serve as a benchmark. Because the resource allocation problem is NP-complete, optimal solutions may only be found when graphs are relatively small. We then present a distributed greedy algorithm, a distributed fair algorithm, and a distributed randomized algorithm, with various complexity and performance.

3.4.1 Optimal Solutions: Benchmark

Given the list coloring problem and a graph $G = (V, E, L)$, we seek the solutions to optimization problem (3.1). As discussed in Section 3.2, the optimization problem is NP-complete. Therefore, in order to find the optimal solution(s), we must search through all valid color assignments, and find the one(s) that maximizes (3.1).

We carry out this search in a breadth-first recursive manner, with the starting node chosen arbitrarily. More specifically, when the node i is visited, we enumerate all combinations of channel allocations for this node permissible by the available channels at i, and iterate through each configuration and tentatively assign it to the node. If there is a conflict between the current assignment attempt and neighboring nodes whose channels have already been selected, we abort this assignment. The complexity of this algorithm is $\prod_i^N 2^{k_i} = O(2^{NK})$. This algorithm is easily modified to find optimal solutions under additional constraints such as fairness, variance of resource allocation, etc. For example, suppose that we seek to maximize (3.1) but with the constraint that each node has at least one channel assigned (if such an assignment exists). We may modify the process so that zero channel assignment is not a permissible option.

The complexity can be reduced if we use the color decoupling property described in Section 3.3.1 and modify our search algorithm so that the color assignment is optimized subsequently for each color. The resulting algorithm has a complexity of $O(K \, 2^N)$. However, this reduction of complexity is not achieved without penalty. For example, since channels are assigned independently of each other, it is no longer straightforward to find a policy that assigns each user at least one channel.

3.4.2 Distributed Greedy Algorithm

Because the resource allocation problem is NP-complete, heuristics are needed to study large graphs. In this section, we present a distributed greedy algorithm with the objective of maximizing the utilization.

Because of the color decoupling property discussed in Section 3.3.1, the distributed greedy algorithm handles colors one by one. For each color, a greedy assignment is calculated to maximize the number of nodes assigned to this color. To elaborate, consider the assignment of the color i. A subgraph $G_i = \{V_i, E_i\}$ is generated, where a node belongs to V_i if and only if it belongs to V and color i can be used at the node. An edge connects two nodes in V_i if and only if these two nodes are connected in the original graph G. For instance, for the case presented in Figure 3.1, the subgraph for color A consists of nodes 1, 2, and 5, and a link between 1 and 2.

The following is the description of the greedy algorithm.

1. Each node looks up the color i in its available color list. It finds the nodes in its neighbor list that also have this color available, denoting this sublist as N_i (including itself).
2. If its link degree is the smallest in N_i, it picks the color i. Note that the link degree here means the number of nodes it connects to that also have this color available. If not, it will look up the next available color and go back to (1).
3. If one or more neighboring nodes has the same smallest link degree (a tie exists), it uses its color degree to break the tie. The color degree means the number of available colors a node has. If its color degree is the smallest among the nodes with the tie, it will also pick color i. Otherwise, it will look up the next available color and go back to (1).
4. If a tie again exists, the nodes involved will generate random numbers independently. If this node has the smallest random number, it will pick the color. Otherwise, it will switch to the next available color and go back to (1).
5. It stops when all of its available colors are processed.

The greedy algorithm performs as if nodes are ranked according to their link degrees from low to high *for each color*. Then the color is assigned to the nodes according to their link degrees from low to high. When a tie exists, the number of assigned colors of each node is used to break the tie. Nodes with fewer assigned colors have higher priority. If the nodes have the same number of assigned colors, ties are broken randomly. The worst-case complexity of the algorithm is $O(KN^2)$, where K is the number of colors and N is the total number of nodes. The algorithm can result in very unfair allocation. Nodes with lower link degrees will obtain more resources in general.

3.4.3 Distributed Fair Algorithm

As discussed in the previous section, the greedy algorithm can result in very unfair allocation. In this section, we discuss a distributed algorithm with fairness considerations. The algorithm has three steps.

Step 1 is to build an acyclic directional graph. The building of the acyclic directional graph is motivated by Ref. [9], although the link degrees of nodes are not taken into account in Ref. [9] and no iteration is required due to the difference in objectives. All nodes exchange information about their link degree, color degree (color degree is the number of available colors at each node), and a random number. The edges are oriented from higher color degree to lower; i.e., nodes with a smaller number of colors are the receivers. If two connected nodes have the same number of colors, the edge is oriented from higher link degree to lower link degree. If there is a tie again, the edge is oriented from the node with the larger random number. Thus, an acyclic graph is generated. Figure 3.2 shows the graph for the case presented in Figure 3.1. A node is a sink node if there is no edge oriented from it. A node is a source node if there is no edge oriented to it. Note that it is possible that there are multiple source and sink nodes. Node x is node y's upstream neighbor if the edge is oriented from x to y, and downstream otherwise. Coloring starts from sink nodes to source nodes as described in Step 2.

The heuristics of the direction is as follows: nodes with more available colors should yield to nodes with fewer colors to maintain a balance. When this ties, link degrees are used to break the tie. A node with lower link degree has higher priority so that fewer nodes are affected by its color selection. If it ties again, then the random numbers chosen by the nodes are used to break ties. Note that each node has one fixed random number in each iteration to avoid loops. In comparison, the fair algorithm takes into account the color degree first, and link degree second, while the greedy algorithm takes the opposite order. This results in different performance in terms of fairness and throughput.

Step 2 is to assign colors. At most one color is assigned to each node in one iteration. The color assignment starts with sink nodes. If a node is

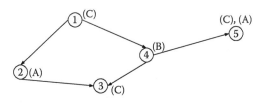

Figure 3.2 Graph.

a sink node, it picks a color that can be used by the minimum number of neighbors. It updates this information to all neighbors, who remove the color from their available color lists. In addition, a set-color token is passed to the neighbors to enable their color searching. If a nonsink node obtains set-color tokens from all its downstream neighbors, the node becomes a sink node and repeats the same process. Thus, the color selection is performed from sink to source nodes step by step.

Step 3 is to start the next iteration if needed. After a source node performs the process, it generates a reset token and passes it to all its neighbors. If a node receives reset tokens from all its upstream neighbors, it sends a reset token to all its downstream neighbors. After all nodes receive a reset token, the system resets and goes to step 1 with the remaining colors. If a node is out of available colors, it quits the operation. When no nodes have available colors remaining, the operation stops.

As shown in Figure 3.2, in the first iteration, sink nodes 3 and 5 select colors first (both select color C). Then the tokens are passed to nodes 2 and 4, who select colors A and B respectively. Last, the source node 1 selects color C. After the reset process, explained in Step 3, the second iteration begins, where node 5 picks color A and ends the process. Note that in each iteration, a new graph is generated to reflect the remaining color availability.

In each iteration, at least one color will be assigned to a node. Thus, there are at most $O(NK)$ iterations. Assume the maximum link degree is Δ. The complexity of the scheme in each iteration is $O(\Delta NK \log K)$. Thus, the worst-case complexity is $O(\Delta N^2 K^2 \log K)$. In general, in each iteration, more than one color is assigned and the average complexity is better. The communication cost per iteration is $O(N)$ where broadcast is assumed.

3.4.4 Randomized Distributed Algorithm

The above distributed algorithms may need a large number of iterations when a large number of nodes and colors are involved. When the size of the network is large, large communication overhead will occur. So the distributed randomized algorithm is proposed to reduce the delay and communication cost. The algorithm is inspired by the IEEE 802.11 backoff algorithms in MAC protocols, although the objective is different. In 802.11, the station doubles its contention window to reduce the probability of collision when its transmission fails. In our algorithm, a node will increase its chance to win the next color when it fails to get a color contending with its neighbors. The description of the randomized algorithm is as follows:

(1) Each node first generates a random number for each of its available colors uniformly distributed from [0, *window*]. At the beginning, window is set to START, which is a relatively small value (e.g., 1).

Then, each random number is divided by the color degree of a node.

(2) Nodes exchange the random numbers and the available colors with their neighbors.

(3) Within one round, the node goes through all its available colors. If it has the highest random number among all its neighbors for one color, it will win the contention and be able to use the color from then on.

(4) At the beginning of the next round, each node exchanges information with its neighbors on what colors it gets. Each node deletes the colors that are obtained by its neighbors from its available color list. Each node then updates its window by the following rule: if it loses one color, it doubles its window (i.e., $window = 2 * window$); if it wins one color, it divides its window by 2 (i.e., $window = window/2$).

(5) There is also a STOP value for the window evolution. The window value cannot exceed the STOP value even if a node keeps losing contentions. STOP can be several times the value of START. For example, STOP = 8 * START.

(6) After recomputing the window, each node regenerates the random numbers for its remaining colors and goes into the next round.

The randomized distributed algorithm has a small communication overhead. It also converges faster compared with the distributed fair algorithm, especially when the number of nodes and colors are large. For the first set of random numbers, we make it inversely proportional to the color degree of each node so that the node with a lower color degree should have a higher chance to pick colors in the first round. In the following rounds, the intuition is that the node that loses the contention should have a higher chance to win in the next round. On the other hand, the node that wins the contention should lower its chance to win in the next round. We may want to limit the window evolution within a certain range to avoid the situation that a node may have a very high chance to win its remaining colors because it loses many colors in the previous round. Thus, a certain level of utilization and fairness can be achieved.

3.5 Numerical Results

In this section, numerical results are used to illustrate the impact of location-dependency and time-variance on the channel availability, and evaluate the performance of the proposed algorithms. We focus on the following aspects: (1) the potential of opportunistic resource availabilities vs. fixed resource availability, (2) the impact of the characteristics of primary users, and (3) the effect of time-varying channel availability and limited information exchange on the proposed algorithms. We first study a snapshot of the network. This means that the channel availability is fixed during the execution of the algorithms. Thus, the impact of the location-dependency on the channel availability is illustrated. Then we will introduce the time-variance into the channel availability and evaluate its impact.

We measure utilization by computing the average number of channels (colors) assigned to each vertex, which is also referred to as utility. The fairness metric is the variance of channel allocations in each assignment. The smaller the variance, the fairer the assignment. If all nodes obtain the same number of channels in an assignment, then the fairness metric is zero. In general, the fairness metric is not zero even if the assignment is max-min fair, because nodes have different available channels and it may be impossible for all nodes to obtain the same number of channels. The third metric used is the effective nonopportunistic bandwidth discussed in Section 3.2, which is defined as the equivalent nonopportunistic bandwidth required to achieve the same performance as the opportunistic spectrum band use.

3.5.1 Simulation Setup

Given a topology geographically, we use the following model to generate a snapshot of the channel availability for the nodes of interest. Each channel has N_I primary users (also referred to as interferers) that are uniformly distributed in a unit square area. With probability p_I, each of the primary users is active, independently of other users. A channel is available at a vertex (the secondary user) if and only if it is not within the interference range (R_I) of any active primary users of the channel. We generate the availability of K different channels independently. This is similar to the scenarios shown in Figure 3.1, except that all primary users have the same interference range, R_I. The purpose here is to generate correlated channel availability profiles; i.e., nodes closer to each other are likely to have similar channel availabilities. When a node in the graph represents a cell with a positive geographical size, such a generation scheme is less precise because the size of the cell is not taken into account. Nonetheless, the main purpose is to generate correlations among the channel availabilities of nodes, and thus similar results should be observed. We will present the way to generate the time-varying channel availability in a later part of the discussion.

We illustrate the results in two fixed topologies. In the fixed topology, unless otherwise specified, we set $R_I = 0.1$, $p_I = 0.2$ and K, the total number of channels in the network, to be 15. The first topology is a simple six-node ring as shown in Figure 3.3. The objective here is to study symmetric topologies, where all nodes are identical: each node has two

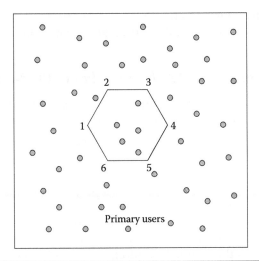

Figure 3.3   Ring topology.

neighbors and the same characteristics of available colors (in a statistical sense). Although not shown here, we observe that all six users obtain roughly the same amount of bandwidth averaged over simulations.

The second topology shows a combination of symmetric and asymmetric nodes, as in Figure 3.4. In particular, node 4 is in the worst position, as it interferes with all other nodes. Nodes 1–3 and 5–6 are symmetric, as are nodes 7 and 8.

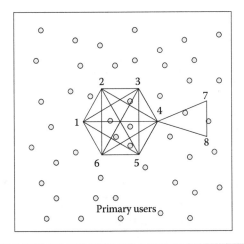

Figure 3.4   Hexagon-triangle topology.

We note that simulations have also been performed using a large number of random topologies. Similar observations and conclusions have been yielded and thus are not included.

3.5.2 Performance under the Snapshot of the Network

We first show the performance of the proposed algorithms under the snapshot of the network. We generate 100 different snapshots of the network for each topology. The performance is averaged over all nodes and over all snapshots.

Ring: Figure 3.5 compares the average performance (average over all nodes and over all simulations) of the four algorithms with different sets of parameters in the ring topology. The first subplot compares the average utility (number of channels assigned per node) of the four algorithms. We observe that in all cases, the greedy algorithm can always achieve the same utilization as the optimal solution, followed by the randomized and fair algorithms. With the increase of N_I, each node observes fewer available channels ('avgcolor'), and thus lower average utility. As an illustration of fairness, the second subplot shows the variance among different nodes

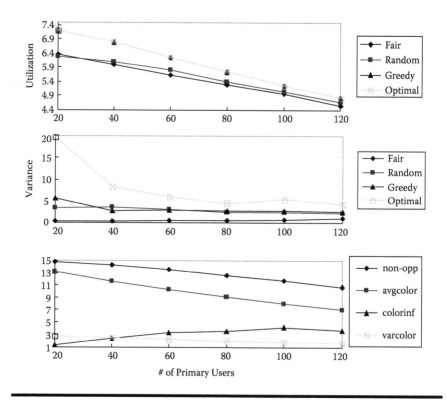

Figure 3.5 Ring topology performance.

of the four algorithms, which is the variance of each allocation scheme averaged over 100 simulations. The fair algorithm has the lowest average variance value and thus claim to being the most fair. The random and greedy algorithm have similar performance. In summary, in the ring topology, the greedy algorithm can achieve the best utilization at the cost of relatively high variance. The fair algorithm can achieve the smallest variance, but also with the smallest utilization. The randomized algorithm has the most moderate performance among the three. The advantage of the randomized algorithm is that it has the smallest communication overhead and convergence time. Such performance reflects the differences in the design principles of the three algorithms.

The last subplot includes the average number of channels available per node, its variance, the average number of neighbors sharing the color, and the effective nonopportunistic channel. They are "avgcolor," "varcolor," "colorinf," and "non-opp," respectively, in the legend. The first three metrics reflect the characteristics of the channel availabilities. In particular, "avgcolor" is the average number of channels available per node. It is equivalent to 3×0.62GHz in the case where there is 62% white space under 3GHz. It is a measure of channel availability. Its comparison with effective nonopportunistic bandwidth, denoted by "non-opp," is to help us understand the question of whether or not that 62% of white space under 3GHz is equivalent to an additional band of 3×0.62GHz. The second metric, "varcolor," is the variance of color availability in each snapshot. The third metric "colorinf" is related to the likelihood a node obtains a channel. For instance, if a channel is available to nodes 1–3, then the value at node 2 is 3 (including itself, three nodes will share the channel). The larger the value is, the less the average utilization (for a given topology) is due to higher-level competitions. The metric drawn in the figure is averaged over all nodes and all colors.

In this topology, the effective bandwidth is calculated as follows: one nonopportunistic frequency that is available to all nodes can be used by every other node. Thus, to assign an average of m channels to each node, the required number of effective nonopportunistic channels is $2m$. When the number of interferers is small (e.g., $N_I = 20$, the leftmost node), a color is likely to be available to most nodes, and thus the average channels per node is 13 with a small variance. Recall that the total number of channels is 15. In this case, the value of the effective nonopportunistic channel is close to that of average available channels. As the number of interferers increases, the channel availability to different nodes varies more, and the difference between the effective bandwidth and the average number of channels available per node is larger. This indicates that *the impact of the opportunistic bandwidth allocation is more significant when the variance of the channel availability is higher.* This is intuitive and is also observed in other simulation results.

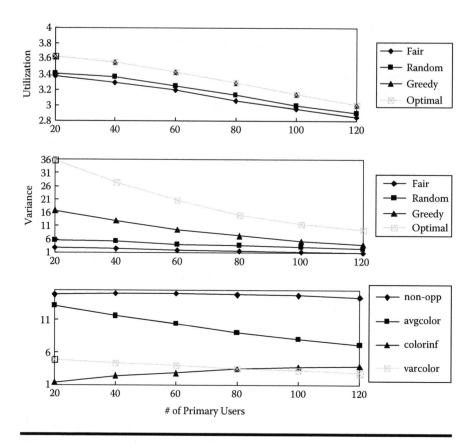

Figure 3.6 Hexagon-triangle topology performance.

Hexagon–triangle: Figure 3.6 shows the result of the simulation of the hexagon–triangle (H–T) topology using the same set of parameters as in Figure 3.5. The figure is drawn using the same notations. The effective bandwidth in the H–T topology is calculated as follows:

$$B_e = B_4 + \max\left(\sum_{i=1,2,3,5,6} B_i, \sum_{i=7,8} B_i\right),\qquad(3.4)$$

where B_i is the average number of channels of node i. The rationale is that when node 4 is using a channel, no one else can use it. Similarly, only one of nodes 1, 2, 3, 5, 6 can use the channel, and one of nodes 7 and 8 can use the channel. In this simulation, the value of the utility of the fair algorithm is used to calculate the effective bandwidth shown in the third subplot, which is in fact a lower bound of the effective bandwidth because we cannot claim the optimality of the fair algorithm.

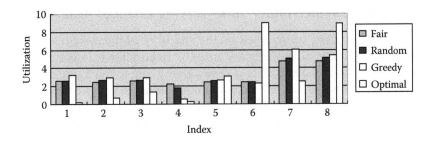

Figure 3.7 Hexagon–triangle topology bar graph.

For the utilization and variance performance, compared with the ring topology, the relative order of the three algorithms is more obvious. The greedy algorithm can achieve the best utilization at the price of the highest variance. The fair algorithm has the lowest variance, but also the lowest utilization. The randomized algorithm performs in between.

To further understand the principle of the algorithms, let us take a look at Figure 3.7. It shows the performance of different algorithms at different nodes. The four bars from left to right are the results of the fair, randomized, greedy, and optimal algorithms. Since node 4 is the node that interferes with the greatest number of other nodes, the optimal and greedy algorithms allocate the least amount of the spectrum to it in order to maximize the total utilization. In contrast, the fair and randomized algorithms allocate much more of the spectrum to node 4 so that these algorighms are more fair than the optimal and greedy algorithms. For the fair algorithm, node 4 has a slightly lower allocation compared with other nodes because it is more likely to run out of colors first. A similar result is also observed for the randomized algorithm, where node 4 needs to compete with 8 other nodes.

For the last subplot in Figure 3.6, we can see that the effective bandwidth decreases much slower than that of the average utility per node as N_I increases. This is due to the fact that the term $\sum_{i=1,2,3,5,6} B_i$ decreases slowly as N_I increases because there is always a good chance a channel is available to at least one of these five nodes. This illustrates that the impact of opportunistic channel availability is more significant on dense topologies (the clique of nodes 1–6) than on sparse topologies in terms of the difference between the effective bandwidth and the average channel availability.*

---

* It is arguable that the effective bandwidth should be calculated using the total utility and should not distinguish among nodes. Our choice is to emphasize the variability of the topology.

3.5.3 Performance under Time-Varying Channel Availability

In the above studies, we focused on the snapshot of the system where the channel availability of the primary users is fixed after it is generated. In practice, the channel availability is time-varying due to the traffic activity of the primary users. So it is essential to study the impact of the time-varying channel property on the performance of the proposed algorithms.

We introduce the time-varying channel availability at secondary users by varying the usage of primary users as discussed in Section 3.2. In the simulation, we set $p_{11} = 0.8$ and $p_{01} = 0.2$. Such a policy can be considered as an approximate exponential ON/OFF traffic model for the primary users with an average ON/OFF period of 5 time slots.

We assume that in each time slot, secondary users can only exchange information once. This can happen when the changes in channel availability are relatively fast and the information exchange channel between secondary users has limited bandwidth and/or experiences delay(s). This constraint is imposed to test the "real-time" performance of the proposed algorithms that may not converge in one iteration. That is, after one time slot, not all the available channels will be assigned. We call this partial assignment real-time performance. We are interested in the difference between the real-time performance and offline converged performance, i.e., the difference between the partial channel allocation done in one time slot and the converged result, where the latter is what we get under the snapshot of the network. Such a difference reveals the impact of the time-varying channel availability on the performance of the proposed algorithms. The smaller the difference, the better the performance when the proposed algorithms are used under time-varying channel availabilities.

Figure 3.8 and Figure 3.9 show the performance comparisons among three distributed algorithms in the ring topology and H-T topology, respectively. We do not show the performance of the optimal algorithm because it only serves as a benchmark under the snapshot of the network. In the simulation, the total number of channels is still 15, and the number of primary users is 80.

As shown in the figures, the randomized algorithm has the smallest difference in utilization between the real-time and converged performance, followed by the greedy algorithm. The fair algorithm has the largest difference between the real-time and converged performance. Recall that the randomized algorithm is designed with the objective of small communication overhead and short convergence time. It means that the randomized algorithm allocates most of the available channels in the first iteration. On the other hand, due to the complexity of the fair algorithm, it can only allocate a small portion of its available channels in one iteration, leading to the large difference. The small variance of the fair algorithm is caused by the small value of the assigned colors. Since the variance is defined as

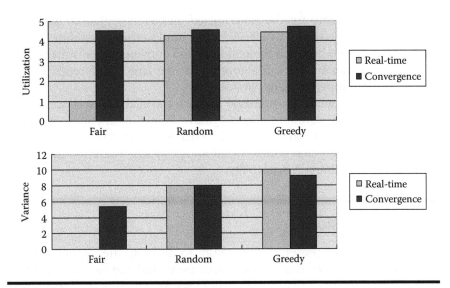

Figure 3.8   Real-time vs. convergence in the ring topology.

the standard deviation, the variance could also be small when the value of the assigned colors is small.

From the above results, we can conclude that the time-varying channel availability does affect the performance of the proposed algorithms. But the difference between real-time and converged performance is not

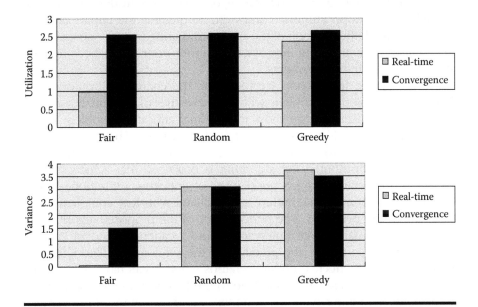

Figure 3.9   Real-time vs. convergence in the H-T topology.

very large, especially for the randomized and greedy algorithms. Under the time-varying channel availability, the randomized algorithm has the smallest difference between the real-time and converged performance due to its low complexity. The fair algorithm has the largest difference and the greedy algorithm performs in between.

In a network with opportunistic spectrum availability and access, the channel allocation algorithm may only have limited information exchange before the current channel availability changes. Therefore, the algorithm with low complexity and communication overhead is preferred due to the time-variance in channel availability.

3.6 Conclusions

In this chapter, we present a framework to illustrate the relationship between channel availability of secondary users and the usage of primary users. We pointed out that certain properties are inherent to systems with opportunistic spectrum availability and are independent of resource allocation schemes. The first of these properties is location-dependency. As the channel availability observed by the secondary users is determined by the primary users, neighboring secondary users tend to have similar channel availability. The second is the time-variance. Due to the traffic load of the primary users, a secondary user may observe different channel availability over time. These two characteristics are unique to cognitive-radio systems and are captured in this framework. We also introduce a new metric, effective nonopportunistic bandwidth, to quantify the impact of white space. Although the intuition of the spectrum sharing is clear—spatial reuse and statistical multiplexing improve throughput—our work sheds light on *quantifying* such improvements.

Based on the framework, we formulate the channel allocation problem as a list coloring problem with the objective of maximizing the total spectrum utilization. We show the optimal allocation problem is NP-complete, and develop several distributed algorithms. The proposed distributed greedy algorithm achieves close to optimal resource utilization. In addition, a distributed fair algorithm is proposed that achieves better fairness while maintaining a good level of spectrum utilization. Last, a distributed randomized algorithm is introduced with low complexity and communication overhead. The performance of the proposed algorithms is validated in scenarios with time-varying and location-dependent channel availability. In particular, the greedy and the randomized algorithms are shown to remain effective under scenarios where users have only limited information exchange capability. Due to the time-variance of the channel availability, algorithms with low complexity and communication overhead are preferred.

The chapter is an attempt to understand the impact of dynamic spectrum availability and the opportunistic exploration of it. Further studies include performance analysis, general definitions of utilization and fairness metrics, and experimental tests.

References

[1] M. McHenry. Spectrum white space measurements. Presented to New America Foundation BroadBand Forum (June, 2003). http://www.new america.net/Download_Docs/pdfs/DocFile_185_1.pdf.

[2] FCC. In the matter of promoting efficient use of spectrum through elimination of barriers to the development of secondary markets, WT Docket No. 00-230, FCC Rcd 20604.

[3] FCC. Facilitating opportunities for flexible, efficient, and reliable spectrum use employing cognitive radio technologies, notice of proposed rule making and order, FCC 03-322.

[4] FCC. Establishment of an interference temperature metric to quantify and manage interference and to expand available unlicensed operation in certain fixed, mobile and satellite frequency bands, ET Docket No. 03-237, FCC 03-289.

[5] FCC. Promoting efficient use of spectrum through elimination of barriers to the development of secondary markets, WT Docket No. 00-230, report and order and further notice of proposed rulemaking, FCC 03-113 (secondary markets report and order and FNPRM).

[6] DARPA. Defense advanced research projects agency (DARPA) neXt Generation (XG) communications program. http://www.darpa.mil/ato/programs/XG/.

[7] I. Katzela and M. Naghshineh, Channel assignment schemes for cellular mobile telecommunication systems, *IEEE Personal Communications.* 3 (3), 10–31 (Jun., 1996).

[8] L. Narayanan, Channel assignment and graph multicoloring, *Handbook of Wireless Networks and Mobile Computing.* pp. 71–94, (2002).

[9] N. Garg, M. Papatriantafilou, and P. Tsigas, Distributed long-lived list colouring: how to dynamically allocate frequencies in cellular networks, *Wireless Networks* 8(1), 49–60, Springer (2002).

[10] W. K. Hale, Frequency assignment: theory and applications, *Proceedings of the IEEE* 68, 1497–1514 (Dec., 1980).

[11] J. Janssen and K. Kilakos, Adaptive multicolourings, *Combinatorica* 20(1), 87–102 (2000).

[12] J. Janssen, D. Krizanc, L. Narayanan, and S. Shende, Distributed online frequency assignment in cellular networks, *J. Algorithms* 36(2), 119–151 (2000). http://dx.doi.org/10.1006/jagm.1999.1068.

[13] R. Prakash, N. G. Shivaratri, and M. Singhal. Distributed dynamic channel allocation for mobile computing. In *Proceedings of the Fourteenth Annual ACM Symposium on Principles of Distributed Computing*, pp. 47–56. ACM Press (1995). http://doi.acm.org/10.1145/224964.224970.

[14] F. P. Kelly, Blocking probabilities in large circuit-switched networks, *Applied Probability* 18, 473–505 (1986).

[15] A. Nasipuri, J. Zhuang, and S. R. Das. A multichannel CSMA MAC protocol for multihop wireless networks. In *Proceedings of IEEE Wireless Communication and Networking Conference*, 1999.

[16] A. Adya, P. Bahl, J. Padhye, A. Wolman, and L. Zhou. A multi-radio unification protocol for IEEE 802.11 wireless networks. In *Broadnets*, IEEE, (2004).

[17] J. So and N. Vaidya. Multi-channel MAC for ad hoc networks: handling multi-channel hidden terminals using a single transceiver. In *MobiHoc*, ACM (2004).

[18] P. Bahl, R. Chandra, and J. Dunagan. SSCH: Slotted seeded channel hopping for capacity improvement in IEEE 802.11 ad-hoc wireless networks. In *Proceedings of MobiCom*, ACM (2004).

Chapter 4

Rendezvous and Coordination in Cognitive Networks

Paul D. Sutton and Linda E. Doyle

Contents

4.1 Cognitive Networks and Dynamic Spectrum Access

Haykin[1] defines a cognitive radio as "an intelligent wireless communication system that is aware of its environment and uses the methodology of understanding-by-building to learn from the environment and adapt to statistical variations in the input stimuli" with the overarching aim of providing reliable communication whenever and wherever needed while efficiently using the resources available to it. The cognitive radio is built upon software-defined radio (SDR) technology, and a great deal of research effort is currently devoted to the development of radio devices capable of *cognitive* operation. Cognition in this sense includes making observations of the operating environment, using these observations to arrive at optimization decisions, and reconfiguring device operating parameters while building upon past experience to learn from mistakes and proactively adapt to the environment.

Cognitive *network* research recognizes that, while highly capable devices are essential in achieving the goals of reliable "anytime, anywhere" communications, cognitive operation is required not just at the level of the individual device but on an end-to-end basis, across the entire network. This vision of a cognitive network is examined by Thomas et al.,[2] where the scope of cognitive behavior is extended to encompass the entire network and influence all levels of the protocol stack. Within the network, environmental observations can be made locally and shared among devices to improve reliability. On the basis of these observations, optimization decisions that take into consideration all members of the network must be made. In the case of an infrastructure-based network, these decisions may be made unilaterally by a device with a global view of the network; however, in a distributed network, agreement must be reached cooperatively. Similarly, reconfiguration of device operating parameters must be performed in coordination with other member devices of the network in order to maintain network connectivity.

A key application for which cognitive network behavior is particularly suited is that of dynamic spectrum access (DSA). Recent reports[3] have shown spectrum to be severely underutilized across almost all frequency bands and in both rural and urban settings. This inefficiency stems from the *command-and-control* approach to spectrum regulation traditionally taken by regulatory bodies worldwide. Under the command-and-control model, interference between wireless systems is avoided through the assignment of licenses. Licensees are strictly limited in terms of the wireless technology to be deployed, locations of transceivers, and maximum permissible power levels. While this approach has successfully mitigated the threat of harmful interference between systems, it has led to serious underutilization of spectrum resources, inflated costs of access to spectrum, and severely stunted innovation. Conversely, the potential benefits of a more liberal approach to spectrum regulation have been demonstrated over the past decade by the remarkable success of unlicensed systems within the industrial, scientific, and medical (ISM) spectrum bands, particularly that of the IEEE 802.11 standard for wireless local area networks (WLANs) within the 2.4 GHz band.

Recognizing the shortcomings of current spectrum management approaches, a number of regulatory bodies including the Federal Communications Commission (FCC)[4] in the US and the Office of Communications (OfCom)[5] in the UK have begun to explore the potential for more liberal approaches. In considering these alternatives, the potential has been recognized for increasingly adaptive wireless systems to coordinate and autonomously use spectrum resources while avoiding the creation of harmful interference. Such cognitive networks are often termed *dynamic spectrum access networks* (DySPANs). Although a wide range of alternative spectrum approaches are currently being considered, they typically fall into one of three key categories:

■ Exclusive usage rights
■ Spectrum commons
■ Opportunistic use

The *exclusive usage* or property rights model[6,7] of spectrum management involves the assignment of exclusive licenses under which the wireless technology used, services provided, and business models adopted are at the discretion of the licensee. The creation of harmful interference is avoided through the use of technology-neutral restrictions. Licenses may be traded in an open market, reducing barriers to entry, promoting innovation, and encouraging efficient use of spectrum.

In contrast, the *commons* model[8,9] of spectrum management promotes shared access to available spectrum resources for a number of users. Responsibility for interference avoidance is thus devolved from the regulator to the networks themselves, and an agreed protocol or etiquette is used to allocate spectrum in times of congestion and to resolve conflicts. Only networks conforming to the adopted protocols are granted access to the spectrum. Cognitive networks are key to the concept of a spectrum commons in providing the awareness and adaptability required for successful operation under such a model.

Cognitive network technology also forms the basis for the third alternative model of spectrum management under consideration, the *opportunistic access* model.[1,10] Under this model, cognitive networks are permitted to independently identify spectrum resources that are unused at a given time and place and utilize them while protecting other users from the effects of harmful interference. In this way, existing spectrum users should remain unaffected by the operation of the cognitive network. Opportunistic spectrum access is also referred to as *easement* usage and may take the form of either underlay or overlay access. Underlay access involves the use of signal powers below that of the noise floor experienced by existing spectrum users. This is the approach adopted by technologies such as ultra-wideband (UWB), commonly used for short-range, high-data-rate applications such as wireless USB.[11] On the other hand, overlay access involves the detection and use of spectrum *white spaces*. These unutilized spectrum bands are temporarily occupied by the network and immediately relinquished upon the reappearance of higher priority primary users. Opportunistic overlay spectrum access is the approach taken by the proposed IEEE 802.22 standard for wireless regional area networks, discussed in Section 4.2.3 below.

A more liberal approach to spectrum management is essential to the development and implementation of systems based on cognitive network technology. However, a number of key technical challenges must first be overcome before such regulatory change can take place. These challenges include the design of robust yet flexible radio platforms, reliable spectrum scanning techniques for identification of unutilized resources, and coordination mechanisms for cognitive network creation, synchronization, and maintenance. In the following sections, we outline the challenge of cognitive network rendezvous and coordination, we examine some existing techniques used to achieve network coordination, and we look at a number of new techniques which may be applied to cognitive networks.

4.2 The Challenge of Cognitive Network Rendezvous and Coordination

As the demand for wireless systems and services has grown over the past two decades, so too have the flexibility and sophistication of the wireless networks developed to meet that demand. In order to provide robust, broadband communication links, wireless devices must make optimal use of the limited power, processing, and spectrum resources available to them. Cognitive network devices achieve this by dynamically choosing operating parameters which are best suited to the conditions under which they operate. This ability to adapt to suit the operating environment is improved by broadening the range of operating parameters available to a given device. However, as the range of possible operating parameters broadens, the role of network coordination becomes increasingly important. In order to establish and maintain network connectivity, devices within the network must choose compatible parameters with which to communicate. Key among these are the carrier frequencies, signal bandwidths, and signal waveforms chosen.

Coordinated choice of operating parameters is especially important in the initial establishment of communication links within a network. The process of network rendezvous refers to this initial creation of network links and may take place during network creation, during network entry by a new device, or during network reestablishment following a disruptive event. In the case of a dynamic spectrum access network, such events might include the appearance of a primary user within the spectrum band currently occupied.

We will examine the challenge of network rendezvous by first looking at approaches taken within a number of existing wireless systems.

4.2.1 Static Control Channels—Global System for Mobile (GSM)

A popular approach for achieving network rendezvous in cellular systems is the use of fixed-frequency control channels. Equipped with a prior knowledge of the operating frequencies and bandwidths of these control channels, network devices can perform time and frequency synchronization, obtain network configuration data, and join the network.

GSM is the world's most popular second-generation cellular technology. Developed in the mid-1980s by a special working group of the European Conference of Post and Telecommunications (CEPT), GSM was introduced to the European market in 1991 and today accounts for 82% of the global mobile market with over 2 billion subscribers.[12]

Like many conventional networks, GSM uses fixed-frequency control channels which are known in advance by all subscriber devices. Due to this static configuration, network rendezvous is achieved in a straightforward manner. Channels of 200 KHz are defined within two 25 MHz bands, centered at 902.5 MHz and 947.5 MHz. Frequency division duplexing (FDD) is used with base-subscriber transmissions taking place in the upper band and subscriber-base transmissions occurring in the lower band. Forward and reverse link channel pairs are separated by 45 MHz and are identified using ARFCNs (absolute radio frequency channel numbers). Time division multiplex access (TDMA) is used within each channel to permit sharing between up to 8 subscribers. A physical channel is thus described using an ARFCN and a unique time slot (TS).

There are 34 specific ARFCNs that are defined as broadcast channels (BCH), used for periodic control signaling by base stations. Within the forward link of each BCH, TS 0 is reserved for control signaling, while TS 1–7 are used for regular data traffic. The control data broadcast is determined by a repetitive 51-frame sequence known as a control multiframe. The GSM control multiframe is illustrated in Figure 4.1.

Within the 51-frame sequence, frames 0, 10, 20, 30, and 40 are used for frequency correction, allowing subscribers to synchronize their local clocks to the frequency of the base station. Synchronization bursts follow each frequency correction burst in frames 1, 11, 21, 31, and 41. Synchronization bursts contain a *frame number* and a *base station identity code* and are used to adjust subscriber timing according to the distance from the base station. Frames 2–5 are used to broadcast control information including cell and network identification, control channel configuration, channels

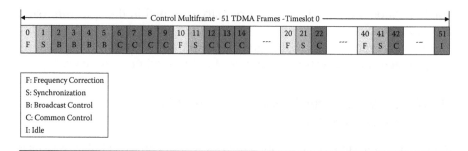

Figure 4.1 **GSM control multiframe. Reproduced from** *Wireless Communications,* **Theodore S. Rappaport.**[13]

used within the cell, and congestion. The remaining broadcast time slots are used for paging and grant access on the forward link and for random access on the reverse link.

Thus by scanning a known control channel, a GSM subscriber can achieve time and frequency synchronization and receive control messages about the network configuration prior to network entry. In this way, the use of periodic signaling on fixed-frequency control channels provides a straightforward technique for achieving network rendezvous.

4.2.2 In-Band Control Signaling—WiMAX

An alternative approach for achieving network rendezvous is the use of in-band signaling. In this case, control signals are not only broadcast on fixed-frequency control channels but rather are transmitted together with data traffic on all used channels. This is the approach adopted for WiMAX, a broadband solution for wireless metropolitan area networks based on the IEEE 802.16 family of standards.

Although the IEEE 802.16 family of standards describe a wide range of physical (PHY) and medium access control (MAC) layer options for fixed and mobile access in both the 2–11 GHz and the 10-66 GHz ranges, these options have been reduced to a set of specific profiles for implementation by the Worldwide Interoperability for Microwave Access (WiMAX) Forum. Profiles are defined for fixed and mobile access using a range of operating frequencies, channel bandwidths, and duplexing modes. Network rendezvous is achieved within each profile through use of in-band signaling.

WiMAX signals are orthogonal frequency division multiplexing (OFDM)-based, and equivalent uplink and downlink frame structures are specified for both time-division duplexing (TDD) and frequency-division duplexing (FDD) implementations. With TDD implementations, uplink and downlink subframes are transmitted sequentially within a single channnel, while FDD operation specifies simultaneous transmission on separate uplink and

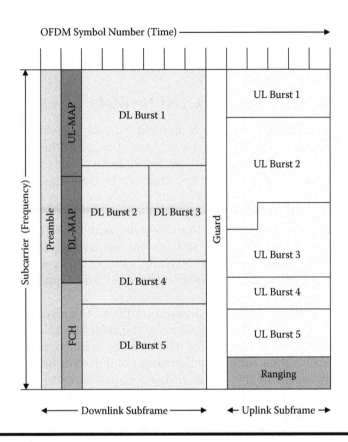

Figure 4.2 WiMAX frame structure. (Reproduced from *Fundamentals of WiMax*, Jeffrey G. Andrews, Arunabha Ghosh, and Rias Muhamed.[14] With permission.)

downlink channels. The TDD subframe structures are outlined in Figure 4.2. Each frame begins with a preamble, used for time and frequency synchronization. This is followed by a frame control header (FCH) which includes information on frame configuration, usable OFDM subcarriers, and modulation and coding schemes, as well as the structure of the MAP messages used to specify uplink and downlink bandwidth allocations for individual subscribers. This is followed by the uplink and downlink MAP messages themselves. WiMAX uses a wide range of modulation and coding schemes that may be tailored to suit channel conditions for individual subscribers; however, control signals are transmitted using a single, highly reliable scheme such as BPSK with 1/2 rate coding and repetition coding.

This downlink frame structure effectively facilitates subscribers performing network rendezvous. Upon commencing operation, a subscriber scans each of the available WiMAX channels in turn for frame preambles. Upon detection of a preamble, timing and frequency synchronization is performed, after which control messages may be received and used to configure the device prior to network entry. Subscribers then enter the network using the contention-based ranging channel of the uplink subframe.

4.2.3 Dynamic Spectrum Access Networks

Network rendezvous using both in-band control signaling and fixed-frequency control channels depends upon the use of known channels which may be scanned for control information. Dynamic spectrum access networks achieve highly efficient spectrum use by detecting and occupying spectrum white space—frequency bands which are unoccupied by other networks at given times and places. As the availability of this white space spectrum fluctuates over time, the channels used by a DySPAN system may not be known in advance and can change unpredictably over time as other primary and secondary networks occupy and vacate them. Under these conditions, more complex techniques are required to perform network rendezvous.

The first networking standard based on the use of dynamic spectrum access is currently under development by the IEEE 802.22 working group on wireless regional area networks. The standard seeks to outline the PHY, MAC, and air interface for use by license-exempt devices on a noninterfering basis in broadcast TV spectrum between 41 and 910 MHz. A fixed point-to-multipoint (P-MP) topology is specified with a minimum peak downstream rate of 1.5 Mbps per subscriber. A critical requirement for the standard is the protection of legacy networks operating in the same spectrum bands, including broadcast TV networks as well as other services such as wireless microphones. In order to provide legacy network protection, a system of distributed sensing is proposed whereby spectrum sensing is performed by each subscriber device and observations within a cell are pooled by the base station in order to reach spectrum occupancy decisions.

The draft standard specifies use of OFDM-based modulation for both the uplink and downlink, and network rendezvous is facilitated through use of a specified superframe structure, illustrated in Figure 4.3. At the start of each superframe is a preamble which is used for frame detection and synchronization. The preamble is followed by a superframe control header (SCH) which provides all information needed to associate with the base station and enter the network. A key challenge for 802.22 devices commencing operation, however, is the detection of those channels currently in use by the network.

Upon starting operation, an 802.22 subscriber will perform a scan of the TV channels in order to build a spectrum occupancy map and identify vacant channels. Each of these channels may potentially be used by the network, and so each is scanned for superframe preambles. Upon detection

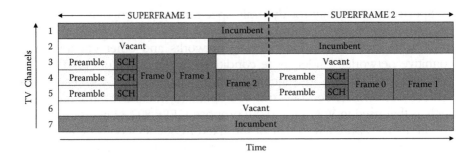

Figure 4.3 IEEE 802.22 superframe structure. (Reproduced from *Evolution of Spectrum-Agile Cognitive Radios: First Wireless Internet Standard and Beyond*, Kiran Challapali, Carlos Cordeiro, and Dagnachew Birru.[15] With permission.)

of a superframe, synchronization is performed and the SCH is used to obtain control information for the network. Once network rendezvous is achieved, the spectrum occupancy map is provided to the base station for use in the distributed sensing process.

Network rendezvous is also complicated by the use of channel bonding in the 802.22 draft standard. Channel bonding is used to acheive the high data rates required by the 802.22 specifications and involves the simultaneous use of up to three unoccupied TV channels, providing an overall bandwidth of 18 MHz for 6 MHz channels. In performing network rendezvous, however, a subscriber does not have prior knowledge of the number of bonded channels in use by the network. For this reason, the superframe preamble and SCH used for network rendezvous are transmitted independently on each channel used (see Figure 4.3). Subscriber devices may thus detect transmitted superframes and receive SCH control messages by scanning individual channels.

4.3 Cognitive Network Rendezvous

The highly flexible nature of dynamic spectrum access cognitive networks means that the challenge of network rendezvous and coordination may not be successfully overcome using the approaches adopted in more conventional, static networks. Highly efficient use of spectrum resources may be achieved where cognitive networks are leveraged to opportunistically access spectrum white spaces while protecting other network devices from the effects of harmful interference. However, in the case of such opportunistic spectrum access, network devices must tailor carrier frequencies, signal bandwidths, and waveforms to suit the white space spectrum available at a given time and place.

Spectrum availability for dynamic spectrum access cognitive networks depends upon the occupancy of other systems operating in the same spectrum band, both high-priority legacy networks and other equal-priority cognitive networks. Under these conditions, a static channel may not be allocated, and so a fixed-frequency control channel may not be used to achieve network rendezvous and coordination. In-band control signaling such as that used in WiMAX and the proposed IEEE 802.22 networks offers an alternative approach to that of the fixed-frequency control channel by including control traffic together with data traffic in uplink and downlink frames. However, this approach relies on the use of known waveforms and fixed-bandwidth channels which can be individually scanned for frames containing control data. In the case of IEEE 802.22, these channels are determined by the main wireless system present in the same spectrum band, terrestrial broadcast TV. Under more general dynamic spectrum access conditions, however, a wide range of heterogeneous networks may be required to utilize the same spectrum band. These networks may use a range of differing channel schemes and may need to dynamically adapt channel bandwidths according to system requirements. Thus, in order to achieve highly efficient spectrum use, cognitive networks must take advantage of any unoccupied spectrum which can be reliably detected and must not be constrained by fixed channel sizes and carrier frequencies. A solution to the challenge of cognitive network rendezvous and coordination must therefore facilitate device discovery, network creation, and network entry while supporting high degrees of flexibility in the carrier frequencies, signal bandwidths, and waveforms utilized by the network.

A significant advantage of such a highly flexible mechanism for cognitive network rendezvous would be the abililty to apply the same approach to a number of different networks. Cognitive networks may comprise a number of highly adaptable heterogeneous devices, each capable of adopting a wide range of operating parameters, including carrier frequencies and signal bandwidths as well as waveform-specific parameters. Such devices are not confined to a single network but may join a range of networks, both cognitive and legacy, through careful choice of operating parameters. A common network approach for rendezvous and coordination would permit cognitive devices to quickly detect the presence of all compatible networks in a given location, choose that most suited to its requirements, and perform rendezvous.

4.3.1 A New Approach

A promising approach for achieving cognitive network rendezvous is that of physical-layer signaling. The approach involves manipulating key fundamental properties of a wireless signal in order to permit detection, analysis, and identification without prior knowledge of signal parameters, including carrier frequency and bandwidth. In this way physical-layer signaling permits network rendezvous to be performed in conjunction with the spectrum sensing required for opportunistic spectrum access.

One such physical-layer signaling approach proposed by Horine et al.[16] makes use of a beaconing signal with an easily detected power spectrum. The beaconing signal uses a number of tones transmitted at discrete relative frequency offsets and with discrete relative amplitudes. A ratiometric decision statistic is then derived from the power spectrum at the receiver, and detection is performed. The use of such a distinctive beacon signal facilitates device discovery and link establishment where operating frequencies are not known *a priori*. However, a key drawback of the proposed approach is its sensitivity to frequency-selective fading. A deep fade at one or more of the distinctive tones would severely distort the ratiometric decision statistic used. In addition, the use of a dedicated beaconing signal requires coordinated transmission and scanning by members of the network. If these processes do not occur simultaneously, rendezvous cannot be achieved.

A more robust approach to physical-layer signaling involves the use of signal cyclostationarity. Many of the communications signals in use today may be modeled as cyclostationary signals due to the presence of one or more underlying periodicities that arise from the coupling of stationary message signals with periodic sinusoidal carriers, pulse trains, or repeating codes. These underlying periodicities may also occur as a result of other processes used in the generation of the signal, including sampling and multiplexing.

The cyclostationary properties of communications signals can be leveraged to accomplish a number of key tasks, including signal detection,[17] classification,[18] synchronization[19,20] and equalization.[21] Gardner and Spooner[22] identify a number of advantages provided by cyclostationary signal analysis over alternative radiometric approaches. Among these are a reduced sensitivity to noise and interfering signals as well as the ability to extract key signal parameters, including carrier frequencies and symbol rates.

Using signal cyclostationarity, it is possible to detect the presence of a signal of interest and determine key signal parameters, including carrier frequency and bandwidth. These capabilities make cyclostationary analysis a promising approach for overcoming the challenge of cognitive network rendezvous. However, a number of significant drawbacks exist. Reliable cyclostationary signal analysis often requires the use of highly complex receiver architectures and long signal observation times. Additionally, the inherent cyclostationary properties of a communications signal depend upon fundamental properties of the signal such as modulation type or symbol keying rate. These properties are not easily manipulated without directly impacting system performance.

In order to overcome these drawbacks, a number of authors[23,24] have proposed the intentional embedding of cyclostationary properties in communications signals in order to facilitate detection and analysis. These induced cyclostationary features can often be manipulated independently of the key signal parameters adopted, and hence used to generate unique signatures which can be tailored to the specific requirements of the system. Additionally, induced features may permit reliable detection and analysis using low-complexity receiver architectures and short observation times.

4.4   Cyclostationary Signatures

One approach for achieving cognitive network rendezvous and coordination proposed by the authors involves the use of *cyclostationary signatures*[25], intentionally embedded in orthogonal frequency division multiplexing (OFDM) waveforms.

OFDM waveforms involve the conversion of a high-rate serial data stream to a number of parallel low-rate streams, which are mapped onto a set of closely spaced orthogonal carriers for transmission. The use of multiple low-rate data streams results in longer symbol durations and provides robustness against the intersymbol interference that can arise from multipath propagation in non-line-of-sight applications. A cyclic prefix may be appended to each OFDM symbol in order to improve this robustness against multipath fading. In the context of dynamic spectrum access, OFDM waveforms offer significant benefits due to the flexible nature of their spectral shape. As the spectrum of an OFDM waveform is composed of multiple subcarriers, the shape of that spectrum may be manipulated by adjusting the data symbols transmitted on each subcarrier. Subcarrier nulling[26], for example, permits the bandwidth of an OFDM waveform to be tailored to the channel that is available.

The spectrum sculpting capabilities afforded by OFDM waveforms are key in the generation of cyclostationary signatures. Cyclostationarity in a wireless communications signal manifests itself as a correlation pattern in the spectrum of that signal. Thus by artificially creating a spectrum correlation in a signal, cyclostationarity is induced and a signature may be embedded. Such a spectrum correlation may be generated in an OFDM waveform using the low-complexity subcarrier mapping technique illustrated in Figure 4.4. Here, F_0 is the carrier frequency of the signal, F_s is the bandwidth, and p is the number of subcarriers between mapped sets. By redundantly transmitting a data symbol on more than one OFDM subcarrier, a correlation is created between those subcarriers and a cyclostationary signature is embedded in the waveform. This process of subcarrier mapping can be described by

$$\gamma_{n,k} = \gamma_{n+p,k}, \qquad n \in M \qquad (4.1)$$

where $\gamma_{n,k}$ is the data symbol transmitted on subcarrier n during OFDM symbol k, and M is the set of subcarrier values to be mapped. The resulting

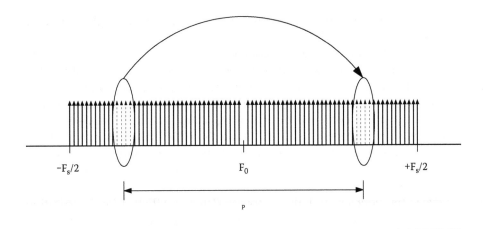

Figure 4.4   Cyclostationary signature generation.

cyclostationary signature occurs at cyclic frequency

$$\alpha_{\text{sig}} = \frac{p}{T_s} \tag{4.2}$$

where T_s is the OFDM symbol period.

A spectral correlation density (SCD) analysis may be used to identify cyclostationary features present within a range of cyclic frequencies of a given waveform. Figure 4.5 shows the SCD of two OFDM waveforms. The first of these (Figure 4.5(a)) illustrates the SCD of an OFDM waveform without an embedded signature. The second (Figure 4.5(b)) illustrates the SCD

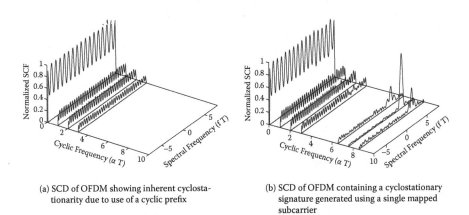

(a) SCD of OFDM showing inherent cyclostationarity due to use of a cyclic prefix

(b) SCD of OFDM containing a cyclostationary signature generated using a single mapped subcarrier

Figure 4.5   Spectral correlation density (SCD) showing inherent and embedded cyclostationary features.

of a waveform containing a signature embedded using a single subcarrier mapping.

For the purpose of achieving cognitive network rendezvous and coordination, cyclostationary signatures can be used to accomplish the three key tasks of signal detection, network identification, and frequency rendezvous.

An OFDM signal containing an embedded cyclostationary signature can be detected by a cognitive device using a time- or frequency-smoothed cyclic cross periodogram (CCP)[27]. Using the CCP, cyclostationary features present at a single cyclic frequency within the signal can be analyzed. This approach permits signature detection without the computational complexity required to perform a full cyclostationary analysis over a wide range of cyclic frequencies, and also accommodates an efficient implementation using the fast Fourier transform (FFT). In this way, a cognitive device can detect a signal present at any carrier frequency within the received band. In addition, using a wide-band receiver, signature detection can be performed over a bandwidth greater than that of the signal of interest. As cyclostationary signatures generated using OFDM subcarrier mapping are present throughout the transmitted signal, they may be detected using any received portion of that signal.

Network identification involves differentiation between signals transmitted by members of different networks. The highly flexible nature of cognitive radio devices permits them to join a wide range of network types through reconfiguration of their operating parameters. In searching for a suitable network at a given time and location, a key advantage is the ability to recognize networks and differentiate between those using similar air interfaces. Cyclostationary signatures facilitate this type of network identification by allowing unique signatures to be embedded in similar OFDM-based signals. The use of subcarrier set mapping provides a high degree of flexibility in the cyclic and spectral frequencies of the embedded signatures. Simply by adjusting the mapped subcarrier set locations, signatures with unique cyclic frequency or spectral frequency properties are generated. This approach is illustrated in Figure 4.6, which shows the *alpha-profiles* of two unique signatures generated at two different cyclic frequencies. In this way, signals transmitted by members of two networks may be differentiated, and network identification is possible.

The third key task that is facilitated using embedded cyclostationary signatures is frequency acquisition of the signal of interest. Following detection of the signal and identification of a suitable network, the cognitive device must synchronize with the signal to establish a communications link prior to joining the network. A vital part of the synchronization process is estimation of the carrier frequency of the signal of interest. Cyclostationary signatures may be generated in such a way that the resulting cyclostationary feature is directly related to the carrier frequency of the signal. Figure 4.7 illustrates a cyclostationary signature generated using mapped subcarrier

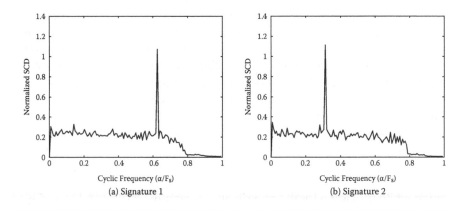

Figure 4.6   Unique cyclostationary signatures generated at different cyclic frequencies (a) Signature 1. (b) Signature 2.

sets equidistant from the signal carrier frequency. The resulting feature is centered at the carrier frequency and so, upon detection, can be used directly to perform coarse carrier frequency acquisition.

Cyclostationary signatures are thus a highly flexible tool for achieving cognitive network rendezvous and coordination by facilitating signal detection, network identification, and frequency rendezvous. Signatures enable coordination without prior knowledge of carrier frequency, signal

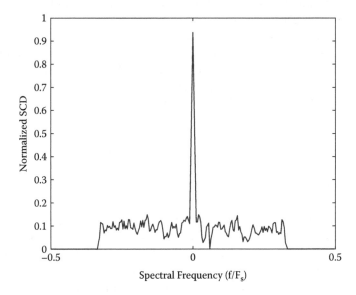

Figure 4.7   Spectral frequency at α_{sig}, the cyclic frequency of the detected signature.

bandwidth, or waveform-specific parameters such as modulation complexity, and so increase the range of operating parameters which may be dynamically adjusted within the network. As the scope for reconfiguration within the network broadens, so the ability of the network to adapt to changing environmental conditions and system requirements improves.

While cyclostationary signatures provide some of the key functions required, they are by no means a complete solution to the challenge of cognitive network rendezvous and coordination. Further techniques are required, for example, to facilitate waveform parameter matching between cognitive network devices and robust medium access control (MAC), and network (NET) layer protocols are needed to ensure reliable end-to-end communication while providing for the coordinated spectrum sensing and reconfiguration required within a dynamic spectrum access cognitive network. However, cyclostationary signatures provide a robust foundation upon which further techniques, protocols, and etiquettes for cognitive networks can be developed.

4.5 Conclusion

Cognitive networks hold the promise of high-data-rate, robust network connectivity whenever and wherever needed while ensuring the efficient use of available resources. Dynamic spectrum access cognitive networks focus upon the highly efficient use of radio spectrum by occupying frequency bands which are unused at a given time and place, and autonomously avoiding the creation of harmful interference.

Before such advanced wireless networks can be realized, a wide range of challenges must be overcome. These challenges are multidisciplinary in nature and encompass all layers of the network protocol stack, as well as such fields of research as machine learning, complex systems science, and game theory. A key principal requirement of any cognitive network capable of dynamic spectrum access, however, is a robust mechanism for achieving network rendezvous and coordination. In this way, network connectivity can be ensured and communication between individual devices established and maintained.

As cognitive networks become more flexible and dynamic, the challenge of network rendezvous and coordination becomes more pressing. Within more conventional, static networks, key operating parameters may be known in advance and may be used effectively to facilitate the initial establishment of communication links between devices. When these operating parameters are no longer fixed but are rather dynamically adjusted to suit the operating environment, conventional approaches for establishing network connectivity may no longer be employed.

An alternative approach for achieving rendezvous and coordination in cognitive networks involves the use of physical-layer signaling. In this way, low-level physical properties of a signal are manipulated in order to provide an easily recognizable *signature* in order to facilitate signal detection and assist in the process of establishing a communications link. Thus, a single mechanism for network rendezvous and coordination can be employed, though a wide range of signal parameters may be dynamically adjusted. One such physical-layer signaling approach proposed by the authors involves the intentional embedding of cyclostationary features in data-carrying OFDM-based signals. These *cyclostationary signatures* are generated, detected, and analyzed using low-complexity receive and transmit architectures and facilitate detection of a signal of interest, network identification, and frequency rendezvous.

Robust techniques such as the use of cyclostationary signatures are key in overcoming the technical challenges for cognitive network technology while facilitating the flexibility needed to fulfill its full potential.

References

[1] S. Haykin, Cognitive radio: Brain-empowered wireless communications, *IEEE J. Select. Areas Commun.* 23(2), 201–220 (2005).

[2] R. W. Thomas, D. H. Friend, L. A. Dasilva, and A. B. Mackenzie, Cognitive networks: Adaptation and learning to achieve end-to-end performance objectives, *IEEE Communications Magazine* 44 (12), 51–57 (2006).

[3] M. McHenry and D. McClosky. New York City spectrum occupancy measurements September 2004. Technical report, Shared Spectrum Company, www.sharedspectrum.com, (2004).

[4] P. Kolodzy. Spectrum policy task force report. Technical report, Office of Engineering and Technology, Federal Communications Commission (Nov. 2002).

[5] M. Cave. Review of radio spectrum management. Technical report, Department of Trade and Industy and HM Treasury (2002).

[6] A. De Vany. Property rights in the electromagnetic spectrum. *The New Palgrave Dictionary of Economics and the Law*, 3rd ed., Peter Newman ed. Palgrave Macmillan (1998).

[7] D. Hatfield and P. Weiser. Property rights in spectrum: Taking the next step. In *2005 First IEEE International Symposium on New Frontiers in Dynamic Spectrum Access Networks, 2005. DySPAN 2005.*, pp. 43–55.

[8] W. Lehr and J. Crowcroft. Managing shared access to a spectrum commons. In *2005 First IEEE International Symposium on New Frontiers in Dynamic Spectrum Access Networks, 2005. DySPAN 2005.* pp. 420–444.

[9] P. Weiser and D. Hatfield, Policing the spectrum commons, *Fordham Law Review* 74, 101–131 (2005).

[10] I. Akyildiz, W. Lee, M. Vuran, and S. Mohanty, NeXt generation/dynamic spectrum access/cognitive radio wireless networks: a survey, *Computer Networks: The International Journal of Computer and Telecommunications Networking* 50(13), 2127–2159 (2006).

[11] Wireless universal serial bus specification revision 1.0. USB Implementers Forum, Inc. (May 2005).

[12] www.wirelessintelligence.com. Last accessed: 27/9/2007.

[13] S. Rappaport, *Wireless Communications: Principles and Practice*, 2nd edition, Prentice Hall (2002).

[14] J. Andrews, A. Ghosh, and R. Muhamed, *Fundamentals of WiMax*. Prentice Hall (2007).

[15] K. Challapali, C. Cordeiro, and D. Birru. Evolution of spectrum-agile cognitive radios: first wireless internet standard and beyond. In *WICON '06: Proceedings of the 2nd Annual International Workshop on Wireless Internet*, p. 27, ACM Press, New York, NY, USA (2006).

[16] B. Horine and D. Turgut. Link rendezvous protocol for cognitive radio networks. In *2nd IEEE International Symposium on New Frontiers in Dynamic Spectrum Access Networks, 2007. DySPAN 2007*. pp. 444–447 (2007).

[17] W. Gardner, Signal interception: a unifying theoretical framework for feature detection, *IEEE Trans. Commun.* 36(8), 897–906 (1988).

[18] A. Fehske, J. Gaeddert, and J. Reed. A new approach to signal classification using spectral correlation and neural networks. In *2005 First IEEE International Symposium on New Frontiers in Dynamic Spectrum Access Networks, 2005. DySPAN 2005*. pp. 144–150.

[19] F. Gini and G. Giannakis, Frequency offset and symbol timing recovery in flat-fading channels: A cyclostationary approach, *IEEE Trans. Commun.* 46(3), 400–411 (1998).

[20] H. Bolcskei, Blind estimation of symbol timing and carrier frequency offset in wireless OFDM systems, *IEEE Transactions on Communications*. 49(6), 988–999 (2001).

[21] R.W. Heath, and G. Giannakis, Exploiting input cyclostationarity for blind channel identification in OFDM systems, *IEEE Transactions on Signal Processing* 47(3), 848–856 (1999).

[22] W. Gardner and C. Spooner, Signal interception: performance advantages of cyclic-feature detectors, *IEEE Transactions on Communications*. 40 (1), 149–159, (1992).

[23] K. Maeda, A. Benjebbour, T. Asai, T. Furuno, and T. Ohya. Recognition among OFDM-based systems utilizing cyclostationarity-inducing transmission. In *New Frontiers in Dynamic Spectrum Access Networks, 2007. 2nd IEEE International Symposium on DySPAN 2007* pp. 516–523, (2007).

[24] P. D. Sutton, K. E. Nolan, and L. E. Doyle. Cyclostationary signatures for rendezvous in OFDM-based dynamic spectrum access networks. In *New Frontiers in Dynamic Spectrum Access Networks, 2007. DySPAN 2007. 2nd IEEE International Symposium on*, pp. 220–231 (17-20 April, 2007).

[25] P. Sutton, K. Nolan, and L. Doyle, Cyclostationary signatures in practical cognitive radio applications, IEEE J. Select. Areas Commun. Special Issue— Cognitive Radio: Theory and Applications 26(1) 13–24 (2008).

[26] T. Weiss and F. Jondral, Spectrum pooling: an innovative strategy for the enhancement of spectrum efficiency, IEEE *Communications Magazine.* 42(3), S8–14, (2004).

[27] B. Sadler and A. Dandawate, Nonparametric estimation of the cyclic cross spectrum, *IEEE Transactions on Information Theory* 44(1), 351–358, (1998).

MEDIUM ACCESS CONTROL (MAC) LAYER OF COGNITIVE RADIO NETWORKS

II

Chapter 5

Resource Allocation in Cognitive Radio Networks

S. Anand and R. Chandramouli

Contents

5.1 Introduction

The development of software-defined-radio-(SDR) enabled[1] cognitive radio networks has facilitated dynamic sharing by users of the available spectrum on an opportunistic basis. Users belonging to one network can sense spectrum opportunities in another network and contend for the unused spectrum in the new network. These users thereby become *secondary* users in the new network. Users that originally subscribed to the second network are called *primary* users. The combined interference caused by all the secondary users to existing primary users should be below a specified threshold. Applications of this idea to ultra-wideband wireless networks are presented by Lansford[2]. Liu and Wang[3] present the characteristics of spectrum-agile networks and study them using two metrics: (1) the effective nonopportunistic bandwidth and (2) the space-bandwidth utilization. Akyildiz et al.[4] present a comprehensive survey of cognitive radio networks.

Resource allocation is a topic that has been widely studied for cellular and wireless networks. The resources commonly referred to in cellular networks are

- Timeslots in systems deploying time division multiple access (TDMA) and orthogonal frequency division multiplexing (OFDM)
- Carrier frequencies in systems deploying frequency division multiple access (FDMA)
- Subcarriers in systems deploying orthogonal frequency division multiple access (OFDMA)
- Transmission power and transmission rate most commonly in systems deploying code division multiple access (CDMA)

In Section 5.2 we present a survey of the different types of research problems and solution techniques for channel assignment in cellular networks.

Resource allocation and other topics in cognitive radio networks are also being studied by standards committees. The IEEE Standards Coordinating Committee 41[5] for Dynamic Spectrum Access Networks (DySPAN) was founded in the first quarter of 2005 jointly by the IEEE Communications Society (ComSoc) and the IEEE Electromagnetic Compatibility (EMC) Society. This committee is responsible for the development of standards in the area of dynamic spectrum access. The committee has four associated working groups and one study group:

- The 1900.1 working group is responsible for the definitions and terminologies for radio spectrum management.
- The 1900.2 working group is responsible for the recommended practice on interference and coexistence analysis.
- The 1900.3 working group is responsible for the conformance evaluation of SDR software modules.
- The 1900.4 working group deals with distributed resource allocation for devices in heterogeneous wireless access environments.
- The 1900.A study group focuses on dependability and regulatory compliance for radio systems with dynamic spectrums access (DSA) capabilities.

5.2 Channel Assignment in Cellular Networks

Channel allocation in cellular networks is a classical problem typically studied using integer programming and graph coloring techniques. Several studies have been reported on the capacity of cellular systems with fixed channel allocation (FCA)[6-12] and dynamic channel allocation (DCA).[13-17] Gamst[6] studied algorithms for FCA and presented a lower bound on the required number of channels. McEliece and Sivarajan[7] studied the asymptotic behavior of channelized cellular systems with FCA. Bounds on the offered traffic capacity (Erlangs per channel) were presented. The blocking probability was computed from the carried traffic by using the Erlang-B loss formula. Sarkar and Sivarajan[8] presented channel assignment algorithms for cellular systems with FCA to achieve the bounds given by McEliece and Sivarajan[7]. Sarkar and Sivarajan[9] also presented channel assignment algorithms to satisfy reuse constraints based on co-channel as well as adjacent channel interference. Gupta[10] obtained improved bounds relative to those obtained by Sarkar and Sivarajan[9] by including co-site constraints. Jayateertha and Sivarajan[11] studied the performance of FCA algorithms for different traffic conditions and obtained the optimal traffic distribution to maximize the erlang capacity and minimize the number of cells. Sidi and Starobinski[12] devised an n-dimensional Markov chain model (where n is the total number of available channels), and arrived at product-form solutions to compute the blocking probability for cellular systems with FCA, with low mobility, and with high mobility. The number of states in this chain was $N^{(n+1)}$ where N was the number of cells.

Performance of cellular systems with DCA has also been studied in detail.[13,15,16] Everitt and Macfadyen[13] derived expressions for blocking probability in cellular systems with maximum packing assignment (MPA). The approach was very complex, and approximations based on the approach by Kelly[14] were used. Varghese[15] studied the performance of channelized cellular systems with DCA and showed that DCA offers less blocking for the same traffic when compared to FCA. The set of channels that were feasible for allocation to a user were called *available channels*. It was shown that the system performance when any one of the available channels was allocated to a new user at random was similar to the performance when other algorithms for DCA were used. However, no analytical study of the blocking performance was made. Cimini et al.[16] studied cellular systems with DCA, and performed the analysis for computing blocking probability with an ad hoc Erlang-B approximation for each cell. The expressions obtained did not take into account the interference conditions on the channels. A more accurate approximation was suggested by Sidi and Starobinski[12] as an extension to their analysis of systems with FCA. However, the model used was computationally very complex and required solving a Markov chain with $N^{(n+1)}$, where N was the number of cells and n was the number of available channels. It is observed that in the above references, the analysis of blocking probability did not take the interference conditions on the different channels into account. Anand et al.[17] presented a two-dimensional Markov chain approach to evaluate the blocking probability of cellular systems with DCA. Each state in the Markov chain was a two-tuple (m, k), where m represented the number of channels in use in a cell, and k represented the number of channels that cannot be used due to violation of interference constraints. Linear and two-dimensional circular cellular systems were studied. In all the above-mentioned literature, the studies and analysis took into account systems with voice-only traffic.

Channel assignments for orthogonal frequency division multiplexing/multiple-access (OFDM/OFDMA) systems with data traffic have also been studied.[19,20] Chuang and Sollenberger introduced the concept of packetized DCA or dynamic packet assignment (DPA)[19], in which channels were allocated to data sessions only during the on periods and released during the off periods. Jayaparvathy et al.[20] analyzed such a system (which is also applicable to IEEE 802.16d fixed broadband wireless access (FBWA) systems[21]) by extending the two-dimensional Markov chain approach by Anand et al.[17] and computed the mean delay and throughput for such systems. A comparison of DCA and DPA was made and DPA was shown to obtain about four times the system capacity as compared to DCA.

The channel assignment problem in cognitive radio networks has similarities to that in cellular networks. Cellular networks take into account co-channel and adjacent channel interference while assigning channels to users. The criterion for reusing a channel in cellular networks is that the combined interference received from all co-channel users is below a specified threshold. Constraints on interference from users using the adjacent channels can be added to result in a channel assignment resulting in a better quality of received signal at the cost of added complexity. Similarly, it is also desired that the interference caused by a new user to existing co-channel users be below a specified threshold. Also, it is possible to add constraints on the interference received by a primary user from secondary users using adjacent channels. One can specify different thresholds of co-channel interference for primary and secondary users and treat the channel assignment problem in cognitive radio networks similarly to that in cellular networks. Hence, it would be tempting to conclude that most approaches for resource allocation in cellular networks can also be applied to cognitive radio networks. However, it is essential to take into account the heterogeneity of the networks, and hence the allocated resources. For example, the topologies in cellular networks tend to be less time-varying, and hence the network graph model, while using graph coloring techniques, would in most cases result in a graph with well-defined nodes and edges. However, when considering cognitive radio networks, it is essential to note that the availability of the spectrum and the reuse constraints vary more dynamically due to the arrival of primary users. More importantly, when a user can choose two networks that are inherently different (e.g., a channelized network vs. a CDMA-based network, or a cellular network instead of an ad hoc network like a wireless LAN), then the types of resources and the constraints related to allocating them would differ. Some of the issues related to dynamic changes in reuse constraints have been addressed in cellular systems with DCA. Yet, in those systems, most of the parameters, like the number of cells and number of available channels, would remain fixed. Also, given the homogeneity of the system, it is easier to deal with newly arriving calls and calls that are handed off. However, in cognitive radio networks, it is likely that a user detects different types of heterogeneous networks and has to be allocated resources from one or more of them. In such cases, it is essential to modify the mechanisms that have been proposed for cellular networks. While these could be simple adaptations in some cases, in most cases it is quite complex. It is therefore essential to devise resource allocation mechanisms specifically for cognitive radio networks.

5.3   Channel Assignment in Cognitive Radio Networks

There have been few studies on channel assignment in cognitive radio[22–25] using graph-theoretic[22] and integer programming approaches[25]. Zheng and Peng[22] provided heuristics for spectrum assignment for different fairness criteria and evaluated them. Each node in the cognitive radio system was modeled as a vertex in a graph, and vertices joined by colored edges (color corresponding to the available channels in available networks). The channel assignment problem was then modeled as a graph coloring problem and heuristics were provided for different criteria for fairness. Xin et al.[23] studied assignment of channels to nodes to form topologies depending on the parameter that is desired to be optimized. The channels available for each node were modeled as a layered graph. Streenstrup[24] studied three types of frequency assignment problems: common broadcast frequencies, non-interfering frequencies, and direct source-destination communication frequencies, and modeled each one of them as a generalized graph coloring problem, presenting centralized and distributed heuristics for efficient channel assignment. Hou et al.[25] appled ad hoc routing protocols and modeled the channel allocation problem with capacity constraints as a mixed integer nonlinear programming problem to obtain tight lower bounds using linear programming relaxations. Heuristics for a distributed approach based on Lagrangian relaxation and gradient search methods were proposed.

Game-theoretic approaches[26–28] and learning-automata-based approach[29] for spectrum access in cognitive radio networks have also been studied. Larcher et al.[26] presented an *n*-player noncooperative game-theoretic approach for secondary user spectrum access. A utility function based on the access delay and collision probability was proposed, and the existence and convergence to a Nash equilibrium was shown. Xing et al.[27] modeled the interference caused to the primary users and defined an interference temperature constraint to limit this interference. An algorithm for the optimal supported link subset was proposed. The authors also modeled the spectrum sharing problem as a potential game and showed the existence of a deterministic pure strategy Nash equilibrium. Xing et al.[28] presented the homo-egualis game-theoretic model for secondary user spectrum access in which users were modeled to behave like a human society to reduce the unfairness in spectrum access. Xing et al.[29] presented a learning-automata-based approach where users were classified as quality sensitive and price sensitive users, and the price dynamics were studied. The authors also considered a model in which the networks could cooperate to attain better spectrum utilization. Achievable capacity in cognitive radio networks was studied by Devroye et al.[30] The authors considered cases of a genie-aided cognitive radio channel, where the receiver is noncausally provided the message from the transmitter in an interference channel, and then compared this with the case of a causal cognitive radio channel to obtain an achievable capacity region.

5.4 Some Research Problems in Cognitive Radio Networks

The developments in cognitive radio networks being few and the field being in its nascent stages, there is great opportunity for research and development in both academia and the industry. The current topics of research include development of test procedures and test beds for conformance testing, in addition to numerous theoretical, as well as practical, issues related to spectrum management. We provide details on problems specific to complexity analysis of channel assignment algorithms, network selection, and keyless security in Sections 5.4.1, 5.4.2, and 5.4.3, respectively.

5.4.1 Complexity Analysis of Channel Assignment Algorithms

It was mentioned in Section 5.2 that the channel assignment algorithms used for cellular networks cannot be directly reused for cognitive radio networks. The optimum channel assignment problem in cognitive radio networks taking into account all constraints on primary and secondary users is in general an NP-complete[31] problem. However, the impacts of fairness and link quality constraints on the hardness of the problem have not been well studied. The link quality constraints refer to the signal-to-noise ratio or the distance ratio for using a channel, while fairness refers to evenly distributing the number of channels to users. The cognitive radio network can be mapped to a graph $G = (V, E)$, where the vertices, V, represent the secondary users (nodes), and the edges, E, represent the secondary transmissions (links) between these users. Some of the notations used in this subsection are summarized below:

V	Set of all nodes (vertices of G)
$e_{i,j}$	Link formed when node i transmits to node j (note that the existence of $e_{i,j}$ does not imply $e_{j,i}$)
E	Set of all possible links in the network (edges in G)
E_t	Set of links that are active in subchannel t, where $t > 0$
$SIR(e_{i,j}, t)$	Signal-to-interference ratio of edge $e_{i,j}$ in subchannel t
T	Total number of active/available subchannels
S_i	Set of subchannels assigned to link e_i

The signal-to-interference ratio (SIR) constraint on the quality of the signal requires that a channel assignment should ensure that all the links experience an SIR above a particular threshold, γ_{th}. If E_t is the set of all links assigned to a subchannel t, then channel assignment is said to satisfy the minimum SIR constraint if

$$SIR(e_{ij}, t) \geq \gamma_{th}, \forall t \in \{1, \ldots, T\}, \forall e_{ij} \in E_t. \tag{5.1}$$

If link e_i is assigned subchannel t, i.e., $e_i \in E_t$, then it is said to satisfy the minimum distance ratio constraint if $\forall e_j \in E_t - e_i$,

$$d(Tx(e_j), Rx(e_i)) \geq (1 + \delta_{th})R_{max}, \tag{5.2}$$

where R_{max} is the maximum transmission radius in the network, $Tx(e_j)$ denotes the transmitter of link e_j, and $Rx(e_i)$ is the receiver of link e_i. That is, the distance between the receiver of the given link and the transmitters of other active links sharing the same subchannel t should be larger by a factor of $1 + \delta_{th}$ compared to the maximum transmission radius of the network. It can be shown that the SIR constraint can be mapped to an equivalent distance ratio constraint.

Let S_i be the be the set of subchannels assigned to link e_i. Then S_i is represented as

$$S_i \triangleq \{t, e_i \in E_t \, \forall t \in \{1, \ldots, T\}\}. \tag{5.3}$$

Depending on the number of subchannels assigned to each link, channel assignment can be classified as unfair, 1-fair, and fair. A spectrum assignment is called unfair if there is at least one link that is not assigned to any subchannel, i.e., $\exists\, i \in \{1, \ldots, |E|\}$ such that $|S_i| = 0$. Such an assignment could lead to loss of connectivity in a multihop network and should be avoided. To preserve connectivity, it is essential that $|S_i| \geq 1 \; \forall i \in \{1, \ldots, |E|\}$. A channel assignment that ensures connectivity is called a 1-fair assignment. A channel assignment is called fair if $|S_i| = d \; \forall i \in \{1, \ldots, |E|\}$, where $d \geq 1$.

A spectrum assignment problem is represented as a two-tuple $(\mathbf{G}, \mathbf{C})$, where $\mathbf{G}$ is a graph representing the cognitive network, and $\mathbf{C}$ is the set of constraints (i.e., distance and fairness). Depending on the fairness criteria provided by the constraints, the channel assignment problems can be classified as

- distance-constrained unfair assignments
- distance-constrained 1-fair assignments
- distance-constrained fair assignments

Let J be the number of simultaneous transmissions in the network. Then it can be shown that the distance-constrained unfair assignment problem can be solved in polynomial time if $|E| >> J$, while the distance-constrained 1-fair assignment and distance-constrained fair assignment problems can be shown to be NP-complete. If the distance threshold is δ_{th}, the maximum transmission radius of a node is R_{max}, and the area covered by the network is A, then it can be shown that the maximum number of feasible distance-constrained fair transmissions in the network is upper-bounded by $\left(|E| - \pi \delta_{\text{th}}^2 R_{\text{max}}^2 n_0\right)^L$ where $L \triangleq \frac{4A}{\pi \delta_{\text{th}}^2 R_{\text{max}}^2}$.

5.4.2 Network Selection for Secondary Users

Another interesting and important topic for investigation is assignment and reassignment of secondary users to networks. Secondary users can choose networks based on several factors that include signal strength, quality-of-service (QoS), and price. Also, reassignment of users to different networks or different channels in the same network could result in a better spectrum utilization. Each user i enters the system with a minimum rate requirement and is willing to pay a maximum price. There are n networks, and each network contains a specified number of channels that can be used by a secondary user. Any secondary user, when allocated a channel in a network, causes interference to the primary users in the network. It is desired to limit the maximum interference to the primary users below a specified threshold. Users are to be assigned to networks to satisfy all the above constraints.

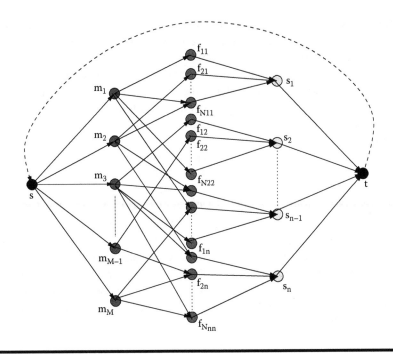

Figure 5.1 A network graph representation of cognitive radio systems.

The concept of multicommodity flows[32] can be applied to network selection and channel assignment to nodes in a cognitive radio network. The set of users, the set of available channels, and the set of available networks form nodes in a 5-partite graph (with additional source and sink nodes) as shown in Figure 5.1. In Figure 5.1, node m_i, $1 \le i \le M$, represents the ith user, node f_{jk} denotes the jth channel in the kth network, and node S_k denotes the kth network. Nodes s and t denote dummy source and sink nodes, respectively. The interference, rate of transmission, and price can be modeled as commodities in the network. This is further explained as follows. A directed edge (s, m_i) exists to all users. A directed edge (m_i, f_{jk}) exists if user m_i can use channel f_{jk}, i.e., user m_i senses channel f_{jk} to be available. A directed edge (f_{jl}, S_k) exists if and only if $l = k$. Edges (S_k, t) exist $\forall\, k$. Finally, there is a dummy directed edge from t to s, which translates the minimum cost flow problem to the max flow problem.

Along any edge e, the flow on the edge is represented as $[x_e^1\ x_e^2\ x_e^3\ x_e^4]$. x_e^1 denotes the price parameter on the edge, x_e^2 denotes the interference parameter, x_e^3 denotes the rate of transmission, and x_e^4 denotes usage of a particular channel in a network. For edges e of the form (s, m_i), the flow term x_e^1 denotes the price paid by user m_i. For any edge e of the form (m_i, f_{jk}), x_e^2 denotes the interference caused by user m_i to a primary user in network S_k using channel f_{jk}, and x_e^3 denotes the rate at which user

m_i transmits on channel f_{jk} in network S_k. For any edge e of the form (f_{jk}, S_k), the flow term x_e^2 denotes the total interference experienced by a primary user on channel f_{jk} in network S_k, and $x_e^4 = 0$ if channel f_{jk} is not assigned to any user in network S_k and $x_e^4 = 1$ otherwise. By assigning appropriate capacities to the edges, the channel assignment and network selection problem can then be modeled as a minimum cost flow problem and can be used for allocation and reassignment of channels and networks to users.

It can be shown that a newly arriving secondary user m_i can be assigned a channel f_{jk} in network S_k if and only if there exists a flow-augmenting directed path from s to t through m_i, f_{jk}, and S_k (i.e., there exists a directed flow-augmenting path $s \rightarrow m_i \rightarrow f_{jk} \rightarrow S_k \rightarrow t$). It can also be shown that the L commodity flow maximization model proposed results in at most $L \lfloor \frac{|A|}{2} \rfloor$ reassignments, where $|A|$ is the number of edges in the network. The computational complexity of the network assignment problem based on the multicommodity flow maximization approach can be shown to be $O(M + N + F)^3 \log (M + N + F)$, where M is the total number of users, N is the total number of available networks, and $|F|$ is the total number of channels in all networks.

5.4.3 Secrecy Capacity of Cognitive Radio Networks

Secrecy capacity was studied for systems with keyless security. A system typically consists of a source (or a transmitter), a destination (or a receiver), and an eavesdropper as shown in Figure 5.2. The source transmits information, which is received both by the receiver and the eavesdropper. Secrecy capacity is roughly the maximum rate at which the source can transmit such that the bit error rate (BER) at the destination approaches zero while that at the eavesdropper approaches 0.5. For some cases, the secrecy capacity is the difference between the Shannon capacity of the channel between the source and destination and that between the source and eavesdropper.

For example, in the system shown in Figure 5.2, let the channel gain between the source and destination be h_d and let the channel gain between

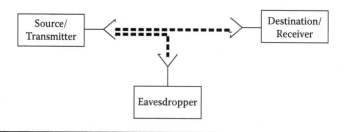

Figure 5.2 A wired/wireless channel with an eavesdropper.

the source and the eavesdropper be h_e. If the channel bandwidth is W and the source transmits at power P_s, then the secrecy capacity, C_S, is given by

$$C_S = W \ln \left(\frac{1 + P_s h_d}{1 + P_s h_e} \right).$$

(5.4)

Wyner[33] first showed that positive secrecy capacity can be achieved without having a secret key of larger entropy than that of the message. Barros and Rodrigues[34] used complete channel state information (CSI) to study the secrecy capacity for wireless channels. Gopala et al. extended Wyner's[33] work for fading wiretap channels[35] and then to fading wireless channels.[36,37] The authors analyzed three cases: (1) full channel state information between source and eavesdropper and source and the destination, (2) full channel state information between source and destination but no channel state information between the source and eavesdropper, and (3) no channel state information between source and destination as well as between source and eavesdropper. The secrecy capacity for each of these cases was obtained with constraints on the average transmission power of the source. Lai and El Gamal[38] studied cooperative secrecy by having a relay node help secure communication. They considered a system with an additional node called a relay node. The relay node decodes the message from the source, encodes it again, and transmits it to the destination. The eavesdropper can listen to the source as well as the relay. The relay node, once in a while, randomly transmits a message which is not in the codebook of the destination. The destination discards such messages but the eavesdropper is "confused." The authors achieve bounds on the rate of equivocation for such a system. Tekin and Yener[39] studied multiple source-destination pairs and cooperative jamming. Multiple source-destination pairs were considered and the sum secrecy capacity of the network was maximized. The authors also studied cooperative jamming, wherein some users transmit to add to the interference of the eavesdropper and thereby improve secrecy capacity. To our knowledge, no work has explicitly addressed the secrecy capacity of cognitive radio networks. The secrecy capacity of cognitive radio networks with underlay or overlay transmissions from secondary users is still unknown.

The concept of transmission power allocation with pricing and its effect on the secrecy capacity of cognitive radio networks is an interesting problem to investigate. In the context of cognitive radio networks, this amounts to allocation of powers to secondary users so as to maximize the individual secrecy capacity or the sum secrecy capacity of primary users. It may be possible to admit some secondary users to improve secrecy capacity of a primary user at the cost of added interference to the primary user.

The problem is formulated as follows: Consider a wireless system with bandwidth W and with N sources/transmitters and N destinations/receivers. Let the channel gain from source i to destination j be h_{ij}. Let source i

transmit at power P_i. The signal-to-interference ratio (SIR) obtained at destination i, x_i is given by

$$x_i = \frac{P_i h_{ii} G_i}{\sum_{j \neq i} P_j h_{ji} + N_0 W},$$ (5.5)

where G_i is the gain of user i, which could be due to either error control coding or spectrum spreading, and N_0 is the power spectral density of white noise. In the absence of an eavesdropper, each source would transmit at a power P_i^* that results in maximizing the satisfaction experienced by destination i. The satisfaction experience by destination i is modeled by a utility function $f(x_i)$, which has the following properties according to microeconomics[40,41] and game theory:[42,43]

- $f(x_i) \geq 0$.
- $f'(x_i) \geq 0$, i.e., the utility function is a nondecreasing function of the SIR.
- $f''(x_i) \leq 0$, i.e., the utility function is a concave function of SIR. This is also termed as the law of diminishing marginal utility.
- $\lim_{x_i \to \infty} f'(x_i) = 0$.

The function $f(x_i) = W \ln (1 + x_i)$ satisfies the above-mentioned properties. The power allocation problem to the sources can be formulated as an optimization problem as mentioned below.

$$\text{Maximize } f(x_i),$$ (5.6)

subject to

$$0 \leq P_i \leq P_{\max} \ \forall i.$$ (5.7)

Another formulation could be

$$\text{Maximize } \sum_{i=1}^{N} f(x_i),$$ (5.8)

subject to the constraints in Equation (5.7). Since the SIR x_i depends on the transmission powers of all the sources (as described in Equation (5.5)), the power allocation problem described by Equations (5.6) and (5.7) can be viewed as an N-player noncooperative game.[42,43] It can be shown that the Pareto optimal solutions occur when $P_i = P_{\max}$ for some i and the Nash equilibrium is $P_i = P_{\max} \ \forall \ i$. The solution for the optimization problem in Equation (5.8) subject to the constraints Equation (5.7) can be obtained as $P_i = P_{\max} \ \forall \ i \in S \subset \{1, 2, \ldots, N\}$ and $P_k = 0$, $k \notin S$. Therefore, it is observed that the allocation results in "unfairness." One possible means of performing a fair allocation is to impose a pricing or a penalty function to all users* such that users experiencing larger SIR pay a larger price. Thus, by defining a pricing function $v(x_i) \; \forall \; i$, the optimal power allocation problem can then be reformulated as

$$\text{Maximize } \hat{f}(x_i) = f(x_i) - \lambda v(x_i), \tag{5.9}$$

subject to constraint (5.7). In Equation (5.9), λ is the pricing parameter.

By choosing the pricing function $v(x_i)$ suitably, the power allocation problem can be solved as an SIR allocation problem and the powers of each source can be obtained from the respective SIR at each destination by solving the matrix equation

$$\mathbf{p} = N_0 W (\mathbf{D}_1 - \mathbf{A})^{-1} \mathbf{D}_2^{-1} \mathbf{1}, \tag{5.10}$$

where $\mathbf{p} = [P_1 \quad P_2 \quad P_3 \quad \cdots \quad P_N]^T$, $\mathbf{1}$ is the column vector of length N with all entries unity, $\mathbf{I}$ is the $N \times N$ identity matrix, and $\mathbf{A}$, $\mathbf{D}_1$, and $\mathbf{D}_2$ are given by

$$\mathbf{A} = \begin{bmatrix} 0 & \frac{b_{21}}{b_{11}} & \frac{b_{31}}{b_{11}} & \cdots & \frac{b_{N1}}{b_{11}} \\ \frac{b_{12}}{b_{22}} & 0 & \frac{b_{32}}{b_{22}} & \cdots & \frac{b_{N2}}{b_{22}} \\ \frac{b_{13}}{b_{33}} & \frac{b_{23}}{b_{33}} & 0 & \cdots & \frac{b_{N3}}{b_{33}} \\ \vdots & \vdots & \vdots & \ddots & \vdots \\ \frac{b_{1N}}{b_{NN}} & \frac{b_{2N}}{b_{NN}} & \frac{b_{3N}}{b_{NN}} & \cdots & 0 \end{bmatrix}, \tag{5.11}$$

$$\mathbf{D}_1 = \begin{bmatrix} \frac{G_1}{x_1} & 0 & 0 & \cdots & 0 \\ 0 & \frac{G_2}{x_2} & 0 & \cdots & 0 \\ 0 & 0 & \frac{G_3}{x_3} & \cdots & 0 \\ \vdots & \vdots & \vdots & \ddots & \vdots \\ 0 & 0 & 0 & \cdots & \frac{G_N}{x_N} \end{bmatrix}, \tag{5.12}$$

---

* The terms "users," "receivers," and "destinations" are used interchangeably here and henceforth in this section.

and

$$
\mathbf{D}_2 = \begin{bmatrix} b_{11} & 0 & 0 & \cdots & 0 \\ 0 & b_{22} & 0 & \cdots & 0 \\ 0 & 0 & b_{33} & \cdots & 0 \\ \vdots & \vdots & \vdots & \ddots & \vdots \\ 0 & 0 & 0 & \cdots & b_{NN} \end{bmatrix}.
\tag{5.13}
$$

It is observed that the matrix $(\mathbf{D}_1 - \mathbf{A})$ is a $\mathcal{Z}$-matrix[44] and it is possible to obtain nonnegative solutions for $\mathbf{p}$ if and only if $(\mathbf{I} - \mathbf{D}_1^{-1}\mathbf{A})$ is an $\mathcal{M}$-matrix[44]. From the properties of $\mathcal{M}$-matrices,[45] it is possible to obtain lower bounds and upper bounds on the pricing parameter λ such that the solution to the power allocation problem is feasible. The obtained solutions can also be used to study the asymptotic behavior of the system, i.e., when $W \to \infty$ and $N \to \infty$ such that $N/W = \rho$.

In the presence of an eavesdropper, it is then possible to evaluate the secrecy capacity at each receiver i if the channel gains from all sources to the eavesdropper are known. It can be shown that maximizing the secrecy capacity in the presence of an eavesdropper results in the same power allocation as that of maximizing the utility (with pricing) in the absence of any eavesdropper.

References

[1] A. Harrington, C. Hong, and T. Piazza, Software defined radio: The revolution of wireless communication, white paper, Ball State University, http://www.bsu.edu/cics/alumni/whitepapers/

[2] J. Lansford, UWB coexistence and cognitive radio, *Proc. IEEE Intl. Workshop on Ultra wide band systems (UWBS 2004),* May 2004.

[3] X. Liu and W. Wang, On the characteristics of spectrum agile radio networks, *Proc. IEEE Intl. Symp. on New Frontiers in Dynamic Spectrum Access (DySPAN 2005),* November 2005.

[4] I. F. Akyildiz, W. Lee, M. C. Vuran, and S. Mohanty, Next generation/ dynamic spectrum access/cognitive radio: A survey, Elsevier J. on *Computer Networks,* vol. 50, pp. 2127–2158, 2006.

[5] IEEE Standards Coordinating Committee 41 for Dynamic Spectrum. Access, http://www.scc41.org

[6] A. Gamst, Some lower bounds for a class of frequency assignment problems, *IEEE Trans. on Vehic. Technol.,* vol. 35, no. 1, pp. 8–14, February 1986.

[7] R. J. McEliece and K. N. Sivarajan, Performance limits of channelized cellular systems, *IEEE Trans. on Info. Theory*, vol. 40, no. 1, pp. 21–34, January 1994.

[8] S. Sarkar and K. N. Sivarajan, Hypergraph models for cellular mobile communication systems, *IEEE Trans. on Veh. Tech.*, vol. 47, no. 3, pp. 460–471, May 1998.

[9] S. Sarkar and K. N. Sivarajan, Channel assignment algorithms satisfying co-channel and adjacent channel reuse constraints in cellular mobile networks, *IEEE Trans. on Veh. Tech.*, vol. 51, no. 5, pp. 954–967, September 2002.

[10] R. K. Gupta, K. N. Sivarajan, and C. Yu, An algorithm for channel assignment schemes in cellular systems using hypergraph models, *Proc. NCC'2000,* New Delhi, January 2000.

[11] M. G. Jayateertha and K. N. Sivarajan, Design of highway cellular systems, *Proc. NCC'1997*, Madras, January 1997.

[12] M. Sidi and D. Starobinski, New call blocking versus handoff blocking in cellular networks, *Proc. IEEE INFOCOM'96*, April 1996.

[13] D. E. Everitt and N. W. Macfadyen, Analysis of multicellular mobile radio-telephone systems: A model and evaluation, *British Telecommun. Tech. Jl.,* vol. 1, pp. 37–45, 1983.

[14] F. P. Kelly, Blocking probability in large circuit switched networks, *Advances in Applied Probability*, vol. 18, pp. 473–505, April-June 1986.

[15] K. Varghese, Decentralized dynamic channel allocation in cellular networks, M.E Project Report, Department of ECE, Indian Institute of Science, Bangalore, January 1999.

[16] L. Cimini, G. Foschini, I. Chih-Lin, and Z. Miljanic, Call blocking performance of distributed algorithms for dynamic channel allocation in micro cells, *IEEE Trans. on Commun.*, vol. 42, no. 8, pp. 2600–2607, August 1994.

[17] S. Anand, A. Sridharan, and K. N. Sivarajan, Performance analysis of channelized cellular systems with dynamic channel allocation, *IEEE Trans. on Vehic. Technol.*, vol. 52, no. 4, pp. 847–859, July 2003.

[18] M. Eriksson, Dynamic single frequency networks, *IEEE Jl. on Sel. Areas in Commun.,* vol. 19, no. 10, pp. 1905–1914, October 2001.

[19] J. C. I. Chuang and N. R. Sollenberger, Spectrum resource allocation for wireless packet access with application to advanced cellular internet service, *IEEE Jl. on Sel. Areas in Commun.,* vol. 16, no. 6, pp. 820–829, June 1998.

[20] R. Jayaparvathy, S. Anand, and S. Srikanth, Performance analysis of dynamic packet assignment in cellular systems with OFDMA, *IEE Proc. on Commun.,* vol. 152, no. 1, pp. 45–52, February 2005.

[21] IEEE P802.16: Draft amendment to IEEE standard on local and metropolitan area networks—Part 16: Air interface for fixed wireless access systems-Medium access control modifications and additional physical layer specifications for 2–11 GHz, December 2004.

[22] H. Zheng and C. Peng, Collaboration and fairness in opportunistic spectrum access, *Microsoft Technical Report, MSR-TR-2005-23,* February 2005.

[23] C. Xin, B. Xie, and C. C. Shen, A novel layered graph model for topology formation and routing in dynamic spectrum access networks, *Proc. IEEE Intl. Symp. on New Frontiers in Dynamic Spectrum Access (DySPAN 2005)*, November 2005.

[24] M. E. Streenstrup, Opportunistic use of radio-frequency spectrum: A network perspective, *Proc. IEEE Intl. Symp. on New Frontiers in Dynamic Spectrum Access (DySPAN 2005)*, November 2005.

[25] Y. T. Hou, Y. Shi, and H. D. Sherali, Optimal spectrum sharing for multi-hop software defined radio networks, *Proc. IEEE INFOCOM'2007*, May 2007.

[26] A. Larcher, H. Sun, M. V. Schaar, and Z. Ding, Decentralized transmission strategies for delay sensitive applications in spectrum agile networks, *Proc. IEEE Pkt. Video Workshop 2004*, December 2004.

[27] Y. Xing, C. Mathur, M. A. Haleem, R. Chandramouli, and K. P. Subbalakshmi, Real-time secondary spectrum sharing with QoS provisioning, *Proc. IEEE Consumer Commun. and Networking Conf. (CCNC 2006)*, January 2006.

[28] Y. Xing, R. Chandramouli, S. Mangold, and N. Sai Shankar, Dynamic spectrum access in open spectrum wireless networks, *IEEE Jl. on Sel. Areas in Commun.*, vol. 24, no. 3, pp. 626–637, March 2006.

[29] Y. Xing, R. Chandramouli, and C. Cordeiro, Price dynamics in competitive agile spectrum access markets, to appear *IEEE Jl. on Sel. Areas in Commun.*

[30] M. Devroye, P. Mitran, and V. Tarokh, Achievable rates in cognitive radio networks, *IEEE Trans. on Info. Theory*, vol. 52, no. 5, pp. 1813–1827, May 2006.

[31] M. R. Garey and D. S. Johnson, *Computers and Intractability: A Guide to the Theory of NP Completeness*, V. Klee Ed. Freedman and Co., New York, 2003.

[32] R. K. Ahuja, T. L. Magnanti, and J. B. Orlin: *Network Flows: Theory, Algorithms and Applications*, Prentice Hall Inc., 1993.

[33] A. D. Wyner, The wire-tap channel, *Bell Systems Tech, Jl.*, vol. 54, no. 8, pp. 1355–1387, 1995.

[34] J. Barros and M. R. D. Rodrigues, Secrecy capacity of wireless channels, *Intl. Symposium on Info. Theory (ISIT 2006)*, July 2006.

[35] P. K. Gopala, L. Lai, and H. Elgamal, Secrecy capacity of fading wiretap channel, *ICASSP 2007*.

[36] P. K. Gopala, L. Lai, H. Elgamal, On the secrecy capacity of fading channels, *Intl. Symposium on Info. Theory (ISIT 2007)*, January 2007.

[37] P. K. Gopala, L. Lai, and H. Elgamal, On the secrecy capacity of fading channels, accepted in *IEEE Trans. on Info. Theory*.

[38] L. Lai and H. El Gamal, The relay eavesdropper channel:Co-operation for secrecy, accepted in *IEEE Trans. on Info. Theory*.

[39] E. Tekin and A. Yener, The general Gaussian multiple access and two-way wire-tap channels: Achievable capacity and cooperative jamming, *IEEE Trans. on Info. Theory*, Spl. Issue on Info. Theoretic Security.

[40] D. M. Kreps, *A Course in Microeconomic Theory*, Prentice Hall, 1990.

[41] H. Varian, *Intermediate Microeconomics: A Modern Approach*, W. W. Norton and Co., 1996.

[42] R. Gibbons, *Game Theory for Applied Economists*, Princeton University Press, 1992.

[43] H. Gintis, *Game Theory Evolving*, Princeton University Press, 1997.

[44] R. Horn and C. Johnson, *Topics in Matrix Analysis*, Cambridge University Press, 1991.

[45] M. Fielder and V. Ptak, On matrices with non-positive off-diagonal elements and positive principal minors, *Jl. of Math., Czechoslovakia,* vol. 12, pp. 382–400, December 1962.

Chapter 6

Development and Trends in Medium Access Control Design for Cognitive Radio Networks

Chao Zou, Chunxiao Chigan, and Zhi Tian

Contents

6.1 Introduction

Cognitive radio (CR) technology offers a much more flexible way to utilize the wireless spectrum than traditional radio technology based on a fixed spectrum access policy. As a result, the medium access control (MAC) layer protocols for CR networks are quite diversified from the perspectives of application scenarios, spectrum sharing mechanisms, control message sets, and the styles of handshakes among CR users. Although the precise definition of cognitive radio is still under debate, it is commonly accepted that MAC protocols for CR networks are meant to possess one or several of the following functionalities:

1. The capability to handle dynamic access over multiple channels. Multichannel access has become a de facto requirement for CR networks. For secondary CRs sharing licensed bands with primary users, the channel availability can be time varying, which requires that medium access be carried out dynamically over multiple channel opportunities. For coexisting CRs in unlicensed bands, multichannel access allows nearby users to share spectrum, which increases the spatial usage efficiency and thus considerably improves the throughput of the entire network. Furthermore, multichannel access makes it feasible to separate the control channel from data channels so as to eliminate collisions between control messages and data packets. When each CR node is equipped with multiple transceivers, the use of multichannel access allows for simultaneous control message exchange and data communication which reduces network latency.

2. The capability to coordinate with other spectrum-sharing CRs. Coordination is another important feature in most CR networks. A collaborative radio can use services of other radios as leverage to further

its goals or the goals of the network. Coordination can help CRs to acquire more accurate information about their external environments and accordingly make more efficient spectrum access decisions. Particularly, coordination is essential for coexistence between primary and secondary users. A CR can only determine whether a primary transmitter is presently emitting in a certain spectrum, but does not know whether the primary receiver is locally present; however, the objects to be protected from the interference of secondary users are primary receivers. Because CRs usually cannot communicate directly with primary users, it is imperative for secondary CR users to perform coordinated negotiation in order to estimate the spectrum opportunities accurately. Coordination, by its nature, consumes medium resources and needs to be done in a controlled manner through MAC.

3. The capability to reason using intelligent decision algorithms. Intelligence is a distinct feature that sets cognitive radios apart from conventional radios. CRs are "smarter" than ordinary radios in several respects. They can not only automatically detect spectrum holes and switch channels accordingly, but also optimize the channel access parameters such as modulation and coding technique, transmission power, and transmission time, and jointly configure multiple layers of the network stack to improve overall performance. Many CR MAC designs have adopted advanced reasoning algorithms based on graph theory, stochastic theory, game theory, and so on. Furthermore, some cognitive radios can predict the channel availability from historical records and observed environment. The input-output information flow that passes through the reasoning engines needs to be controlled by the MAC protocol.

With different levels of hardware capability and software complexity, the sensing ability, adaptability, and intelligence degree of CRs are kaleidoscopic, which gives considerable room for MAC design flexibility in CR networks. Depending on the targeted application scenario and network structure, the challenges for MAC protocol design in CR networks can be quite diverse. Nevertheless, there are several common challenges for most MAC protocol designs, as discussed below.

6.1.1 Coexistence of Heterogeneous Communication Devices

The coexistence issue in CR networks includes two aspects: one is the coexistence of primary users and secondary CR users in licensed bands, and the other is the coexistence of CRs in unlicensed bands.

For the coexistence of primary and secondary users, it is required that transmission by secondary users does not interfere with the communication of primary users. A common approach is for secondary users to first detect the spectrum holes in which the incumbent devices are inactive. These temporarily idle spectrum holes are then allocated among the secondary users. Another approach is that the secondary users control their transmission power so that the interference imposing on primary users does not exceed a tolerable threshold.

Channel allocation is a key function in approaches to solving the problem of coexistence of CRs in both licensed and unlicensed bands. In general, spectrum allocation in a licensed band cannot be as flexible as that in an unlicensed band, due to strict spectrum regulatory limitations in licensed bands.

6.1.2 Common Control Channel Issue

This issue arises when a CR network operates in a licensed band. As spectrum availability for secondary users fluctuates over time and location, it is hard to find a common channel to exchange messages on the scale of the entire network. There are two possible solutions: one is to assign the common control channel locally since nodes in a given neighborhood observe the same or similar spectrum activities; the other is to allow the control messages to be exchanged in any channel. A MAC protocol for CR networks would need to properly handle this issue in order to percolate control messages throughout the network.

6.1.3 Multiple Hidden Terminal Problem

The multiple hidden terminal problem arises when multiple users attempt to access one of multiple channels. The root of this problem is that a user with a single transceiver can use only one channel in each sensing period. Therefore, a control message on one channel between a transmitter and a receiver may collide with a data communication on the same channel between another pair of users who have just finished their handshake on a different channel. Details of the multiple hidden terminal problem will be illustrated in detail in Section 6.2.

6.1.4 Diversified Local Goals and Actions of CRs

The strategy space for the whole network may be multidimensional, i.e., a CR user can dynamically adjust multiple parameters to access spectrum, including transmission power, channel selection, modulation method, antenna parameters, etc. Hence, the spectrum allocation may entail a multivariable optimization problem, which increases the allocation complexity. Also, local goals of different users can differ in terms of the quality of service (QoS) requirements, the required energy consumption rate, the latency constraint, and so on. In light of the heterogeneous local goals among CRs, spectrum should not be allocated in a uniform way, but be allocated to meet each user's individual requirements.

6.1.5   Recursive Adaptation

The adaptation of CR users is a complex process because each CR's adaptation changes the operating environments for all other radios. Collectively, the individual actions of a group of CRs would appear as an interactive recursive decision process, which may negatively impact the network performance if the adaptation process cannot quickly converge.

6.1.6   Coordination and Negotiation

Coordination and negotiation are commonly adopted in distributed negotiation-based MAC protocols. A critical motivation for negotiation is to provide CRs with accurate and adequate information about the operating environment. Due to the scalability issue, negotiation cannot be performed simultaneously across the entire network. It is thus necessary for CR users to autonomously form local clusters to negotiate. The clustering algorithm, including the processes of cluster construction, header selection, and cluster management, needs to be specified in negotiation-based MAC protocols.

6.1.7   Cross-Layer Design

In CR networks, the spectrum sharing mechanism and efficiency depend on the network configuration at networking and other higher layers, while the performance of these higher layers hinges on MAC-layer activities. In such a setting a cross-layer protocol design is a promising approach to enhance the overall network performance. The high flexibility and programmability of CRs lend themselves well to implementing optimal protocols across layers, yet few papers have fully explored the cross-layer design issue for CR networks. In Section 6.4, we will introduce several MAC protocols based on cross-layer design for ad hoc networks and discuss how they can be modified and enhanced for CR networks.

In the ensuing sections, we will provide a systematic description of existing MAC protocols for CR networks and explain how they cope with the aforementioned design challenges. We categorize the MAC protocols into two main types: one is direct-access-based MAC, which refers to protocols with a single-dimensional strategy space and a goal realized by a pure control message handshake without a reasoning algorithm, and the other is a technique based on dynamic spectrum allocation called DSA-driven MAC, which refers to protocols with an embedded reasoning algorithm. The rest of this chapter is organized as follows: Section 6.2 introduces direct-access-based MAC protocols for CR networks, including distributed protocols for infrastructureless networks and centralized protocols for infrastructure networks. The distributed protocols will be further divided into contention-based and coordination-based types for elaboration. Section 6.3 organizes and presents DSA-driven MAC protocols according to their adopted DSA algorithms, and analyzes their merits and drawbacks. Section 6.4 is dedicated to MAC protocols with cross-layer design, along with recommendations on how to extend these protocols to CR networks. Concluding remarks are given in Section 6.5.

6.2 Direct-Access-Based Cognitive Radio MAC Protocol

Direct access has two meanings: First, the access strategy of a CR user is direct—the CR user either selects a channel to access or gives up accessing any channel. Second, the goal of the CR user is also direct—each user only tries to reach its local goal without any consideration of others or the global network, i.e., there is no global optimization issue. Owing to these design principles, in a direct-access-based MAC protocol, the strategy of CR user i can be represented by a one-dimensional variable S_i that represents the channel selection. If $S_i = k$, $k = 1, \ldots, K$, then the CR user decides to communicate over the kth channel out of the K available channels; if $S_i = 0$, then the CR user is either idle or waiting for vacant channels. Meanwhile, the goal of CR user i at the MAC layer can be expressed as a single-variable, single-objective optimization problem: $\max U_i(S_i)$. Protocols of this type are generally inherited from IEEE 802.11 with improvements to adapt to multichannel environments. Based on the structure of the network, existing direct-access-based multichannel MAC protocols can be classified as centralized MAC, which is used for infrastructure networks, and distributed MAC, for infrastructureless networks. For distributed MAC protocols, there are generally two classes: one is the contention-based MAC protocol, in which the negotiation (handshake) only happens between the sender and its intended receiver; the other is the coordination-based MAC protocol, in which each node coordinates with other nodes besides its intended sender or receiver to make decisions. We present these two classes of MAC below.

6.2.1 Direct Contention-Based MAC Protocols

Direct contention-based MAC protocols revolve around two core issues. First, as this type of protocol is based on sender-receiver handshaking, the sender and its receiver should find a control channel on which to exchange control messages. Second, the sender and receiver should choose an unoccupied channel on which both of them may perform duplex data communication.

In [14], distributed multichannel MAC protocols for ad hoc wireless LANs are classified into three general categories: (i) common control channel with a dedicated control radio, (ii) split-phase, and (iii) common hopping sequence. This taxonomy also makes sense for direct contention-based CR MAC protocols for CR networks. Next, we introduce each of the three types of distributed CR MAC protocols and also give brief comparisons among them.

6.2.1.1 Common Control Channel Approach

In the common control channel approach, it is preferred that each CR be equipped with one control radio and one data radio. The control radio is always tuned to the common control channel (or code) where all devices contend for access to the data channels (or codes). Since each device always listens to the control channel, it can overhear control messages from its neighbors and thus easily keep track of the status of all channels and neighbors. It is also possible for the CR to have a single transceiver. In this case, the data sender must be tuned to the control channel to handshake with its intended receiver before sending data. Since CRs might not acquire the correct specification of the data channel due to the lack of consecutive monitoring over the control channel, they may need to handshake again on the data channel.

This approach is used by references [1-6], including RBCS-MAC (receiver-based channel selection MAC) [1], DCA (dynamic channel assignment) [2], DCA-PC (dynamic channel assignment with power control) [3], DPC (dynamic private channel) [4,5], and DCSS-MAC (distributed coordinated spectrum sharing MAC) [6]. The sender–receiver handshake follows the common wisdom that, upon receipt of the request packet, which includes the available channel list from the sender, over a certain common control channel, the receiver replies with the selected data channel based on a certain spectrum selection policy. This handshake is referred to as the CFSR (channel filtering sender-receiver) handshake mechanism in this chapter.

Early protocols of this type were not intended for CR networks; nevertheless, they are suitable for CR networks working in unlicensed bands. The CFSR handshake mechanism has been widely adopted by many recent MAC protocols designed for CR networks. We will first introduce MAC protocols that can be implemented in CR networks in unlicensed bands, followed by the presentation of MAC protocols designed for CR networks in licensed bands.

6.2.1.1.1 MAC Protocols in Unlicensed Bands

A typical MAC protocol based on the common control channel approach is RBCS-MAC [1], which was developed to address the issue of collision between control messages and data packets through the multi-channel access method. The network chooses a dedicated channel as the control channel and other channels as data channels. The control channel is used to exchange control messages such as Request to Send (RTS), Clear to Send (CTS), Reservation (RES), and so on. Before sending an RTS, the data sender first scans all the data channels and builds a list of free data channels available for transmission. If the free data channel list is empty, the data sender enters backoff and re-senses the data channels after the expiry of the backoff time. The list of nonempty free data channels will be appended to the RTS and sent to the data receiver. Except for the data receiver, other nodes overhearing the RTS on the control channel would defer their transmissions until the completion of the CTS. This is in contrast to IEEE 802.11, in which other nodes wait until the completion of the Acknowledgment (ACK) packet. The rationale is that both the data and ACK packets are transmitted in a data channel and thus cannot interfere with other RTS/CTS handshakes.

After the data receiver receives the RTS successfully, it also checks all the data channels and creates its own free data channel list. If its own list has some common free channels with the list in RTS, then the data receiver selects the best one among them (the channel requiring the minimum received power) and sends the selected channel information in the CTS packet. Otherwise, the data receiver refuses to send a CTS and the data sender retries sending RTS after the backoff time runs out.

After the data sender receives the CTS, it uses the channel indicated in the CTS to transmit data. Other nodes in the vicinity of the data receiver refrain from transmitting on the data channel indicated by the CTS for the duration of the entire data transmission (including ACK). If the data receiver successfully receives the data packet, it sends ACK back to the data sender on the same data channel. If the data sender fails to receive an ACK, it enters backoff and retry after timeout.

The DCA MAC protocol in [2] alleviates the sensing burden in RBCS-MAC by designing a channel usage list (CUL) into each node, which enables it to set up a free channel list (FCL) and calculate the network allocation vector (NAV). DCA assumes that each node is equipped with two transceivers, one of which monitors the common control channel to collect neighboring node information, while the other is used for data communication. Each node X stores a CUL, which keeps records of when the neighboring hosts access channels. To do so, entry i of CUL ($CUL[i]$) of node X has three fields: (1) *host*, a neighbor host ID of X; (2) *ch*, the data channel number used by *host*; and (3) *rel_time*, the time when channel *ch* will be released by *host*. The CUL is built by listening to control messages on the control channel. With the aid of the CUL, the FCL can be set up and the NAV can be calculated as well. The entire process for RTS/CTS handshake can be detailed in ten steps. Here we only focus on how the FCLs are built by the data sender and receiver and how the NAVs are calculated.

To build an FCL for a data sender, the data sender checks its CUL to find a channel D_j that will be free after the RTS/CTS handshake. This condition can be checked as follows: for all the entries in CUL, if $CUL[i].cb = D_j$, then

$$CUL[i].rel\_time \leq T_{curr} + (T_{DIFS} + T_{RTS} + T_{SIFS} + T_{CTS}). \qquad (6.1)$$

Here T_{curr} is the current clock time of the sender, T_{RTS} is the time to transmit RTS, T_{CTS} is the time to transmit CTS, T_{SIFS} is the time length of interframe spacing, and T_{DIFS} is the length of distributed interframe spacing. All D_js that satisfy the above condition are added into the FCL.

For other unintended hosts receiving the RTS, they have to inhibit themselves from using the control channel for a period of NAV_{RTS}:

$$NAV_{RTS} = 2T_{SIFS} + T_{CTS} + T_{RES} + 2\tau. \qquad (6.2)$$

where τ is the maximum propagation time for the control message. This is to avoid having the unintended hosts interrupt the RTS/CTS/RES dialogue between the data sender and receiver.

The data receiver, upon receiving the RTS, checks all the entries in its CUL to determine whether there is any data channel D_j satisfying $D_j \in FCL$ and

$$\text{if } CUL[i].cb = D_j, \text{ then } CUL[i].rel\_time \leq T_{curr} + (T_{SIFS} + T_{CTS}). \qquad (6.3)$$

If there exists such a channel, then this channel is a free channel that can be used for data communication. The data receiver picks any such D_j and put its ID in the CTS. Another item contained in the CTS is NAV_{CTS}, satisfying $NAV_{CTS} = L_d/B_d + T_{ACK} + 2\tau$, where L_d is the length of data that will be received, B_d is the bandwidth of the data channel, and T_{ACK} is time to send ACK. Any node receiving the CTS will update the entry in its CUL that satisfies $CUL[i].host = ID_{data\_receiver}$ to $CUL[k].cb = D_j$ and $CUL[k].rel\_time = T_{curr} + NAV_{CTS}$.

The DCA is more advantageous than the RBCS-MAC in terms of the sensing cost, because each node does not need to keep sensing both control and data channels. Indeed, in DCA each node only needs to sense the control channel to build the CULs, while the data-transmission time of each node can be calculated from the NAV derived from its CUL.

The DCA-PC MAC protocol in [3] further improves the DCA by incorporating the functionality of power control, which brings several advantages: (1) it can reduce co-channel interference with neighboring hosts, (2) it may increase channel reuse in a physical area (cf. Figure 6.1), and (3) it can save the energy for transmission.

Specifically, in DCA-PC, each mobile host A keeps an array $POWER[N]$, where N is the number of neighboring nodes of A. $POWER[id]$ registers

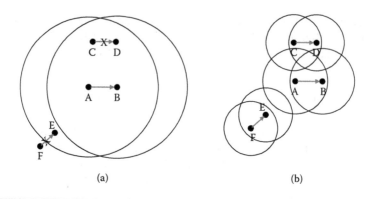

Figure 6.1 Transmission scenarios with (a) no power control and (b) power control. In (a), node A transmits at a high power that covers both the intended receiver B, and the unintended nodes D and E, which prevents transmitters C and E from using the same channel. In (b), all transmitters A, C and F use the same channel, but control their transmission powers so that there are no mutual interferences among the three links.

the level of power that should be used by A to send a data packet to the *i*th neighbor.

Suppose that node Y wants to send a packet to node X at a transmission power P_{Y_T}, such that X receives the packet with the desired received power P_{X_R}. In a wireless channel, the signal power falls off with the distance d between the transmitter and receiver, according to the path-loss model

$$P_{X\_R} = cP_{Y\_T}d^{-n}, \tag{6.4}$$

where $n \geq 2$ is the path-loss exponent dependent on the channel characteristic, and c is a constant dependent on the antenna gains and signal wavelength.

Note that the control message is always transmitted at the maximum power $P_{T,\max}$. Thus, the power of the control message received by X is

$$P_{X\_R,\max} = cP_{T,\max}d^{-n}. \tag{6.5}$$

With Equations (6.4) and (6.5), we have $P_{Y\_T} = P_{X\_R}P_{T,\max}P_{X\_R,\max}^{-1}$. In DCA-PC, the designated power P_{X_R} for all nodes is set to $P_{R,\min}$, which is the minimum power level at which a mobile host can recover signals from noise. Hence, the minimum power consumption required for Y is

$$P_{Y\_T} = P_{R,\min}P_{T,\max}P_{X\_R,\max}^{-1}. \tag{6.6}$$

If node X always monitors the communications from Y on the control channel and tells Y the observed $P_{X\_R,\max}$, then node Y can use (6.6) to control the value of *POWER[id]* to P_{Y_T}.

Because of the added power control functionality, the DCA-PC protocol can enhance the CUL in the DCA protocol by appending a fourth field, *CUL [i].int*. This indicates whether the signal sent by *CUL [i].host* on the data channel *CUL [i].ch* can be overheard by node *i*. Consequently, the process of building the FCL in DCA-PC differs slightly from that in DCA. In DCA-PC, each data sender checks its CUL to find a channel D_j that satisfies

$$CUL[i].ch = D_j \Rightarrow CUL[i].rel\_time \le T_{curr} + (T_{SIFS} + T_{CTS})$$

$$\vee \quad \{(CUL[i].int = 0) \wedge (POWER[CUL[i].host] > POWER[B])\}. \quad (6.7)$$

Compared with DCA, D_j has an additional chance to be selected by the data sender, i.e., the signal from host *CUL[i].host* on channel D_j does not interfere with the data sender, and the yet-to-be-transmitted signals from the data sender to the intended receiver will not interfere with host *CUL[i].host*. The same channel selection criteria are also be used at the data receiver side.

With the aid of multi-channel access, most direct contention-based MAC protocols built on the common control channel approach improve the CR network performance in terms of throughput and average delay compared with conventional networks. Simulation results in [1–3] have quantified the performance gains in throughput achieved by using RBCS-MAC, DCA, and DCA-PC over the conventional IEEE 802.11. Also, compared with IEEE 802.11, these CR MAC protocols incur fewer packet collisions and hence fewer retransmissions, particularly at high loads, which in turn considerably reduces the average packet delay of the network.

6.2.1.1.2 MAC Protocols for Licensed Bands

Traditional multiple-channel MAC protocols based on the common control channel approach are not suitable for CR networks in licensed bands with primary-secondary user hierarchy, since they do not consider the differences between primary and secondary users. For CR networks operating in licensed bands, secondary CR users have to comply with the following constraints: (i) they transmit only on those spectrum segments that are unutilized by the primary users; (ii) their transmissions cause no performance degradation to the primary users; and (iii) there is no exchange of signaling information between the primary and secondary users. With these constraints, CR MAC protocols for licensed bands are more complex than those for unlicensed bands. Some exemplary CR MAC protocols for licensed bands are found in [7–10].

AS-MAC (ad hoc SEC medium access control) in [7] is designed for the coexistence of the GSM cellular network as the primary network, and a multihop ad hoc network as the secondary CR network. AS-MAC performs three basic tasks: (1) it detects the frequency bands utilized by the entities of primary users, i.e., base stations and mobile stations in the GSM cellular network; (2) it creates and maintains a dynamic picture of the portion of primary resources that are unutilized; and (3) it provides a flexible facility for CR nodes to access primary resources and perform their flash channel filter sender-receiver (CFSR) handshakes. Each CR sender–receiver pair uses the CFSR handshake mechanism to select a channel, i.e., the sender provides candidate channels, and the receiver selects a preferred channel. Simulations show that when the spectrum utilization by primary users is low, the total spectrum utilization by both primary and secondary users can be improved 4 to 5 times via AS-MAC, compared with a network only allowing primary users.

The MAC protocol in [12] emphasizes spectrum utilization efficiency, user fairness, and QoS provisioning for CR devices coexisting with primary users on both licensed and unlicensed bands. The entire medium access process is divided into four stages: observe, plan, decide, and act.

In the *observe* stage, a PIT (primary user information table), an RIT (reservation information table), and a CIT (contention information table) are established to help a CR recognize the spectrum opportunities.

In the *plan* stage, a CR transmitter first checks the PIT to see whether a primary user occupies the channel. If so, the CR defers its transmission; otherwise, it uses the *p*-persistent algorithm to determine whether its data frame can be transmitted or not. For fairness, the contention window (CW) sizes for a real-time (RT) flow and a non-real-time (NRT) flow are set differently, so that a delay-sensitive RT traffic flow can access the channel faster than NRT data flows. Among NRT nodes, fairness is achieved by assigning smaller CW sizes for nodes that have already accumulated many unsent packets.

In the *decide* stage, a contention-free period is assigned during which the channel can be accessed only by RT nodes, thus satisfying the delay-QoS requirements.

Finally, in the *act* stage, a distributed frame synchronization mechanism is developed to coordinate frame transmissions without a centralized controller.

Simulations in [10] demonstrate that in a hidden terminal environment, the throughput performance of this CR MAC protocol is at least 50% better than that of the legacy CSMA/CA MAC protocol. Furthermore, the mean and variance of the packet access delays are reduced to be only one-sixth of those of the CSMA/CA.

A common drawback in the above MAC protocols is that each CR sender and receiver pair relies on its own spectrum sensing outcomes to decide the transmission opportunities. This gives rise to possible missed detections of primary transmitters, either when the primary transmitter is too far away or as a result of wireless channel fades. However, the primary receiver could be within the transmission range of the CR sender, resulting in interference to the licensed primary user. To address this issue, [6] proposes a cognitive MAC protocol that enables CR nodes to choose proper data channels that do not interfere with licensed users. This is done by having CRs in the network collaborate and exchange information about the channel status: a time slot for listening is added during the RTS and CTS exchanging period to get reports on the licensed users from neighboring nodes. This strategy circumvents missed detection of active primary transmitters, because a primary transmitter that cannot be detected by one CR user may be detectable by some other CRs at different geographic locations.

So far, we have introduced the principle of the common control channel approach and discussed several typical CR MAC protocols for both licensed and unlicensed bands. In this approach, all the control messages are conveyed over one common channel, which makes it easy for each node to know the information about channel usage through monitoring this control channel. The handshake on the control channel also guarantees collision-free data communication if the network allocation vector (NAV) contained in control packets is reliable. However, control messages may still collide on this channel due to the hidden terminal problem. In addition, a message sender might refrain from sending a control message if it overhears other control messages. Both situations may render the precalculated NAV inaccurate, which in turn may result in packet collisions on data channels.

The performance of a common-control-channel-based CR MAC protocol relies heavily on the efficiency of the control message exchange. To improve the efficiency, future research may take several directions: (1) dynamically adjusting the bandwidth of the control channel based on traffic load, or (2) allowing migration of the common control channel. This is especially important for CR networks running in licensed bands, where it is hard to find a common control channel for a multihop CR network due to the fluctuation of channel availability over time and location.

Finally, the common control channel approach is not confined to direct access based MAC protocols. If a CR user is equipped with multiple transceivers, it is possible to have the CR access multiple channels simultaneously. In this situation, the optimization of resource allocation (with respect to transmission channels, powers, rates, etc.) is another issue worth studying. Because the optimization inevitably entails intelligent reasoning, this issue in fact sets the stage for research on DSA-driven MAC protocols.

6.2.1.2 Split-Phase Approach

In the split-phase approach, time is divided into fixed intervals, each consisting of a control phase and a data phase. In the control phase, all CR devices meet on a default channel to exchange control messages and decide how to allocate the available channels among themselves during the next data phase [13]. In essence, control messages and data packets access the network in a time-division-multiple-access (TDMA) fashion.

The MMAC protocol in [11–12] is a typical split-phase-based MAC protocol. The control phase of each interval is used as an ATIM (announcement traffic indication message) window, during which devices ready to send data packets broadcast their ATIM messages on the control channel and each proposes to meet one of its receivers on one of the available channels. The ATIM message includes the channel preferences of the sender. The prospective receiver either accepts or denies the proposal. Since each device can only choose one channel during each period, the receiver cannot accept a proposal if it has already committed to meeting another device on a different channel. If the proposal is accepted, the receiver returns a confirmation message, ATIM-ACK. Neighbors who overhear the proposal and confirmation can record the number of contending device pairs on each channel, which will help them choose a channel with the least number of device pairs when they handshake in the ATIM window. The traffic load is thus balanced among all channels. This control phase goes on for a fixed amount of time, after which the devices switch to their agreed channels for the data communications. In the data phase, nodes contend for medium access using a traditional RTS/CTS/Data/ACK mechanism.

Some variants of the MMAC mechanism exist, such as the multichannel access protocol (MAP) in [13]. The main difference is that the duration of the data phase is fixed in MMAC, whereas it varies in MAP depending on the agreements made during the control phase.

Multi-channel MAC protocols are easily subject to multi-channel hidden terminal problems, a fact first elaborated in MMAC and illustrated here in Figure 6.2. Suppose that node A has a packet for B, and A sends an RTS on channel 1, which is the control channel. Node B selects channel 2 for data communication and sends a CTS back to A. The RTS and CTS messages should reserve channel 2 in the transmission ranges of both A and B to avoid collisions. However, when node B sent the CTS to A, its neighboring node C could have been busy receiving on another channel, e.g. Channel 3, thus missing the CTS message. At the same time, node C might initiate a handshake with D, and end up selecting channel 2 for communication. The process of control message exchanges is illustrated in Figure 6.2, and evidently results in collisions at node B.

The above multichannel hidden-terminal problem persists in the MMAC protocol for multihop scenarios. In MMAC, only those nodes which are

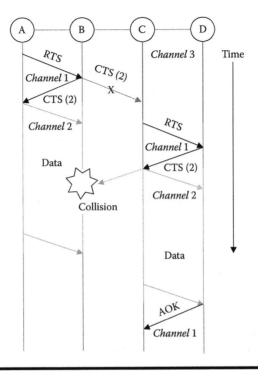

Figure 6.2   A scenario illustrating the multichannel hidden terminal problem.

engaged in data communications actually exchange ATIM control messages. Because not all nearby transmissions can be heard by any node in a multihop scenario, MMAC may fail. As illustrated in Figure 6.3, nodes A and B continuously perform the three-way handshake through ATIM packets, while nodes C and D remain silent since they do not have packets to send. Because C and D are not participating in the ATIM exchanges, after some time node D's clock (and perhaps node C's also) will drift away and hence will lose superframe synchronization with the rest of the network. In this case, node D or C may not be able to switch to the control channel in time. Once this happens, MMAC becomes vulnerable to the hidden terminal problem and will suffer when, for example, node D decides to communicate with node C [14].

Another drawback of MMAC is its inefficient bandwidth utilization. In each ATIM window, only the control channel is used for exchanging ATIM messages, while other channels stay idle. When the total number of channels is large, the wastage makes MMAC extremely bandwidth-inefficient [14].

MMAC also incurs a high communication overhead. For any data packet to be sent, a pair of nodes has to perform a three-way handshake in the control channel and an RTS/CTS handshake in the data channel. Furthermore,

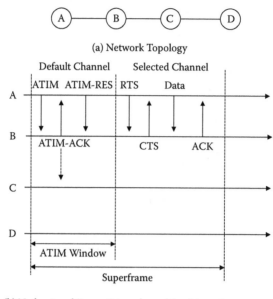

(a) Network Topology

(b) Nodes A and B negotiate a channel for data exchange

Figure 6.3   Procedure of handshaking in MMAC.

MMAC does not support any group communication mechanism such as broadcast or multicast. Because each device relies on ATIM control packets for channel negotiation with a dedicated receiver, it cannot transmit a broadcast/multicast data packet to multiple receivers. Finally, the load-balancing mechanism in MMAC can be quite ineffective when nodes have data packets of different lengths [14].

The drawbacks of MMAC are overcome by the cognitive MAC (C-MAC) protocol in [14], which modifies the superframe structure. Developed for licensed bands, C-MAC provides robust incumbent-protection mechanisms. It assumes that CR terminals are equipped with a single half-duplex radio.

In the C-MAC protocol each channel has its own superframe structure; in contrast, the MMAC protocol only employs a single superframe for the common control channel. With this channel structure, C-MAC does not require all network devices to switch back to the control channel upon the start of every new superframe. Furthermore the use of multiple superframes helps C-MAC to overcome the bandwidth wastage of MMAC, since beacons can be transmitted and recorded on any channel instead of being limited to the common control channel. It also leads to better load balancing and allows more time for actual data transfer.

Similar to MMAC, each superframe in C-MAC is composed of two consecutive parts: the beacon period (BP) and the data transfer period (DTP). A distinctive feature of this multichannel structure is that BPs across channels are nonoverlapping through interchannel synchronization. This allows CR devices to quickly gather information about channels by simply switching channels in ascending order of their BP start time (BPST) and listening for beacon frames during the BP.

Another very important concept in C-MAC is the so-called rendezvous channel (RC). Out of all the channels in use, one is uniquely identified as the RC, which serves as the backbone of C-MAC. The RC is similar to a common control channel, which all the nodes can access to exchange control messages. The difference between an RC and a common control channel is that the RC is composed of BP and DTP, but the common control channel is not involved with data communication. All the nodes should periodically visit the RC. In this case, nodes that do not send beacons on the RC can also get resynchronized through overhearing beacons on the RCs. By periodicly visiting, nodes can also quickly figure out information about network connectivity. Another function of the RC is that it can be used as the vehicle to send group communication (GC) packets.

The RC acts a conduit for sharing load-balancing information of different channels. In C-MAC, nodes can easily figure out the load on a particular channel by analyzing the beacon frames transmitted during a BP. At the time the node goes back to the RC to regain synchronization, it also advertises this channel load. Therefore, nodes can decide which channels to switch to based on the load statistics of that channel. They can also decide to change their current channel.

An important feature of C-MAC is its protection for primary users. This is done through a multistage process that involves detection, notification, and recovery. Once a primary user is detected by a CR, the detection information is included in the beacon frame in every superframe to notify other devices. Since the beacon frame is sent using the most robust modulation and coding, most likely it will be received reliably despite the presence of the incumbent signal from a primary user. Still, it might be possible that the beacons are smeared by very strong incumbent signals (e.g., a TV station). In this case, there is a high probability that if one device detects the presence of a high power signal, so will its neighbors. As a last resort, timeouts are specified so that a device leaves a channel after hearing no beacons for a pre-specified amount of time.

Generally, if a CR user is equipped with only a single transceiver, the split-phase approach outperforms the common control channel approach in terms of throughput and packet delay in most cases. This is because a CR user equipped with a single transceiver in the common control approach may miss the control messages, as they are sent by all users at arbitrary time, whereas in the split-phase approach, all the users must switch to the control channel during the control phase to send and obtain channel status information. With more correct information about the channel status, the split-phase approach is subject to fewer data collisions, thus improving the performance of network throughput and packet delay.

These advantages of the split-phase approach come at a low cost of only one radio transceiver per CR device. On the other hand, this approach requires time synchronization among all devices, which incurs network overhead. Nevertheless, synchronization can be looser than in the frequency-hopping sequence approach that we will introduce in the next subsection, because devices hop less frequently in the split-phase approach.

To further improve the spectrum utilization efficiency of the split-phase approach, it is worthwhile to investigate the time allotment between the control phase and the data phase. In existing protocols, the time lengths of the control and data phases are fixed, regardless of the traffic loads. When the traffic load is low, the use of an unnecessarily long control phase period would waste the spectrum during the control phase, because there are only a few nodes requesting to negotiate for data communication. When the traffic load is high, on the other hand, the use of an inappropriately short control phase period would result in inadequate negotiations, leading to spectrum wastage during the data phase. For example, if there are N_1 transmitter-receiver pairs competing for N_2 channels and the short control phase only permits N_3 times handshaking, where $N_1 \geq N_2 \geq N_3$, then there will be $N_2 - N_3$ channels wasted in the data phase. To alleviate the spectrum wastage, one possible solution is to dynamically set the control phase and data phase periods according to the traffic load. However, as traffic loads vary in different local areas, dynamic control/data phase periods may result in loss of synchronization when neighboring nodes adopt different control phase and data phase periods. For a multihop network, it is challenging to solve the dynamic phase configuration issue in a distributed manner.

6.2.1.3 *Frequency-Hopping Sequence Approach*

In frequency-hopping-sequence-based MAC protocols, each node keeps altering the operation frequency unless the node is in data communication. As idle channels can be quickly occupied by transmission pairs, the channel utility is improved. The frequency-hopping approach outperforms the common control channel method in terms of negotiation efficiency, since it can use any channel for control message exchange and it does not require incessantly monitoring the common control channel to acquire correct channel status.

According to the frequency-hopping sequence that CR users follow, these protocols can be classified into common-frequency-hopping-sequence-based protocols and dynamic-hopping-sequence-based protocols.

6.2.1.3.1   Common Frequency-Hopping Sequence Mechanism

With the common frequency-hopping sequence approach, all nodes in a network are required to follow the same channel-hopping sequence. If a successful handshake between a transmitter and its receiver takes place on one frequency hop, the pair of nodes will stop hopping and continue the data communication on this frequency hop. If the data transfer does not finish by the time the common hopping sequence wraps around and visits the same channel, the pair may broadcast control messages again to reassert their ownership of the channel. After the data communication, they should switch to a common frequency hop to resynchronize to the hopping sequence. By doing so, the CR users do not have to explicitly keep track of the status of the other CR devices, with the result that each CR user can be equipped with only one transceiver. This common frequency-hopping sequence technique is employed by hop-reservation multiple access (HRMA) [15], channel hopping multiple access (CHMA) [16], channel hopping access with trains (CHAT) [17], and receiver-initiated channel hopping with dual polling (RICH-DP) [18].

HRMA [15] permits a pair of nodes to reserve a frequency hop over which they can communicate without interference. A frequency hop is reserved by contention through an RTS/CTS exchange between a sender and a receiver. A successful exchange leads to a reservation of a frequency hop, and each reserved frequency hop can remain reserved by a reservation packet from the receiver to the sender, which prevents those nodes that can cause interference from attempting to use the reserved frequency hop. Once a frequency hop is reserved, a sender is able to transmit data beyond the normal frequency hop dwell time on the reserved frequency hop.

A common frequency hop f_0 is used to permit nodes to synchronize with one another. All idle nodes must dwell on the synchronizing frequency f_0 during the synchronizing slot to exchange synchronization messages. The exchange of synchronization messages on f_0 in the synchronizing period allows nodes to synchronize with one another, i.e., to agree on the beginning of a frequency hop in the common hopping sequence.

CHMA [16] is similar to HRMA. In CHMA, at any given time, all nodes that are not sending or receiving data listen on the common channel hop. A difference from HRMA is that in CHMA, the RTS and CTS for the same handshake can be sent over different channel hops, and therefore the dwell time for a frequency hop can be shorter: it only needs to be long enough to transmit a pair of MAC addresses, a CRC, and framing. The nodes that succeed in a collision-avoidance handshake remain in the same channel hop for the duration of their data transfer, while the rest of the nodes continue to follow the common channel-hopping sequence.

CHAT [17] improves CHMA to allow packets multicast and broadcast. The RTS packet in CHAT defines a bit vector containing 32 bits, with each bit specifying a neighbor node. If a node receives an RTS with its corresponding bit set, it remains in the same channel hop to get ready for incoming data communication; otherwise the node moves to the next channel hop. In the following time slot, the source node negotiates with its intended receivers via single SRTS (specialized RTS) and multiple CTS handshake: if all intended receivers return a CTS successfully, then only one data packet is broadcast; when one or more nodes do not reply with a CTS the source node still broadcasts a data packet to those nodes that have successfully replied with a CTS. This helps the source node avoid involving itself in successive handshakes with all of its neighbors before the broadcast data packet is transmitted. However, this may introduce another problem: if the nodes do not return CTS because they are receiving data from other users, the enforced broadcast from the source node will interfere with these nodes.

Simulation comparisons among CHAT, CHMA, and MACA-CT [19] (a common-control-channel-based MAC protocol for networks using the spread spectrum technique) show the superiority of the frequency-hopping approach over some protocols based on the common control channel approach.

The common frequency-hopping sequence approach still has some drawbacks: CR users dwelling on one channel for data communication are not aware of the status of other channels during the period of data communication. Hence, it is possible that after a transmitter and receiver finish their pairwise communication, when they send control messages on another frequency hop they may interfere with other devices that are currently using this channel hop. Besides, this approach requires tight synchronization over a common control channel.

6.2.1.3.2 Dynamic Hopping Sequence Approach

In the dynamic hopping sequence approach, each node employs a pseudorandom hopping sequence, and node pairs can make agreements on different channels. There is no common hop channel for synchronization, and nodes can rendezvous on different channels.

Slotted seeded channel hopping (SSCH) [20] is a typical dynamic-hopping-sequence-approach-based MAC protocol. It requires only a single radio per node. In SSCH, the hopping sequence, called the *channel schedule* in [20], is a list of channels that the node plans to switch to in subsequent slots, and the time at which it plans to make each switch. The channel schedule can be represented by a current channel ID with a rule for updating the channel. In particular, the schedule is represented as a (channel, seed) pair and the rule for updating the channel is

$$x_i \leftarrow (x_i + a_i) \quad \text{mod total channel number,} \qquad (6.8)$$

where a_i is the seed and x_i is the channel ID. Here $1 \leq i \leq N$, N is the number of seeds. According to (6.8), we can see that the number of seeds represents the number of hopping sequences a node can possess.

Each node maintains a list of the channel schedules for all other nodes it is aware of. When device A wants to talk to B, A waits until it is on the same channel as B. If A frequently wants to talk to B, A adopts one or more of B's sequences, thereby increasing the time they spend on the same channel. This mechanism requires that the sender learn the receiver's current sequences, which can be achieved via a seed broadcast mechanism.

In MCMAC [21], each node picks a seed to generate an independent pseudorandom hopping sequence. Each node has a default seed and thus has a default hopping sequence. When a node is idle, it follows its default hopping sequence. Each node puts its seed into every packet it sends so its neighbors can learn its hopping sequence. For simplicity, nodes are assumed to hop synchronously. When node A has data to send to B, A flips a coin and makes a decision on transmitting with some probability p during each time slot. Once it decides to transmit, it tunes to the current channel of B and sends RTS to B. If B does not reply with CTS, due to a transmission collision or that B is busy, A tries later by coin flips until it receives a CTS reply from B. Once B replies with CTS, both A and B stop hopping to exchange data. Data exchanges normally take several time slots. After the data exchange is over, both A and B return to their original hopping sequence as if there had been no pause in hopping.

Both the common frequency-hopping sequence and dynamic hopping sequence approaches can achieve high network throughput and low packet delay for CR networks working in unlicensed bands. However, there is no way that the common frequency-hopping approach can be used for CR networks operated in licensed bands, because (1) the available channel sets alter over time in the licensed band, so it is impossible to have a fixed channel set for hopping; and (2) the channel availability varies with location, hence it is hard to find a common frequency-hopping sequence for nodes in the whole network. In the case of the dynamic hopping sequence approach, there is the potential for it to be used for CR networks running in licensed bands. However, existing dynamic-hopping-sequence-based MAC protocols cannot be directly implemented for CR networks in licensed bands, as their available channel sets are still fixed. Adopting the dynamic hopping sequence approach for CR networks in licensed bands, however, can be a prospective research field.

6.2.2 Direct Coordination-Based Cognitive Radio MAC Protocol

Compared with direct contention-based MAC protocols, group-negotiation-based (or direct coordination-based) MAC protocols should have better performance. This is because all the nodes within a group are involved in negotiation, and thus the channel usage information of channel competitors can be acquired accurately.

Common spectrum coordination channel (CSCC) MAC protocol [22] is a direct coordination-based MAC protocol proposed to address the coexistence of IEEE 802.11b and 802.16a networks in unlicensed bands. Different types though they are, the CR users negotiate over the same common spectrum coordination channel (CSCC), which is a narrow-band subchannel at the edge of the unlicensed band. Each user periodically broadcasts its spectrum usage information using CSCC and makes decisions upon receiving spectrum usage information. Decisions include frequency selection and power adjustment, if frequency selection alone cannot be used to avoid interference between contending users.

Simulations in [22] show that the CSCC protocol can significantly improve the network throughput compared with the approach of direct contention without coordination, especially when in-band interference between IEEE 802.11b and 802.16a is high. However, there are two limitations for CSCC: (1) as the messages may collide over CSCC, the users may not be able to acquire accurate channel usage information, which consequently results in data-packet collisions on the data channel; and (2) periodic broadcast of channel usage information is redundant if there is no new user joining in the negotiation.

The MAC protocol in [23] is another direct coordination-based protocol wherein each node repeats a MAC-layer configuration protocol every T time units. There are two operation modes in each time period: (1) control-channel-based MAC-layer configuration and normal operation, and (2) normal operation only. During the MAC-layer configuration stage, time is split into intervals referred to as frames. Each frame is further divided into N time slots of equal length. Only node i is allowed to transmit a control message during the ith slot in each frame, and all other nodes listen to the control channel during the ith slot. This ensures that every node in the network gets one chance to transmit within each frame without collisions.

The proposed MAC-layer configuration algorithm enables nodes to identify their neighbors and further dynamically discover the global network topology and physical location of each node in the network. The configuration algorithm also identifies the set of available channels for each node in the network. Based on this information, each CR user can optimize channel and routing selection to transmit data in the normal operation period.

In contrast to the above protocols [22, 23] relying on a common control channel for negotiation, heterogeneous distributed MAC (HD-MAC) [25] is a distributed coordination-based approach that handles dynamic spectrum access without assuming the existence of a common control channel throughout the network. HD-MAC relies on the observation that nearby users share similar spectrum availability, so it is possible to find a control channel in the local area over which users in the vicinity can negotiate for channel assignment. In HD-MAC, each user obtains the available-channel information of its neighbors through channel scanning and beacon broadcast, and sets up a neighbor information list constituted of its neighbors, a list of available channels for each of them, and a schedule of time to connect to each channel. Based on this information, users in the neighborhood select the channel that has the largest connectivity as the coordination channel and negotiate the use of other channels over it.

As the direct coordination approach requires information exchange among nodes throughout the network for negotiation, the scalability issue is a bottleneck for these MAC protocols. The existing solutions can be classified into two types: One is based on time splitting (e.g., [23]). Over the time scale, a specific time interval is divided for negotiation. Each node is assigned a time slot for negotiation so that it can periodically broadcast negotiation packets in its time slot. The other type is based on cluster negotiation, wherein nodes are grouped into different clusters and only nodes within the same cluster coordinate with each other (e.g., [24]).

The time-division-based method requires time synchronization among nodes in the whole network. As traffic load varies with time and location, the design for the negotiation period is a complex issue that is ignored by most of this type of MAC protocols. If all the nodes use the global maximum period, which can provide every node in the network a time slot for negotiation, the efficiency of the control channel will be quite low if the traffic load is low, and the latency for data channel access will be subsequently very long, as much time is divided for negotiation. A possible solution is to dynamically assign the negotiation period according to the connectivity of each node. For example, as shown in Figure 6.4, as node A has five neighbors, the period of negotiation for node A should hold at least six time slots to gather all its neighbor information, while for node B, which has only three neighbors, the period of negotiation only needs to hold four time slots. However, as node A and node B are neighbors, the difference in negotiation periods between the two nodes will result in asynchrony of the negotiation and data communication phases between the two nodes. That is, when B is ready for data communication, A is still in negotiation, which may cause collisions. Therefore the issue of dynamic negotiation period assignment needs further investigation to make the time-division-based method practical for CR network MAC design.

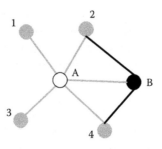

Figure 6.4 Dynamic negotiation period.

For existing cluster-negotiation-based MAC protocols, a common problem is that the cluster is not stable due to the dynamic connectivity in CR networks, caused by mobility or unstable channel availability. Therefore the cluster construction process should be frequently carried out, which causes much communication overhead. Another problem is that the contentions in cluster-based negotiation mechanisms may result in collisions on the control channel within clusters or among clusters. The clustering algorithm and corresponding negotiation mechanism will be the major future research tasks for cluster-negotiation-based MAC protocols.

6.2.3 Centralized MAC Protocol

Up to now we have been focusing on the discussion of distributed MAC protocols. Free-for-all, opportunistic spectrum access for distributed ad hoc CR networks in licensed bands has to resort to spectral bandwidth brokering at the individual node level; thus, the complexity of the MAC protocol and the sensing and agility requirements of individual radios in this case are quite high [28]. Therefore, if conditions allow, the centralized MAC protocol is preferred due to its low complexity. In this section, we will introduce several typical centralized MAC protocols for CR networks. As the typical channel access approaches for CR networks, such as the common control channel approach, split-phase approach, frequency-hopping approach, etc., have been introduced before, we will not put much effort into discussing their implementations in centralized MAC protocols; instead, we will focus on the structure of the CR networks and how the centralized server manages CR clients.

The most well-known CR network with infrastructure is the ongoing IEEE 802.22 project for WRAN (wireless regional area network), which targets wireless broadband access in rural and remote areas [25]. The 802.22 system operates on TV bands, so it has to protect TV sets (which are the primary users) near CR users. The 802.22 system specifies a fixed point-to-multipoint wireless air interface whereby a base station (BS) manages its own cell and all associated consumer premise equipment (CPE) devices. The BS controls the medium access in its cell and transmits in the downstream direction to various CPEs, which respond back to the BS in the upstream direction. No CPE is allowed to transmit before receiving proper authorization from the BS, for the protection of the primary users.

Both BSs and CPEs are responsible for licensed-transmission sensing and incumbent protection. The BS instructs its associated CPEs to perform periodic measurement activities, which may be either in-band or out-of-band. In-band measurement relates to the channel used by the BS to communicate with the CPEs, while out-of-band corresponds to all other unaffected channels. With in-band measurements, the BS needs to quiet the data transmission on the channel so that measurements can be carried out (this period is called the quiet period). To achieve the best operation of measurements, the BS may not need to require every CPE to conduct the same measurement activities. Rather, it may incorporate algorithms that distribute measurement load across CPEs.

Frequency agility, an integral part of design for primary user protection, is achieved through adjusting frequency of operation in time to coexist with primary users. The 802.22 MAC incorporates a vast set of functions that allow the CR devices to efficiently manage the spectrum. Operations such as switching channels, suspending/resuming and adding/removing channels are among the actions in the design in order to guarantee primary user protection and effective coexistence.

During the development process of the IEEE 802.22 protocol, many methodologies were proposed to ameliorate its performance, e.g., [16–27]. In the early version of the 802.22 protocol, during data communication on one channel, it was required to sense the channel periodically, which interrupted data transmission and finally impaired the QoS of WRAN. This critical issue can be addressed by an alternative operation mode proposed in 802.22 called dynamic frequency-hopping (DFH) [26]. In the DFH mode, WRAN data transmission is performed in parallel with spectrum sensing without interruptions, so that efficient frequency usage and reliable channel sensing are achieved. This implies that each CPE should be equipped with multiple transceivers.

In the DFH mode, the time is divided into consecutive operation periods. In each operation period, during data communication on a working channel (called an in-band channel in [26]), sensing is performed simultaneously on the intended next working channels (called out-of-band channels in [26]). It should be noted that the notions of in-band and out-of-band in [26] are different from those in [25]. An out-of-band channel sensed to be vacant is considered to be validated for the next data transmission. A WRAN system thus dynamically selects one of the channels validated in the current operation period for data transmission in the next operation

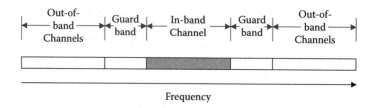

Figure 6.5 Simultaneous sensing and data transmission.

period. This channel can be used for data transmission for up to 2 seconds after the time it was validated. Guard bands between the in-band and out-of-band channels are allocated to mitigate adjacent interference from data transmission to the out-of-band sensing, as shown in Figure 6.5.

In DFH mode, each cell requires at least two channels to perform data transmission and reliable sensing in parallel. If spatially overlapping N cells coordinate their hopping behavior, it can be proven that $N+1$ vacant channels are sufficient for them to coexist as long as the length assigned for a single data transmission is larger than $N*QT$. The QT of one channel is the quiet time for sensing between two adjacent data transmissions, during which no WRAN is allowed to transmit on that channel. A group of nearby WRANs can form a DFHC (DFH community) that supports a coordinated DFH operation to ensure mutual-interference-free channel sensing and maximize the channel utilization efficiency.

Ref. [27] is also dedicated to improving the performance of IEEE 802.22. It adopts the local bargaining mechanism proposed in [30] for the WRAN system. The spectrum allocation process is divided into two stages: the pre-allocation stage and the local bargaining stage. In the pre-allocation stage, each user takes its turn to try to achieve the maximum spectrum bandwidth it requires, in the order of their labeling. The labeling of a node is calculated by the number of its competitive neighbors over its available channel number. In the local bargaining stage, the users who are not able to satisfy the minimum desired bandwidth try to borrow channels from neighbors who have enough channels to satisfy their own maximum requirements. The base station is in charge of checking this situation and judging who should lend channels to these users.

DIMSUMNet (dynamic intelligent management of spectrum for ubiquitous mobile-access network) [28] is also a centralized CR network, but with an infrastructure different from WRAN. A typical application scenario of DIMSUMNet is shown in Figure 6.6, where the centralized entity called a *spectrum broker* is in charge of managing all the spectrum access activities in a region. The spectrum broker shoulders the responsibility of allocating

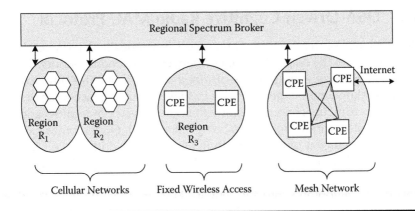

Figure 6.6   Structure of DIMSUMNet.

the coordinated access band (CAB) spectrum to individual networks or users. The CAB is a contiguous chunk of spectrum reserved by regulating authorities such as the FCC for controlled dynamic access. The CAB resembles a licensed band in that the spectrum lease is a short duration license. The difference between the two is that current licenses signify long-term (or permanent) and sole ownership, while CAB leases are awarded by an automated machine-driven protocol, which is timebound and may contain additional constraints such as the extent of spatial region for spectrum use, maximum power, and the exclusive or nonexclusive nature of the lease. The difference between CAB and unlicensed bands is that there is no lease required to use unlicensed bands and no coordination mandated in them. The spectrum broker permanently owns the CAB spectrum and only grants a timebound lease to the requesters. A mechanism named *statistically multiplexed access* improves the spectrum utilization efficiency in the CAB, as it allows a network service provider to use more spectrum and offers better QoS to end users by dynamically adding capacity during peak loads.

In general, most centralized CR network MAC protocols are proposed to solve the coexistence of CR networks with licensed networks, since distributed solutions for primary user protection are very complicated. In such protocols, a central spectrum access manager can possess detailed information about the wireless network so as to implement a highly efficient network configuration and apply far-reaching and flexible policies. The future research directions for centralized CR MAC protocols include optimizing resource allocation algorithms on the central server, and improving the efficiency of negotiation between the central server and clients. These future investigations always require DSA algorithms, which will be discussed in the next section.

6.3 DSA-Driven Cognitive Radio MAC Protocol

The efficiency of dynamic spectrum access is essentially determined by the effectiveness of the negotiation policy between contending senders and receivers. The negotiation policies of *direct-access-based MAC protocols* are based on simple non-cognitive rules lacking the intelligence to achieve efficient spectrum utilization, let alone to address issues of fairness and QoS requirements for individual users being considered. On the other hand, research on intelligent optimal spectrum allocation algorithms, i.e., dynamic spectrum allocation (DSA) algorithms, has always been pursued in a standalone context until recently, when some researchers have made a thrust to embed DSA algorithms into CR network MAC protocol designs. We therefore propose the unified *DSA-driven MAC framework*, and believe it will be the most promising MAC solution for CR networks [57].

DSA-driven MAC mechanisms differ from direct-access-based MAC mechanisms in the dimensions of each CR user's strategy, as well as the goal each user pursues. As mentioned before, in direct-access-based MAC protocols, the strategy S_i for CR user i is one-dimensional and the goal each CR user pursues is to obtain ownership of the desired channel. For DSA driven MAC protocols, the strategy for each CR user is usually multidimensional, including choosing a channel, picking a time slot, and determining transmission power, modulation method, antenna configuration, and so on. In contrast with the standalone goal of the CR user in direct-access-based MAC protocols, DSA-driven MAC protocols can provide a global optimization goals which, however, are always achieved through distributed DSA algorithms.

In DSA-driven MAC protocols, advanced optimization algorithms are adopted to make the spectrum allocation mechanism more efficient, fair, and intelligent. According to the DSA algorithms the MAC protocol employs, DSA-driven MAC protocols can be classified as graph DSA-driven MAC, Markov DSA-driven MAC, and game DSA-driven MAC. It should be noted that this taxonomy is not exclusive, since one DSA algorithm may be based on multiple theories. In this section, we focus on how these theories are applied in DSA-driven MAC protocols; how CRS coordinately negotiate with each other to provide adequate information (including spectrum sensing information and spectrum access strategies) to execute DSA algorithms, while the detailed mathematical optimization process is not our emphasis here.

6.3.1 Graph DSA-Driven MAC Protocol

In graph DSA-driven MAC protocols, the whole CR network is modeled as a directed or undirected graph $G = (V, E)$, where V denotes the vertex set and E denotes the edge set. Generally, there are two types of graph models. One is the *node contention graph* (NCG) model, where each vertex represents a CR user in the network, and each edge connecting two nodes represents that the associated nodes are within each other's transmission range; The other graph model is the *link contention graph* (LCG), or *flow contention* model, where vertices represent active link flows and each edge between two vertices denotes that there is contention between the two flows, i.e., the two flows cannot be active at the same time.

Refs. [30–33] discuss representative graph DSA-driven MAC protocols. Ref. [31] adopts the NCG model and proposes to combine it with a local group negotiation-based MAC protocol in order to provide both efficiency and fairness in spectrum allocation. The system goal is to maximize total logarithmic user throughput, expressed by

$$U(A) = \log \sum_{n=0}^{N-1} \sum_{m=0}^{M-1} a_{m,n}, b_{m,n} \tag{6.9}$$

where M is the number of channels, N is the number of CR nodes $a_{m,n} = 1$ denotes that the spectrum channel m is assigned to user n, (otherwise $a_{m,n} = 0$); and $b_{m,n}$ represents the maximum throughput user n can acquire by using the spectrum channel m. The NCG is represented as an $M \times N$ graph matrix $A_{M \times N}$. Each user updates the graph matrix to maximize the system utility $U(A)$, which implies each user should construct the complete graph model of the local bargaining group through message exchanging in the bargaining group formation stage. The update is completed through iterative local bargaining.

The whole bargaining process has four stages:

1. *Bargaining request:* A node that wants to get more spectrum assignment broadcasts a bargaining request to its k-hop neighbors, where k is the ratio of interference range and transmission range. These neighbors are connected to the node with edges in the NCG.

2. *Bargaining request acknowledgement:* Neighbors that are willing to participate in the bargaining reply to the sender of the bargaining request.

3. *Bargain group formation:* The requester and the neighbors that replied to the request form a bargaining group. For each bargaining group, the requester becomes the group coordinator and performs the computation during the bargaining.

4. *Bargaining:* Each CR user updates its local channel assignment to satisfy the request and maximize system utility in Equation (6.9). In the bargaining process, a poverty line is set to ensure a minimum spectrum allocation to each user, and hence guarantee a certain degree of fairness among users.

In this local bargaining approach, once the bargaining groups are constructed, the bargaining inside each group should not disturb the spectrum assignment of nodes outside the group. To achieve this goal, two conditions should be satisfied: (1) once a node gets one channel from its neighbor, the assignment does not conflict with its neighbors outside the bargaining group; and (2) members belonging to different bargaining groups cannot be directly connected. This is to prevent conflict between different bargaining groups.

Ref. [33] is also based on the NCG model, but the NCG is a directed graph. The directed edge e_i from node $t(i)$ to $r(i)$ denotes that $t(i)$ is intending to transmit data to $r(i)$. In [33], time is divided into fixed length time slots and the DSA algorithm schedules the spectrum access activities in each time slot. The schedule for the time slot TS_k can be represented as a set of two tuples (e_i, Cb_j), which means edge e_i is scheduled to communicate using channel Cb_j during the time slot TS_k. Each link can occupy the allocated channel only in the assigned time slot. An edge is said to be covered if it has been assigned a (time slot, channel) combination. A node is said to be covered if each outgoing link from the node is covered.

The design goal of [33] can be understood from two perspectives: (1) to minimize the number of time slots that have transmission occurring, which signifies that in one time slot, the number of concurrent transmissions should be maximized; and (2) to cover the whole network as soon as possible, which signifies the time slot during which transmission occurrs should be as early as possible.

According to the global goal and system constraints, the spectrum allocation can be formulated as an ILP (integer linear programming) problem. This paper presents a two-round distributed protocol to solve the ILP problem. In the first phase, each node decides on the (time-slot, channel) combination for its entire roster of outgoing links by its rank in the network in order to achieve the global goal. Nodes in the network are ranked based on parameters including the degree of the node, the cardinality of the channel set available at the node, and the node ID. When all neighbors of a node are covered, the node is said to have completed the first phase.

In the second phase, the schedule length (the operation period of a node, i.e., the number of time slots of each operation cycle) of each node is propagated throughout the network, and all the nodes should adjust their schedule lengths so that the whole network has a uniform schedule length. Since in the first phase every node determines its schedule based on its local topology information, the length of the schedule at different nodes could possibly vary. After schedule propagations and adjustments, all the nodes finally should have the same schedule length, which is the maximum schedule length created in the first phase.

Although [33] provides a successful solution to achieve the system goal in its model, the model itself may need some modifications. This is because not all nodes want to use the channel in every working period, and thus the system goal to cover the whole network may not be realistic. Besides, since the communication load on each edge may vary, the ILP constraint that guarantees each edge is covered exactly once may not be reasonable.

For graph-DSA-driven MAC protocols, as the overall network status is represented by a single contention graph, each CR user is required to build such a graph to run the DSA algorithm for CR channel access. This implies that each CR node should master the status of the whole network via message exchange. Therefore, the scalability issue arises when the network becomes large. A possible solution is to partition the whole network into small subnetworks, and within each subnetwork to let each CR user set up its local graph. For this solution, there are two aspects requiring further investigation: one is the graph partition algorithm, including criteria and procedures for graph partitioning in distributed networks; and the other is the coordination mechanism to harmonize the subnetworks to address contentions amongst themselves.

6.3.2 Stochastic DSA-Driven MAC Protocol

In stochastic DSA-driven MAC models [34–43] the process of dynamic spectrum access is depicted as a stochastic process. Through designing the action probability, the spectrum can be automatically allocated optimally and fairly. Among various stochastic models, the Markov model is applied the most. Stochastic models can appropriately reflect the dynamic and uncertainty features of channel availability in CR networks. The design of a stochastic DSA-driven MAC protocol should strike a balance between statistical accuracy and complexity [34].

In [37–43], dynamic spectrum access in CR networks running in a licensed band is modeled as a partially observable Markov decision process (POMDP). In the POMDP network model, each node predicts the spectrum usage based on the statistics of local spectrum observation and historical accessing experience so that it can make an optimal future decision.

A POMDP is represented as five-tuple (S, A, O, P, R), where S is the state space, A is the action space, O is the observation space, $P_a(s, o, s')$ is the probability that action a in state s at time t will give observation o and lead to state s' at time $t + 1$, and $R(s, a, s')$ is the immediate reward that is issued for the transition from s to s' under action a. The goal for the POMDP system is to find a policy that, when followed, can be made to maximize some cumulative function of the rewards.

For the CR networks running on N orthogonal channels, each with bandwidth of B_i, the state of the network is represented as an N dimension vector $[S_1(t), \ldots, S_N(t)]$, $S_i(t) \in \{0, 1\}$, where $S_i(t) = 0$ stands for channel

i being occupied, and $S_i(t) = 1$ stands for channel i being idle. The action space is $\{[A1(t), A2(t)]\}$, where $A1(t)$ is the sensing action at time t, while $A2(t)$ is the accessing action: $A1 = \{a_1, \ldots a_k\}$, $(|A1| = k \leq N)$ is the set of channels a CR user chooses to sense at time t; $A2$ is the set of channels the user chooses to access based on the sensing result and it satisfies $A2 \subseteq A1$. The observation of network state at time t is $\Theta_{j, A_1} = [S_{a1}, \ldots S_{ak}]$, where j is the current network state. For the chosen action $[A1, A2]$, the user can receive a reward γ_{j, A_1, A_2}, defined as $\gamma_{j, A_1, A_2}(t) = \sum_{i \in A_2} S_i(t) B_i$, at the end of this time slot. A simplified procedure for running POMDP is described as follows.

At the beginning of slot t, the knowledge of the internal state of the network based on all past decisions and observations can be summarized by a belief vector $\Lambda(t) = [\lambda_1(t), \ldots, \lambda_M(t)]$, where $\lambda_j(t)$ is the conditional probability (for a given decision and observation history) that the network state is j at the beginning of slot t prior to the state transition. A policy π for the POMDP model is thus given by a sequence of functions mapping from the current belief vector $\Lambda(t)$ to the sensing and access action $[A_1(t), A_2(t)]$ to be taken in slot t.

The global objective of the POMDP MAC protocol is to choose the sensing and access action $[A1, A2]$ sequentially in each slot so that the total expected reward accumulated over T slots is maximized. Under this formulation, a spectrum sensing and accessing strategy is in fact a policy of this POMDP over a finite horizon.

To show the effectiveness of the POMDP-based DSA algorithm, a performance comparison among the optimal approach, the suboptimal greedy approach, and the non-DSA random access approach is conducted in [37]. Simulation results show that the DSA-driven approach outdistances the non-DSA-driven random access approach in network throughput, and the performance of the suboptimal greedy approach is close to the performance of the optimal approach.

Effective as it is, the POMDP framework demands improvements from the following aspects: (1) in the POMDP framework, the state transition probabilities $\{p_{i,j}\}$ are assumed to be fixed and known by all the nodes, which may be not true in practice; and (2) the sensing error model, specified by the receive operating characteristics (ROC) curve, which gives the function of false alarm and miss detection, is fragile, since in practice, only the signal from the transmitter can be detected, and not that from the receiver.

Since there are considerable number of models for the stochastic method, stochastic DSA can be very flexible. The flexible stochastic DSA algorithms provide an opportunity to address various goals for different CR network systems. It is very important to choose the appropriate model based on the specific goal when designing stochastic DSA-driven MAC protocols. The immediate challenges for stochastic DSA-driven MAC protocols include (1) how to determine the channel or user state transition probability based on local historical data, local observation, or exchanged information; and (2) how to balance the accuracy of statistics and the complexity of the model.

6.3.3 Game-Theoretic DSA-Driven MAC Protocol

Game theory has been extensively applied in other areas of network communications. In recent years researchers have started to apply it to dynamic spectrum allocation in CR networks. In game-theoretic DSAs, the interactive behaviors among CRs are modeled as a game model $\Gamma = \langle N, \{S_i\}, \{u_i\}\rangle$. Here N is the set of CR players, and one player often consists of one transmitter and one receiver; S_i denotes the strategy (on transmission powers, accessing channels, modulation methods, etc.) space of player i; and u_i is the local utility function of player i, to be maximized by player i updating its strategy. If a CR has to iteratively adjust its strategies to adapt to the adjustments of other CRs before the whole network converges to a steady state, then the CR network should be modeled as a repeated game model $\Gamma = \langle N, \{S_i\}, \{u_i\}, T\rangle$, where T is the decision timing for the game, determining the time at which CRs can update their strategies by methods including synchronous decision, asynchronous decision, round-robin decision, and random decision [44], among which round-robin decision processes are used most in game-theoretic DSA.

Research results presented in [45–51] are typical game-theoretic DSA algorithms for CR networks. Most theoretical DSA algorithms intend to transfer a global optimization problem into a distributed local optimization problem by introducing pricing in the local utility function. Hence the solution of the global optimization problem can reach the Nash equilibrium of the distributed game. Single-channel multi-channel asynchronous distributed pricing (SC-/MC-ADP) algorithms [47–49] set up the game model based on the Karush–Kuhn–Tucker (KKT) principle to achieve full spectrum utilization. The interference price scheme is introduced in SC-/MC-ADP to reflect the interference levels on available channels at different locations. The SC-/MC-ADP algorithms, which exchange information on users' interference prices during spectrum sharing, are shown to outperform their counterparts ignoring interference prices.

Game-theoretic DSA is especially fit for CR networks due to its inherent advantages: (1) Cognitive radios' actions can be very complex. By modeling the intricate CR network into a game, these complicated elements can be studied in a uniform way. (2) The local goal of each CR can be quite different from that of others, and game theory enables various optimality criteria fit for each player's local goal. (3) Adaptation is an important feature of CRs; nevertheless, each cognitive radio's adaptation changes the environment of the outside world for all the other CRs, and the actions of CRs would then appear as a recursive interactive decision process, which can be naturally represented by a repeated game. (4) Another merit of game-theoretic DSA is the fairness provision. In game-theoretic DSA, the local utility function is often composed of two parts: one is its payoff for the strategy it makes; the other is the price it should pay for its strategy. The utility function may be the payoff minus the price, or other forms that consider both payoff and price. The price mechanism enables fairness as an intrinsic value of the game-theoretic DSA.

Given these salient merits, here we will introduce several typical game-theoretic DSA-driven MAC protocols [52–57] to handle the complex spectrum utilization requirements in CR networks. We will mainly discuss their system goals, local utility functions, and negotiation mechanisms.

Ref. [54] presents a game-theoretic framework for distributed adaptive channel allocation in CR networks. The channel allocation problem is modeled as a normal-form game. The local utility function of the game model is

$$U2_i(s_i, s_{-i}) = - \sum_{j \neq i, j=1}^{N} p_j G_{ij} f(s_j, s_i) - \sum_{j \neq i, j=1}^{N} p_i G_{ji} f(s_i, s_j) \qquad (6.10)$$

where N is the number of CR users in the network, p_j is the transmission power for user j, G_{ij} is the link gain between transmitter i and receiver j, and $f(s_i, s_j) = 1$ if $s_j = s_i$, 0 otherwise. The first item $- \sum_{j \neq i, j=1}^{N} p_j G_{ij} f(s_j, s_i)$ is the payoff for user i, and the second item $\sum_{j \neq i, j=1}^{N} p_i G_{ji} f(s_i, s_j)$ is the price user i should pay for its strategy. It can be proved that the channel allocation game with utility function $U2$ is a potential game.

The MAC protocol in [55] implements a mechanism similar to the channel filtering sender-receiver (CFSR) handshake mechanism to finish the game process, which is elaborated as follows:

(1) At the beginning of each time slot, each user decides to listen to the common control channel with probability $p_b = (N-1)/N$, or go to step (2) with probability $p_a = 1/N$.

(2) Transmitter sends START packet. The packet includes current estimation of the interference created for neighboring users on all possible frequencies, denoted by $I_o(f)$. This information is computed based on information stored in the channel status table.

(3) Receiver computes current interference estimation for itself, $I_d(f)$. Based on $I_o(f)$ received in the START packet, it can determine the utility function $U2_i(f)$ for all channels, and it selects the channel with the highest U.

(4) Receiver piggybacks the newly selected channel information on a signaling packet START_CH, which is transmitted on the common channel.

(5) Transmitter sends ACK_START_CH, which acknowledges the decision of transmitting on the newly selected frequency, and starts transmitting on that channel.

(6) All the other users (transmitters and receivers) that heard the START_CH and ACK_START_CH packets update their channel status tables (CST) accordingly.

This game-theoretical MAC protocol [55] relies on the fact that all game players can communicate with each other so as to negotiate by information exchange. This is subject to the scalability issue. Indeed, most game-theoretical MAC protocols (including those for non-CR networks [31, 52]) suffer from the scalability issue.

In fact, although it is expected that game-theoretic DSA-driven MAC can achieve attractive performance, it faces several challenges before it can be adopted in CR networks. The first challenge is the scalability problem. In a large CR network, the overhead and delay of the game process will be unbearable if all the users play a single game; on the other hand, it is not reasonable to let nodes that are set far apart (and thus have little direct mutual impact) play the same game. Therefore, the protocol should group the CR users into multiple clusters and let the members in the same cluster play the same game. The second challenge is the problem of game timing synchronization. In a repeated game, the players should update their strategies according to a certain timing scheme, e.g., synchronous timing, round-robin timing, and so forth [44]. Hence, the users have to coordinate with each other to play the game in the right order based on the game timing. The last challenge is the game collision problem. The information exchange in one game may interfere with other games in different clusters. Therefore, collision avoidance should be considered in the game process.

Considering these challenges, we propose a unified game-theoretic DSA-driven MAC framework [57] for multichannel CR networks. Our framework includes four components (see Figure 6.7): (1) the game theoretic DSA, (2) the clustering algorithm, (3) the collision avoidance mechanism, and (4) the negotiation mechanism. These are indispensable building blocks for a complete game-theoretic DSA-driven MAC protocol.

The cluster structure is the basis for the negotiation mechanism, since negotiation must be constrained to a single cluster. The collision avoidance mechanism design should guarantee reliability for negotiation and adapt to cluster structure, since intercluster collision is the major collision challenge. The negotiation mechanism is in charge of exchanging the game information required by the DSA algorithm. The concrete content of the game information is dependent on the game policy.

To provide a robust framework compatible for various kinds of DSA algorithms, the collision avoidance mechanism and cluster structure should

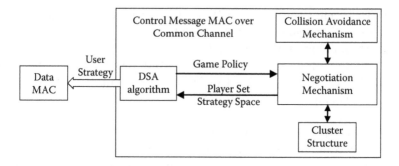

Figure 6.7   Structure of unified DSA-driven MAC framework.

be independent of the DSA algorithms. However, as a player is a pair composed of a data sender and a receiver, once the cluster structure is independent of the DSA algorithm, it may happen that the sender and the receiver are in different clusters. For this intercluster communication, the receiver participates in the game on behalf of the sender, as the sender may not be able to directly communicate with other players in the receiver's cluster. Once the sender gets involved with the game process in the receiver's cluster, it cannot participate in the game process in its own cluster, as one player cannot participate in two games simultaneously. After the negotiation, the receiver should feed the final strategy back to the sender.

The strategies in data communication are established by the game-theoretic DSA algorithm. To prevent the game process from interrupting the ongoing data communication, a common control channel is chosen for negotiation among game players.

Desirable clustering algorithms for the game-theoretic DSA-driven MAC framework should satisfy the following properties: (1) the criterion to form a cluster should be based on the strength of mutual impacts between CR users, (2) the clustering algorithm should be scalable to fit any size (and topology) of networks, (3) the cluster should be stable to avoid frequent reconstruction, and (4) the cluster formation process should be as simple as possible to reduce the overhead for cluster construction.

As the strength of mutual impacts between two CR users is mainly determined by the Euclidean distance between them, a *geographical position-based clustering algorithm* is preferred. In addition, to avoid frequent cluster reconstruction, it is desirable that a cluster be independent from the behavior of the nodes, and such independence can be easily realized if the cluster is only determined by its geographical position.

Therefore in our clustering algorithm, the identity of each cluster is exclusively determined by its location. In this case, each cluster's identity is not affected by the arrival or departure of CRs. Moreover, each CR can

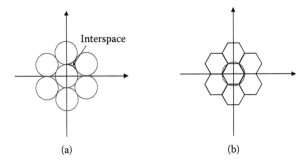

Figure 6.8   Circle and hexagon partition.

independently decide which cluster it belongs to, which simplifies the cluster formation process. Furthermore, the cluster formation is also independent of the network infrastructure. Besides these characteristics, the clustering algorithm is also scalable, as the maximum number of clusters only depends on the network area and is not affected by the number of nodes.

The determination of cluster size is an integral part of a geographical clustering algorithm. To support negotiation mechanisms efficiently, we set the cluster as a fully connected cluster, i.e., the nodes within one cluster can communicate with each other directly. As such, the players within a cluster can negotiate without multihops, so they can avoid using a routing process to exchange control messages. This also entails that the game is only played among nodes that have strong interference with each other.

Supposing the maximum communication distance is d, all the nodes within a fully connected cluster must be located in a circle with a diameter of d. However, the circles cannot cover the whole plane of the network without interspaces (see Figure 6.8(a)), so they should be approximately replaced by isogons with an external diameter of d. To fully cover the whole network plane, it can be proved that the regular hexagon is the optimal option to approach the circle (see Figure 8(b)). Hence, nodes within one hexagon form a cluster. Each cluster is assigned a unique cluster ID, C_{id} (a two-integer pair $<k_i, l_i>$), which is exclusively determined by the hexagon center's coordinates (x_{ci}, y_{ci}):

$$
\begin{cases}
x_{ci} = k_i \cdot \dfrac{3}{4}d, \ k_i = 0, \pm 1, \pm 2 \cdots \\[2mm]
y_{ci} = \begin{cases} l_i d', & k_i \text{ is even} \\[2mm] \dfrac{d'}{2} + l_i d', & k_i \text{ is odd} \end{cases} \quad l_i = 0, \pm 1, \pm 2 \cdots
\end{cases}
\tag{6.11}
$$

where d' is the inner diameter of the hexagon and $d' = \frac{\sqrt{3}}{2}d$. Here the coordinates are all relative coordinates with an arbitrary point (usually the geometric center of the network area) chosen as the reference origin point.

It can be proven that if a node i belongs to the cluster $<k_i, l_i>$, among the distances from node i to all cluster centers, the distance from the node i to its own cluster center (x_{ci}, y_{ci}) is the smallest. Therefore, the process by which one CR node i with coordinates (x_i, y_i) chooses the hexagon is as follows:

1. Using Equation (6.11), it finds cluster ID $< k_j, l_j >$ satisfying

$$
\begin{cases}
x_i - \dfrac{d}{2} \le x_{cj} \le x_i + \dfrac{d}{2}, \\[2mm]
y_i - \dfrac{d}{2} \le y_{cj} \le y_i + \dfrac{d}{2}.
\end{cases}
\tag{6.12}
$$

2. If there is more than one cluster ID satisfying Equation (6.12), it chooses the one that is closest (in Euclidean distance) to itself as its cluster ID.

After CR nodes choose their clusters on their own, all the nodes broadcast the cluster ID and their coordinates using the maximum allowed transmission power and thus obtain the information of other CR nodes within their clusters. For an existing cluster, the newly arriving CR nodes report to the cluster header in order to join the cluster. The cluster header is responsible for handling player arrivals and departures, initiating and ending the game, and processing the control packets within one cluster.

As the roles of all the players in a game are equal and the computation in the game is distributed, it is not desirable to assign one specific node as a header within a cluster. Here we propose a virtual header mechanism, in which the header of a cluster is no longer a node but a cluster-unique packet called Virtual Header (VH), which is initially generated by the node that first joins the cluster. In the VH mechanism, all the jobs that a normal header does are reflected in the VH and its carried token: the player list in the token can reflect the arrival and departure of players; initiation and ending of a game can be reflected by the start and ending of VH propagation; and the control packet processing can only be handled by the VH node, a node which is granted the VH. Details about VH bestowing will be presented later during the discussion of the negotiation mechanism. As the VH keeps changing from one node to another as the game proceeds (refer to the negotiation mechanism discussion in the following part), the VH mechanism can therefore appropriately balance the transmission load on all the nodes.

The collision avoidance mechanism is another core component of our DSA-driven MAC framework, designed to avoid collisions among negotiations of different clusters. Since the busy tone mechanism can completely

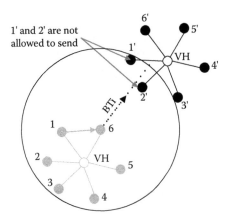

Figure 6.9   Function of *BT_i*.

solve the hidden and exposed terminal problems in ad hoc networks with low overhead [58], we adopt a similar idea and propose a cluster-based busy tone collision avoidance mechanism.

In this mechanism, there are two types of busy tones broadcast by a CR node that is receiving or overhearing: (1) tone BT_i (inside-cluster busy tone) is set up by a node that is receiving messages, serving to prevent nodes out of the cluster from interfering with the negotiation within the cluster, as shown in Figure 6.9; and (2) tone BT_o (outside-cluster busy tone) is set up by a node that is overhearing messages from other clusters, serving to avoid having negotiation among nodes within the cluster be interfered with by nodes from other clusters, as shown in Figure 6.10.

The negotiation mechanism takes charge of managing all the control messages and coordinating CRs to play the game with the right game timing. Before the game process begins, the game player set should be set up and the communication environment parameters (e.g., the noise) required by the game should be collected. Therefore, the negotiation process of the MAC framework should be divided into two stages: the *inquiry stage* and the *formal negotiation stage.* In the inquiry stage, all the cluster members will be queried of their intention regarding data communication by a token packet in the order of their cluster member IDs. Nodes intending to transmit data will become quasi-game players and place the corresponding player information into the token. Next the players (only), conduct the formal game in the subsequent formal negotiation stage, during which time the game information, piggybacked on a negotiation token packet, is passed among players. A typical negotiation process is shown in Figure 6.11, in which node 2, which wants to initiate communication, first reports to the VH node (suppose it's node 1). The VH then carries a token to inquire of all the cluster members in the order of their member IDs, and finally the

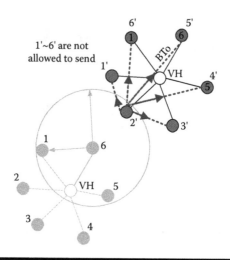

Figure 6.10   Function of BT_o.

VH carries an NG token to coordinate nodes 2 and 4 to process the formal game.

Whenever a node that is not the VH node wants to start a new transmission or stop an ongoing transmission, it will broadcast a REPORT packet intended for the VH node in its cluster. When a VH node not in the inquiry stage receives a REPORT, if there is no report recorded in the VH, it will record the new report. Otherwise, the VH node just discards this redundant report. Once a report is recorded in the VH, the future VH node will initiate the inquiry stage when it is not in negotiation. If a VH

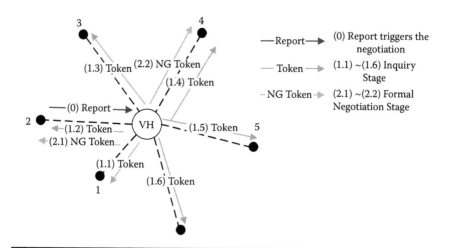

Figure 6.11   Typical negotiation process.

node in the inquiry stage receives a report, the VH node first compares its member ID and the member ID of the reporter. If the member ID of the VH node is smaller than the reporter's, the report is discarded because the reporter will be inquired of in the inquiry stage; otherwise, if there is no report recorded in the VH, the VH node records the report into the VH, and the inquiry stage will be initiated after the current negotiation process.

The entire inquiry stage is guided by a token generated by the VH node at the beginning of the inquiry stage. In the inquiry stage, the token, along with the VH, will be passed around the cluster to collect the information of the quasi game players. The inquiry stage ends when the token returns to the token generator. The advantage of the token-based inquiry is that, since the token awarding only happens between two adjacent members, by adjusting the power for awarding the token, the interference caused by negotiation in the cluster to other clusters can be minimized. Another advantage is that the collisions within the cluster in the inquiry stage can be eliminated, since only one packet (the token) can be transmitted at any time.

At the end of the inquiry stage, the VH node broadcasts the token to the entire cluster, and subsequently each game player can construct the entire player set and strategy space locally. The VH is then passed to the first player in the player set, who will then become the new VH node, and the cluster steps into formal negotiation. Similar to the inquiry stage, the formal negotiation process is guided by a negotiation token (NG token). The NG token, which carries the dynamic game information required by the game players, is passed with the VH around the game players in the order determined by the game timing. Each player updates its local strategy by the information contained in the NG token and replaces the corresponding game information field in the NG token according to the new strategy. It then passes the NG token, as well as the VH, to the next player. The process continues until the game converges to the NE.

To provide an objective evaluation of the performance of the proposed DSA-driven MAC framework, it is compared with another multichannel CR MAC protocol, DCA-PC [3]. DCA-PC is a direct-access-based protocol with transmission power adjustment ability. As QoSe-DSA [50] provides high spectrum utilization efficiency as well as QoS support, we adopt it as the game-theoretic DSA algorithm in our DSA-driven MAC framework. The evaluation metrics are the averaged packet throughput, overhead, and fairness. The throughput of packet i is defined as $P_i/(T_{arrival\_i} - T_{gen\_i})$, where P_i is the size of packet i, T_{gen_i} is the time when packet i is generated, and $T_{arrival\_i}$ is the time when packet i arrives at the packet. Since the transmission rate, queue length, and network congestion all affect the packet throughput, it is a rational criterion to evaluate the overall performance of the CR network. For fairness evaluation, we calculate the average delay of each transmitter to access the data channel, and use the variance of the delays as the benchmark to evaluate fairness.

Simulation results [57] showed that our DSA-driven MAC framework can achieve higher averaged packet throughput than DCA-PC. This is credited to the efficient utilization of the CR spectrum instructed by the intelligent game-theoretic DSA, as well as the low communication overhead generated in our DSA-driven MAC framework. As the game model guarantees fairness among the game players, our DSA-driven MAC also outperforms DCA-PC in the fairness aspect.

The research on the game theoretic DSA-driven MAC protocol is very promising, although still in its infant stage. Several open issues demand further investigation, including

(1) *Common control channel:* For most existing game-theoretic DSA-driven MAC protocols in multichannel networks, it is assumed that there is a common control channel for negotiation. As stated before, this assumption is unlikely to be achieved in a licensed band, since spectrum availability for the secondary user network varies with time and location. Negotiation over an unstable control channel is therefore an open issue for game-theoretic DSA-driven MAC protocols.

(2) *Cluster size:* In principle, grouping CR nodes into clusters will impair the whole network capacity, since a single game is artificially divided into several small games, which cannot achieve global optimization. As the cluster size increases, the performance of scattered games approaches the performance of a single game of the whole network, and therefore the increase of network capacity follows. On the other hand, the cost for negotiation among CR nodes also increases with the increase of cluster size. Determining the optimal cluster size to balance the negotiation cost and the network capacity is another open topic.

(3) *Queueing issue:* To reduce the frequency of game negotiation, it is desirable that after one negotiation, a data transmitter can send as many packets as possible. This implies that the CR node should wait for a period to accumulate more packets and send out these packets in a burst fashion. On the other hand, packet queueing also results in a longer latency. The design of queueing strategy is another topic worthy of investigation.

(4) *Broadcast/multicast support:* In existing DSA-driven MAC protocols, each CR player is constituted of one transmitter and one receiver. There is no way that a CR node can transmit a broadcast/multicast packet with current game models. This issue demands modifying the inner structure of one CR player during the process of player set configuration.

6.4 Cross-Layer-Design-Based MAC Protocol

Another trend for MAC protocol design in CR networks is cross-layer design. This is attractive because the optimal solution for spectrum allocation/access always involves other protocol layers besides the MAC layer. Existing cross-layer-design-based MAC protocols have two main types: one is the routing-oriented MAC protocol, which configures the MAC layer dynamically to adapt to routing information; the other is the flow-oriented MAC protocol, in which medium access takes end-to-end flow competition into account. The key challenge for cross-layer MAC design is how to decouple the joint optimization problem into different layers.

6.4.1 *Routing-Oriented MAC Protocol*

The MAC layer provides the hop-to-hop communication basis for the multihop network layer communication. Thus behaviors in the MAC layer have outright influence on the performance of the routing protocol. The routing protocols designed in association with a specific MAC protocol are often referred to as MAC-oriented routing protocols. On the other hand, the network layer design also affects the performance of the MAC layer, as different routing paths choose different transmitters and receivers in the neighborhood. MAC protocols designed to adapt to different route decisions are called *routing-oriented MAC protocols*, which are discussed in this section.

Link failure caused by fading and interference level changes in a CR network is often sufficient for routing and transport protocols to react, which results in operational inefficiencies. Therefore, there is a need to devise an adaptive mechanism that can suppress the negative impact caused by this type of link failure. Furthermore, intermediate nodes shared by other flows may cause data transmission to be deferred or even to fail, a situation called link blocking. The effect of link blocking as well as link failure can be alleviated by forwarding frames via an alternative path reaching the destination. This is the core idea of [59].

An approach called anycasting is presented in [59], where information on multiple route paths is provided to the MAC layer so that the MAC layer may take advantage of this multiple path routing information. The MAC layer decides the next hop, among the multiple optional next hops given by the link status, to forward the data frame. This MAC protocol, based on the link status information, improves performance but requires certain operational coordination between the routing and MAC layers.

At each hop, the routing layer passes the information on multiple next hops to the MAC layer. The transmitter then broadcasts the RTS to these multiple next hops. The multicast RTS, referred to as MRTS, contains all the MAC addresses of next hop receivers. By positioning the MAC addresses of next hops onto the MRTS frame, a priority order to each next hop is set up. The positioning priority can be determined by information from the network layer (e.g., path length) or a lower layer (e.g., signal strength), or a combination of these two.

When all intended receivers receive the MRTS packet, the receivers respond CTS in the order of their priorities. When the transmitter receives the first successful CTS, it transmits the DATA frame after an *SIFS* interval to the sender of the first CTS. The interval of two adjacent CTSs is *PIFS*. PIFS is longer than SIFS; this is to ensure that other receivers with lower priority hear the DATA before they send CTS and suppress any further CTS transmissions. In this way, the efficiency of the control channel can be improved. If none of the CTSs is received successfully, the transmitter then retries, again following the random backoff algorithm.

Simulation results in [59] show that the MAC protocol based on the routing path diversity improves the network performance in terms of throughput and overhead, especially when the qualities of channels are quite low and the network is highly loaded. By exploiting the multipath routing information, the routing-oriented MAC protocol can improve the performance by reducing interactions between the MAC and network layers, as well as simultaneous routing deployments of multiple traffic sources.

Similar to the idea of [59], for CR networks operating in licensed bands, the procedure for protection of primary users can also make use of the method of alternate route paths by choosing the route path that causes less interference to primary users. However, it may happen that the contention among the CR users will be aggravated when the primary user protection requirement is imposed, as more CR users choose route paths far away from primary users. Therefore, the balance between primary user protection and CR users' contention should be carefully addressed in designs for routing-oriented MAC protocols in licensed-band CR networks.

6.4.2 Flow-Oriented MAC Protocol

Due to the contention for the shared channel, the throughput of each node is limited not only by the channel capacity, but also by the transmissions in its neighborhood. Furthermore, for multihop CR networks, each multihop flow encounters medium access contentions not only from other flows that pass by the neighborhood (i.e., the *interflow contention*), but also its own from its own transmissions because the transmission at each hop must contend for the channel with the upstream and downstream nodes (i.e., the *intraflow contention*). These two kinds of flow contentions can result in severe collisions and congestion, and thus seriously limit the performance of ad hoc networks [60]. As the control for end-to-end flow is the responsibility of the transport layer, MAC protocols concerned with the flow contention issue should be cross-designed with the transport layer.

The flow-oriented MAC protocol in [61] is based on the LCG (link contention graph) model (as discussed in Section 6.3.1) for single-channel access. Given a contention graph, all its maximal cliques can be identified. Maximal cliques are local constructions and capture the local contention relations of the flows. Flows within the same maximal clique cannot transmit simultaneously, but flows in different cliques may do so. A flow may belong to several cliques, and can successfully transmit if and only if it is the only active flow in the clique to which it belongs.

Consider an ad hoc wireless network with a set of nodes V and a set of logical links L. It is assumed that each link l has a fixed finite capacity C_l^0 (packets per second) when active, assuming that a power control algorithm maintains a constant data rate over a realistic fading channel. A successful link transmits at rate C_l^0 for the duration it holds the channel.

Assume the network is shared by a set S of sources indexed by s. Each source s uses a set $L^s \subset L$ of links. The relationship between source models and links is represented by an $L \times S$ routing matrix R:

$$R_{ls} = \begin{cases} 1 & \text{if } l \in L^s, \\ 0 & \text{otherwise.} \end{cases} \tag{6.13}$$

The global objective is to choose source rates $x = \{x_s\}_{S \times 1}$ so as to

$$\underset{x_s \geq 0}{\text{maximum}} \sum_s U_s(x_s) \tag{6.14}$$

subject to

$$FRx \leq \varepsilon \tag{6.15}$$

where $U_s(x_s)$ is the utility source s attains when it transmits at a rate of x_s packets per second, and F is an $N \times L$ contention matrix $\{F_{nl}\}_{N \times L}$. Here N is the number of maximal cliques of the contention graph, each clique n contains a set $L_n \subset L$ of links, and

$$F_{nl} = \begin{cases} 1/c_l^0 & \text{if } l \in L_n, \\ 0 & \text{otherwise.} \end{cases} \tag{6.16}$$

Suppose y is an L-dimensional vector with element y_l being the flow rate on link l. To simplify the analysis, the flow rate y_l is normalized as y_l/c_l^0. In this way, an active link can achieve a rate of 1 and an idle link gets the rate of 0. Since flows within the same clique cannot transmit simultaneously, there is a constraint $Fy \leq \mathbf{1}$, where $\mathbf{1}$ denotes an N-dimensional vector with each component being 1. However, this is only a necessary condition for feasibility of the flow vector y. Considering the complexity of verifying the feasibility of flow vector y, in [62] the constraint imposed by the MAC layer is written as

$$Fy \leq \varepsilon. \tag{6.17}$$

By the definitions of R, y, and x, there is the relation y = Rx. Therefore the constraint (6.17) can be used to derive (6.15).

The global optimization issue can be seen from two complementary perspectives. On one hand, it is a utility-based congestion control problem with the MAC layer constraints. As such, the congestion prices are not decided by the link capacity, but determined by the contention region. In other words, the MAC layer imposes the ultimate constraints to the achievable rates. On the other hand, it is a media access control problem, which is to allocate physical bandwidth to each link with the objective of maximizing aggregate end-user utilities. As such, the resulting MAC protocol is flow-oriented and will allocate more bandwidth to the links with more contention to alleviate flow congestion. Based on this idea, the global optimization problem is decomposed into two subproblems. One can be solved in the transport layer and the other can be solved in the MAC layer.

The mechanism in [61] can be extended to multiple-channel-based CR networks in the future. The multichannel access can relieve the contention among different flows; however, it will also increase the complexity of the cross-layer design, since (1) different flows should be dynamically assigned with different channels, and (2) even in the same flow, the flow may go through different channels to avoid intraflow contention.

These are referred to as "cross-layer" or "joint-layer" designs, yet the major challenge for cross-layer MAC protocol design is actually to *decompose* the intrinsic layer-to-layer mutual impacts that arise from optimization crossing layers. Although cross-layer-based MAC protocol design in CR networks is still in its early stage, it can be predicted there will be more and more work tackling this issue due to its potential advantages.

6.5 Concluding Remarks

In this chapter, we classified existing MAC protocols in CR networks into two categories based on their intelligence level: one is direct-access-based MAC protocols, with single-dimensional local strategy and absence of global optimization; the other is DSA-driven MAC protocols, with the goal to optimize the overall performance of the whole network and usually with multiple-dimensional local strategy. We presented characteristics of different types of CR MAC protocols, analyzed their merits and potential challenges, and identified future research directions. We also briefly introduced cross-layer-design-based MAC protocols at the end of the chapter.

In the section on direct-access-based MAC protocols, we first discussed the direct contention-based MAC protocols. Most early direct contention-based multichannel MAC protocols can work well with CR networks operating in unlicensed bands. However, those protocols cannot be implemented in CR networks running in licensed bands, since they fail to meet the constraints imposed by primary users. Therefore, recent research on direct-access-based MAC protocol design focuses on providing detection and protection of primary users. Most such existing protocols are based on the common control channel approach and the split-phase approach; however, another recent method, the dynamic frequency-hopping sequence approach, also shows potential applicability in licensed-band CR networks.

With the observation that contention-based MAC protocols relying on sender-receiver handshaking may not be capable of ensuring that CR users acquire the correct spectrum usage information, direct coordination-based MAC protocols are proposed to detect the correct spectrum hole and allocate spectrum among CR users more efficiently. Given the complexity and difficulties in constructing infrastructureless CR networks composed of heterogeneous radio users, the first CR network standard, IEEE 802.22, is devised for CR networks with infrastructure. Since infrastructure-based CR networks can provide coordinated spectrum access efficiently with low complexity via a regional spectrum manager, pragmatic centralized MAC protocols are more attractive in this standard.

Compared to simple direct-access-based MAC protocols, DSA-driven MAC protocols utilize CR spectrum more efficiently, since the channel access control is instructed by intelligent DSA algorithms. Most existing DSA-driven MAC protocols only work for CR networks in unlicensed bands. However, with the recent development of DSA techniques, some MAC protocols also adopt DSA algorithms for licensed-band CR networks to efficiently harmonize the coexistence of primary and secondary users.

Cross-layer MAC design is another trend for CR network MAC protocol design to further optimize the overall performance of the whole CR network. Most existing cross-layer-design-based MAC protocols work for unlicensed band access. Cross-layer protocol design for CR networks in licensed bands turns out to be quite complex, involving new technologies in terms of optimization tools, coordination, and signaling protocols. Typical routing-oriented and flow-oriented MAC protocols are introduced at the end of this chapter.

Acknowledgments

This research is partially supported by NSF (CNS-0644056) and U.S. Army, Communications-Electronics Research Development and Engineering Center (CERDEC).

References

[1] N. Jain, Samir R. Das, and Asis Nasipuri, A Multichannel MAC Protocol with Receiver-Based Channel Selection for Multihop Wireless Networks, in *Proc. IEEE International Conference on Computer Communication and Networks (ICCCN 2001)*, October 2001.

[2] S. Wu, C. Lin, Y. Tseng, and J. Sheu, A New Multi-Channel MAC Protocol with On-Demand Channel Assignment for Mobile Ad Hoc Networks, *International Symposium on Parallel Architectures, Algorithms and Networks (I-SPAN)*, 2000.

[3] S. Wu, Y. Tseng, and C. Lin, A Multi-channel MAC Protocol with Power Control for Multihop Mobile Ad Hoc Networks, *The Computer Journal*, vol. 45, no. 1, pp. 101–110, 2002.

[4] W. Hung, K. Law, and A. Leon-Garcia, A Dynamic Multi-Channel MAC for Ad Hoc LAN, in *Proc. 21st Biennial Symposium on Communications*, pp. 31–35, Kingston, Canada, June 2002.

[5] N. Choi, Y. Seok, and Y. Choi, Multi-Channel MAC protocol for mobile ad hoc networks, *IEEE Vehicular Technology Conference*, vol. 2, pp. 1379–1382, 2003.

[6] N. Hyon, T. Yoo, and Sang-Jo, Distributed Coordinated Spectrum Sharing MAC Protocol for Cognitive Radio, in *Proc. IEEE DySPAN 2007*, pp. 240–249, Apr. 17–20, 2007.

[7] A. Mishra, A Multi-Channel MAC for Opportunistic Spectrum Sharing in Cognitive Networks, in *MILCOM 2006*, pp. 1–6, Oct. 23–25, 2006.

[8] V.R. Syrotiuk, M. Cui, S. Ramkumar, and C.J. Colbourn, Dynamic spectrum utilization in ad hoc networks, *Computer Networks: The International Journal of Computer and Telecommunications Networking*, vol. 46, issue 5, pp. 665–678, 2004.

[9] T. Shu, S. Cui, and M. Krunz, Medium Access Control for Multi-Channel Parallel Transmission in Cognitive Radio Networks, in *Proc. Globecom 2006*, Nov. 2006.

[10] L. Wang, A. Chen, and D. Wei, A Cognitive MAC Protocol for QoS Provisioning in Overlaying Ad Hoc Networks, in *Proc. IEEE CCNC 2007*, pp. 1139–1143, Jan. 2007.

[11] J. So and N. Vaidya, Multi-Channel MAC for Ad Hoc Networks: Handling Multi-Channel Hidden Terminals Using a Single Transceiver, ACM MobiHoc, May 2004.

[12] J. So and N. Vaidya, A Multi-Channel MAC Protocol for Ad Hoc Wireless Networks, Technical report, Dept. of Comp. Science, Univ. of Illinois at Urbana-Chamapign, 6, 2003.

[13] J. Chen, S. Sheut, and C. Yangt, A New Multichannel Access Protocol for IEEE 802.11 Ad Hoc Wireless LANs, in *Proc. IEEE Personal Indoor and Mobile Radio Communication 2003*, pp. 2291–2296, 2003.

[14] C. Cordeiro and K. Challapali, C-MAC: A Cognitive MAC Protocol for Multi-Channel Wireless Networks, in *Proc. IEEE DySPAN 2007*, pp. 147–157, Apr. 2007.

[15] Z. Tang and J.J. Garcia-Luna Aceves, Hop-Reservation Multiple Access (HRMA) for Ad Hoc Networks, in *Proc. IEEE INFOCOM*, vol. 1, pp. 194–201, Mar. 1999.

[16] A. Tzamaloukas and J.J. Garcia-Luna-Aceves, Channel-Hopping Multiple Access, in *Proc. IEEE International Conference on Computer Communication and Network (IC3N '00)*, Oct. 2000.

[17] A. Tzamaloukas and J.J. Garcia-Luna-Aceves, Channel-Hopping Multiple Access with Packet Trains for Ad Hoc Networks, In *Proceedings IEEE MoMuC '00*, Oct. 2000.

[18] A. Tzamaloukas and J.J. Garcia-Luna-Aceves, A Receiver-Initiated Collision-Avoidance Protocol for Multi-Channel Networks, in *Proc. IEEE INFOCOM 2001*, pp. 189–198, Vol., Apr. 2001.

[19] M. Joa-Ng and I. Lu, Spread Spectrum Medium Access Protocol with Collision Avoidance in Mobile Ad Hoc Wireless Network, *IEEE INFOCOM '99*, vol. 2, pp. 776–783, Mar. 1999.

[20] P. Bahl, R. Chandra, and J. Dunagan, SSCH: Slotted Seeded Channel Hopping for Capacity Improvement in IEEE 802.11 Ad hoc Wireless Networks, in *Proc. of ACM MobiCom*, 2004.

[21] Hoi-Sheung, W. So, J. Walrand, and M. Jeonghoon, MCMAC: A Parallel Rendezvous Multi-Channel MAC Protocol, in *Proc. IEEE WCNC 2007*, pp. 334–339, Mar. 2007.

[22] X. Jing and D. Raychaudhuri, Spectrum Co-existence of IEEE 802.11b and 802.16a Networks Using Reactive and Proactive Etiquette Policies, *Mobile Networks and Applications*, vol. 11, issue 4, pp. 539–554, Aug. 2006.

[23] S. Krishnamurthy, M. Thoppian, S. Venkatesan, and R. Prakash, Control Channel Based MAC-Layer Configuration, Routing and Situation Awareness for Cognitive Radio Networks, in *Proc. IEEE MILCOM 2005*, Oct. 2005.

[24] J. Zhao, H. Zheng, and G. Yang, Distributed Coordination in Dynamic Spectrum Allocation Networks, in *Proc. IEEE DySPAN 2005*, pp. 259–268, Nov. 2005.

[25] C. Cordeiro, K. Challapali, D. Birru, and N. Sai Shankar, IEEE 802.22: The First Worldwide Wireless Standard Based on Cognitive Radios, in *Proc. IEEE DySPAN 2005*, pp. 328–337, Nov. 2005.

[26] W. Hu, D. Willkomm, M. Abusubaih, et al., Cognitive Radios for Dynamic Spectrum Access—Dynamic Frequency-Hopping Communities for Efficient IEEE 802.22 Operation, *IEEE Communication Magazine*, vol. 45, issue 5, pp. 80–87, May 2007.

[27] Y. Chen, N. Han, S. Shon, and J. Kim, Dynamic Frequency Allocation Based on Graph Coloring and Local Bargaining for Multi-Cell WRAN System, in *Proc. APCC 2006*, pp. 1–5, Aug. 2006.

[28] M.M. Buddhikot, P. Kolodzy, S. Miller, K. Ryan, and J. Evans, DIMSUM-Net: New Directions in Wireless Networking Using Coordinated Dynamic Spectrum Access. *IEEE WoWMoM 2005*, Jun. 2005.

[29] V. Brik, E. Rozner, S. Banerjee, and P. Bahl, DSAP: A Protocol for Coordinated Spectrum Access, in *Proc. IEEE DySPAN 2005*, pp. 611–614, Nov. 2005.

[30] L. Cao and H. Zheng, Distributed Spectrum Allocation via Local Bargaining, in *Proc. IEEE SECON 2005*, pp. 475–486, Sep. 2005.

[31] C. Peng, H. Zheng, and B. Zhao, Utilization and Fairness in Spectrum Assignment for Opportunistic Spectrum Access, *Mobile Networks and Applications*, vol. 11, issue 4, pp. 555–576, Aug. 2006.

[32] H. Zheng and C. Peng, Collaboration and Fairness in Opportunistic Spectrum Access, in *Proc. IEEE ICC 2005*, vol. 5, pp. 3132–3136, May 2005.

[33] M. Thoppian, S. Venkatesan, R. Prakash, and R. Chandrasekaran. MAC-Layer Scheduling in Cognitive Radio Based Multihop Wireless Networks, in *Proc. IEEE WoWMoM 2006*, Jun. 2006

[34] S. Geirhofer, T. Lang, and B. Sadler, Cognitive Radios for Dynamic Spectrum Access—Dynamic Spectrum Access in the Time Domain: Modeling and Exploiting White Space, *IEEE Communication Magazine*, vol. 45, issue 5, pp. 66–72, May 2005.

[35] Y. Xing, R. Chandramouli, S. Mangold, and S. Shanker, Dynamic Spectrum Access in Open Spectrum Wireless Networks, in *IEEE Journal on Selected Areas in Communications*, vol. 24, issue 3, pp. 626–637, Mar. 2006.

[36] Q. Zhao, L. Tong, A. Swami, and Y. Chen, Decentralized cognitive MAC for Opportunistic Spectrum Access in Ad Hoc Networks: A POMDP Framework, in *IEEE Journal on Selected Areas in Communications*, vol. 25, issue 3, pp. 589–600, Mar. 2007.

[37] Q. Zhao, L. Tong, and A. Swami, Decentralized cognitive MAC for dynamic Spectrum Access, in *Proc. IEEE DySPAN 2005*, pp. 224–232, Nov. 2005.

[38] Q. Zhao, L. Tong, and A. Swami, A Cross-Layer Approach to Cognitive MAC for Spectrum Agility, in *Proc. Signals, Systems and Computers Conf. 2005*, pp. 224–232, Nov. 2005.

[39] Q. Zhao, L. Tong, A. Swami, and Y. Chen, Cross-Layer Design of Opportunistic Spectrum Access in the Presence of Sensing Error, in *Proc. Information Sciences and Systems Conf. 2006*, pp. 778–782, Mar. 2006.

[40] Y. Chen, Q. Zhao, and A. Swami, Distributed Cognitive MAC for Energy-Constrained Opportunistic Spectrum Access, in *Proc. MILCOM 2006*, pp. 1–7, Oct. 2006.

[41] A. Swami and Q. Zhao, A Decision-Theoretic Framework for Opportunistic Spectrum Access, *IEEE Wireless Communications*, vol. 14, issue 4, pp. 14–20, Aug. 2007.

[42] Y. Chen, Q. Zhao, and A. Swami, Joint Design and Separation Principle for Opportunistic Spectrum Access, in *Proc. ACSSC 2006*, pp. 696–700, Oct. 2006.

[43] D.V. Djonin, Q. Zhao, and V. Krishnamurthy, Optimality and Complexity of Opportunistic Spectrum Access: A Truncated Markov Decision Process Formulation, in *Proc. ICC 2007*, pp. 5787–5792, Jun. 2007.

[44] J. Neel, Analysis and Design of Cognitive Radio Networks and Distributed Radio Resource Management Algorithms, Thesis of PhD, EE Virginia Tech Sep. 6, 2006.

[45] S. Mathur, S. Lalitha, and N.B. Mandayam, Coalitional Games in Receiver Cooperation for Spectrum Sharing, in *Proc. of Information Sciences and Systems 2006*, pp. 949–954, Mar. 2006.

[46] M.D. Perez-Guirao and K. Jobmann, Cognitive Resource Access Scheme for IR-UWB Autonomous Networks, in *Proc. of WPNC 2007*, pp. 267–271, Mar. 2007.

[47] J. Huang, R. Berry, and M.L. Honig, Distributed Interference Compensation for Wireless Networks, *IEEE Journal on Selected Areas in Communications*, vol. 24, issue 5, May 2006, pp. 1074–1084.

[48] J. Huang, R. Berry, and M.L. Honig, Spectrum Sharing with Distributed Interference Compensation, in *Proc. IEEE DySPAN 2005*, pp. 88–93.

[49] J. Huang, R. Berry, and M.L. Honig, A Game Theoretic Analysis of Distributed Power Control for Spread Spectrum Ad Hoc Networks, *IEEE Intl. Symp. on Information Theory*, Sept. 2005.

[50] T. Jin, C. Chigan, and Z. Tian, Game-theoretic Distributed Spectrum Sharing for Wireless Cognitive Networks with Heterogeneous QoS, *IEEE Globecom 2006*, Nov. 2006.

[51] C. Zou, T. Jin, C. Chigan, and Z. Tian, QoS-aware Distributed Spectrum Sharing for Heterogeneous Wireless Cognitive Networks, *Elsevier Journal of Computer Networks: Special Issue on Cognitive Radio Networks* vol. 52, issue 4, Mar. 2008.

[52] Z. Fang and B. Bensaou, Design and Implementation of A MAC Scheme for Wireless Ad Hoc Networks Based on A Cooperative Game Framework, in *Proc. IEEE ICC 2004*, vol. 7, pp. 4034–4038, Jun. 2004.

[53] R. Sudipta and R.K. Guha, Fair Bandwidth Sharing in Distributed Systems: A Game-Theoretic Approach, *Transaction on Computers*, vol. 54, issue 11, pp. 1384–1493, Oct. 2005.

[54] N. Nie and C. Comaniciu, Adaptive Channel Allocation Spectrum Etiquette for Cognitive Radio Networks, in *Proc. IEEE DySPAN 2005*, pp. 269–278, Nov. 2005.

[55] F. Wang, O. Younis, and M. Krunz, GMAC: A Game-Theoretic MAC Protocol for Mobile Ad Hoc Networks, in *Proc. of Modeling and Optimization in Mobile, Ad Hoc and Wireless Networks 2006*, pp. 1–9, Apr. 2006.

[56] A. Muqattash and M. Krunz, POWMAC: A Single-Channel Power-Control Protocol for Throughput Enhancement in Wireless Ad Hoc Networks, in *IEEE Journal on Selected Areas in Communications*, vol. 23, issue 5, pp. 1067–1084, May 2005.

[57] C. Zou and C. Chigan, A Game Theoretic DSA-Driven MAC Framework for Cognitive Radio Networks, in *Proc. IEEE ICC 2008*, May 2008.

[58] Z. Haas and J. Deng, Dual Busy Tone Multiple Access (DBTMA) —A Multiple Access Control Scheme for Ad Hoc Networks, *IEEE Trans. On Commun.*, vol. 50, pp. 975–985, Jun. 2002.

[59] L. Hyungkeun, L. Jang-Yeon, and C. Jin-Woong, An Adaptive MAC Protocol Based on Path-Diversity Routing in Ad Hoc Networks, in *Proc. of 9th Advanced Communication Technology Conf.*, vol.1, pp. 640–644, Feb. 2007.

[60] H. Zhai, J. Wang, and Y. Fang, DUCHA: A New Dual-Channel MAC Protocol for Multihop Mobile Ad Hoc Networks, in *IEEE Workshop on Wireless Ad Hoc and Sensor Networks*, Nov. 2004.

[61] L. Chen, S. Low, and J. Doyle, Joint Congestion Control and Media Access Control Design for Ad Hoc Wireless Networks, in *Proc. IEEE Infocom 2005*, vol. 3, 2212–2222, Mar. 2005.

Chapter 7

Medium Access Control Protocols for Dynamic Spectrum Access in Cognitive Radio Networks: A Survey

Dusit Niyato and Ekram Hossain

Contents

7.1 Introduction

In a traditional wireless communication system, frequency spectrum is statically allocated to licensed users (i.e., primary users) only. However, since licensed users may not always occupy the allocated radio spectrum, this static spectrum allocation results in spectrum underutilization. This was confirmed in a report from the FCC (Federal Communications Commission)[1] where it was shown that even in a crowded area, more than half of the radio spectrum is not occupied at any given time. This, along with the increasing spectrum demand from emerging wireless applications, is driving the development of new spectrum allocation policies to allow unlicensed users (i.e., secondary users) to access radio spectrum when it is not occupied by primary users. These new spectrum allocation policies, which will be implemented through the *cognitive radio* technology, are expected to improve spectrum utilization while satisfying the spectrum demand for new wireless applications.

Cognitive radio technology can be developed based on the software-defined-radio technique,[2] which was proposed to improve adaptability and flexibility of wireless communication systems. A cognitive radio node will have the ability to observe and learn the spectrum usage activity of the primary users,[3] and subsequently, to make an optimal decision to access the spectrum opportunities without interfering with incumbent services. The emerging IEEE 802.22 standard-based wireless regional area network (WRAN) technology is based on this cognitive radio concept in which the cognitive radio users can opportunistically utilize the TV bands when they are not occupied by incumbent TV services.

Cognitive radio systems can be classified as either *static cognitive radio systems* or *dynamic cognitive radio systems*. In a static cognitive radio system, secondary users observe the activity of the primary users in a fixed spectrum band and access the entire spectrum band if a spectrum opportunity is detected. This type of cognitive radio system can be developed based on IEEE 802.11 and IEEE 802.15.3 technologies. In a dynamic cognitive radio system, secondary users can operate on different spectrum sizes (i.e., transmit using different bandwidths) by changing the transmission parameters in the physical layer (e.g., based on orthogonal frequency division multiplexing (OFDM) or multi-carrier code division multiple access (MC-CDMA)).

Dynamic spectrum access (DSA) is one of the key concepts in cognitive radio. Through DSA, frequency spectrum can be shared among cognitive radio users. There are three models of spectrum sharing in dynamic spectrum access. These models are defined based on the rights of cognitive radio users to access the spectrum.

- *Public commons model:* In this model, the radio spectrum is open to anyone for access. Since all users can access the spectrum, there is no permanent spectrum owner. This access model is currently used in wireless standards (e.g., WiFi and Bluetooth radio) operating in the license-free ISM band.
- *Exclusive usage model:* In this model, radio spectrum is exclusively licensed to a particular user/service. However, to improve spectrum utilization, spectrum can be allocated by the spectrum owner to the cognitive radio users. In other words, permission for the cognitive radio users to access the spectrum must be granted by the spectrum owner.
- *Private commons model:* In this model, unlicensed users (i.e., secondary users) can access the spectrum owned by the licensed users (i.e., primary users) ensuring that the interference to the primary

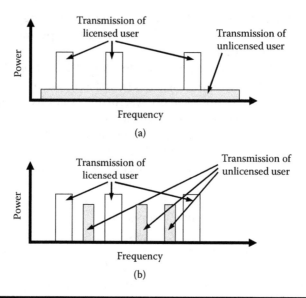

Figure 7.1   (a) Underlay and (b) overlay spectrum access.

users is maintained below the target level. Spectrum access by the secondary users can be performed either in an underlay or an overlay approach (Figure 7.1). In the underlay approach, a secondary user spreads the transmission over a large bandwidth using low transmit power (e.g., ultra-wideband (UWB) transmission). In the overlay approach, a secondary user accesses the spectrum in the frequency or time domain. While power control is crucial for underlay access, spectrum opportunity identification and synchronization are important for overlay access. Note that this private commons model is more flexible than the exclusive usage model, since the spectrum can be opportunistically accessed by secondary users.

To implement dynamic spectrum access for cognitive radio networks, physical and medium access control (MAC) layer communications protocols used in the traditional wireless networks have to be modified accordingly. The primary function of a MAC protocol is to control transmission in the physical layer and to provide access service to the error control and recovery protocols at the link layer. In a cognitive radio network, designing MAC protocols for secondary users is challenging due to the requirement of "peaceful" coexistence with primary users and improved channel utilization by avoiding collisions among users. Synchronization (in time and channel of transmission) between the secondary transmitter and receivers is also required for successful communication in a cognitive radio network.

There are two steps involved in dynamic spectrum access: spectrum exploration (i.e., channel sensing) and spectrum exploitation (i.e., channel

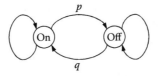

Figure 7.2 On–off activity model of primary user.

access). In the channel sensing step, a secondary user searches for the signal signature of a primary user (i.e., incumbent service). Several techniques for channel sensing in the physical layer have been proposed.[4-8] This sensing is abstracted in the MAC layer to identify whether the channel is occupied by primary users or not.

To abstract spectrum sensing in the MAC layer, secondary users generally consider a two-state activity model (i.e., on–off model*) for a primary user (Figure 7.2). In particular, an "on" state indicates that the primary user is currently accessing the channel. In contrast, an "off" state indicates that the primary user is not accessing the channel. The duration of stay in each state is assumed to be random. For example, it was shown in Ref. [9] that a geometric distribution fits the measurement data very well for wireless LANs operating in the ISM band. However, due to this abstraction of spectrum sensing in the MAC layer, the sensing result in the MAC layer may not be perfect (i.e., sensing error can occur with a nonzero probability). Given this sensing information, a secondary user must make a decision on whether to access the channel or not.

Channel sensing in the MAC layer can be both reactive and proactive. Reactive spectrum sensing relies solely on on-demand sensing. That is, the channel is sensed only when a secondary user has to switch from the current channel (e.g., when a primary user is detected). However, in proactive sensing, periodic channel sensing is combined with on-demand sensing. This proactive sensing can reduce latency of channel access, since channel occupation/availability information is provided through periodic sensing. However, this proactive sensing incurs larger overhead, and consequently, the throughput performance of the secondary users is degraded.

The main MAC layer functions in a cognitive radio can be summarized as follows:

■ To obtain information on channel occupancy. This information will be used by a secondary user to decide whether to transmit data or not and whether to switch to a new channel or not.

---

* The name of the model could be different, for example, busy-idle or occupied-unoccupied.

- To perform negotiation among primary users and secondary users for spectrum allocation, and also among secondary users to perform channel sensing and channel access.
- To synchronize transmission parameters (e.g., channel, time slot) between transmitter and receiver.
- To facilitate spectrum trading functions (e.g., spectrum bidding and spectrum pricing), which involve primary users (or primary service providers) and secondary users (or secondary service providers).

In this chapter, we provide a survey on the MAC protocols for cognitive wireless networks. We review related works on MAC protocols for dynamic spectrum access (Table 7.1). Specifically, the state of the art of

Table 7.1　Summary of Related Work on Cognitive Radio MAC Protocols

Paper	Description	Technique
H. Kim et al.[10]	Optimal sensing period	Optimization
Q. Zhao et al.[11]	Optimal channel sensing and access	POMDP
P. Papadimitratos et al.[12]	AS-MAC for spectrum sharing in GSM network	CSMA protocol
T. Shu et al.[13]	Multichannel cognitive radio MAC with rate and power adaptation	Optimization
L. Ma et al.[15]	Dynamic open spectrum sharing (DOSS) protocol	Busy tone
M. Thoppian et al.[17]	MAC-layer scheduling	Integer linear programming
S. Sengupta et al.[18]	Channel allocation for self-coexistence in IEEE 802.22	Graph coloring
Q. Zhang et al.[19]	Hardware-constrained multichannel cognitive MAC (HC-MAC) protocol	Optimal stopping problem
C. Cordeiro et al.[22]	Cognitive MAC (C-MAC) for synchronization	Rendezvous channel
D. Grandblaise et al.[30]	Dynamic spectrum sharing through leasing	Token based rental protocol
A. C.-C. Hsu et al.[29]	Statistical channel allocation MAC (SCA-MAC)	Optimization
V. Brik et al.[31]	Dynamic spectrum access protocol (DSAP) for spectrum assignment	Coordinated protocol
S. N. Shankar et al.[32]	Queueing model for dynamic spectrum access	M/G/1
C. Doerr et al.[38]	Adaptive MAC protocol	Load measurement
W. Hu et al.[40]	Out-of-band channel sensing in IEEE 802.22 network	Frequency hopping

spectrum sensing, synchronization, spectrum access, and spectrum allocation schemes is summarized. Some open issues in designing MAC protocols for dynamic spectrum access in cognitive radio networks are outlined. Then we present on overview of MAC signaling for the IEEE 802.22 standard-based WRANs. To this end, challenges in 802.22-based MAC design and related work are reviewed.

7.2 MAC Protocols for Cognitive Radio Networks: Challenges and Existing Approaches

The major challenges in designing MAC protocols for dynamic spectrum access in cognitive radio networks can be summarized as follows:

- *Coexistence:* The most important function of a MAC protocol for dynamic spectrum access is to support coexistence between primary and secondary users. Also, coexistence among multiple secondary users (i.e., self-coexistence) is important to avoid collisions due to simultaneous channel access. For example, IEEE 802.22-based cognitive networks must coexist with incumbent service (i.e., TV service) and also must self-coexist with other IEEE 802.22 networks operating in the same or overlapping areas.

- *Primary user's time-varying activity:* Since the channel occupancy by the primary users is time-varying, the MAC protocol at a cognitive radio must be able to detect this time-varying activity of primary users and adjust the channel access policy dynamically.

- *Multichannel access:* In a cognitive radio network, multiple channels are available for access by the secondary users. However, two major challenges arise in a multichannel cognitive radio network. First, if the cognitive radio is hardware-limited (e.g., the number of channels which can be sensed and accessed is limited), an optimal scheme is required for selection of the sets of channels to sense and to access. Second, synchronization between transmitter and receiver is necessary so that transmitter and receiver can communicate in the same channel and at the same time slot.

- *Multihop communications:* Transmission of secondary users can span multiple hops to extend the transmission range (e.g., cognitive mesh networks). However, in different locations, secondary nodes can experience different channel conditions (i.e., different sets of primary users with different channel activity). Also, the *hidden and exposed terminal problems* have to be solved to avoid collision and under-utilization of the available channels.

In the hidden terminal problem, some of the secondary nodes are not aware of transmission from other nodes. In this case, collision

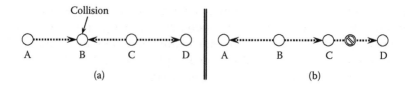

Figure 7.3   Hidden and exposed terminal problems.

can occur. For example, in Figure 7.3(a), node C is not aware of transmission from node A to node B, since node C is out of the transmission range of node A. As a result, node C may start transmission to node D, but this transmission of node C will collide with reception of node B.

In the exposed terminal problem, some of the secondary nodes overhear the transmission of other nodes. In this case, overhearing nodes will defer transmission, which will result in performance degradation. For example, in Figure 7.3(b), node B transmits to node A, but node C overhears this transmission. As a result, node C defers transmission to node D, although transmission from node C would not interfere with the reception at node B.

The research works in the literature related to MAC protocol design for cognitive radio networks can be categorized into the following:

- *Optimal channel sensing and optimal channel access:* Channel sensing is required to gain knowledge of channel availability for a secondary user. Based on the channel sensing information, which could be erroneous, the decision on channel access must be optimally made. For optimal channel sensing, a standard optimization problem can be formulated.[10] Alternatively, joint channel sensing and access decision can be formulated as a Markov decision process.[11] However, channel sensing can be constrained due to hardware limitation (e.g., secondary node cannot sense and access a channel simultaneously). Therefore, the channel sensing strategy needs to be optimized.
- *Synchronization mechanism:* Synchronization and negotiation among secondary users would be required to ensure correct data reception and to avoid collision. Synchronization is important, especially in multihop networks without any centralized controller, to govern transmissions from all nodes. Two major approaches to implementing the negotiation mechanism in a cognitive radio network are based on CSMA/CA[11–14] and busy tone.[15,16]

- *Channel allocation:* Channel allocation is important especially in multihop networks where the channel availability depends on the location and the presence of primary users (i.e., different geographical locations could have different sets of primary users). The idle channels must be optimally allocated to multiple secondary users. Approaches for channel allocation in cognitive radio networks can be designed based on integer linear programming[17] and graph theory.[18]

A summary of the related works on cognitive MAC protocols is provided in Table 7.1.

7.3 Cognitive MAC Protocols for Optimal Channel Sensing and Optimal Channel Access

7.3.1 Cognitive MAC Protocol for Joint Spectrum Sensing and Spectrum Access Optimization

To achieve optimal performance in a distributed cognitive radio network, a MAC protocol must determine a set of channels to sense and a subset of channels to access. Assuming that the activity of the primary users can be modeled by a two-state Markov chain, this problem of channel sensing and access has been formulated as a partially observable Markov decision process (POMDP).[11]

For optimal channel sensing and access, the POMDP was formulated as follows. The state of the system was defined in terms of channel condition (i.e., occupied or idle). The actions referred to sensing and accessing the subset of channels, and an observation referred to the availability of the sensed channel. Reward was defined as the number of transmitted bits. The objective was to maximize the expected total number of transmitted bits in a certain number of time slots. The constraint was to maintain the collision probability with the primary user below a target level. This POMDP can be solved using linear programming. However, when the number of channels increases, the state space of the POMDP formulation becomes large. Therefore, a suboptimal strategy based on a reduced number of states was also considered.

Based on the POMDP formulation, a decentralized cognitive MAC (DC-MAC) was proposed.[11] In DC-MAC, a small number of operations are performed in each time slot (Figure 7.4). First, the channel is sensed. Then, the result of sensing is observed. The secondary user makes a decision on whether to access the channel. If the secondary user accesses the channel, an acknowledgement is used to confirm successful data transmission. The reward is computed at the end of every time slot.

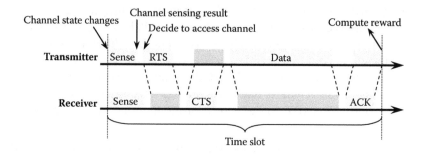

Figure 7.4 DC-MAC: Sequence of operations in a time slot.

When the DC-MAC protocol operates between a sender and a receiver neighboring the same primary user, it can guarantee that sender and receiver will access the same channel without synchronization. However, this may not be true when the sender and the receiver are in different locations. Therefore, handshaking between the sender and the receiver is required. In DC-MAC, RTS and CTS messages are used for synchronization and also to mitigate the hidden and exposed terminal problems (Figure 7.4).

7.3.2 MAC Protocol for Hardware-Constrained Cognitive Radio

With a constraint on radio hardware (e.g., single radio transceiver), spectrum sensing must be optimally performed to achieve the highest performance. With this challenge, a hardware-constrained multichannel cognitive MAC (HC-MAC) protocol was proposed in Ref. [19] Channel sensing in this HC-MAC was formulated as an optimal stopping problem[20] with an objective of maximizing the throughput. The system model was based on OFDM with channel aggregation. The transceiver can spread its signal within a limited number of channels in a given frequency band. Time period T_s is used by a secondary user to sense the channel, while the length of the transmission period is denoted by T_t. This duration of transmission time T_t depends on the acceptable level of interference to primary users.

The problem of optimal channel sensing arises because when a secondary user spends a longer amount of time in sensing, a shorter amount of time is available to that secondary user for channel access. An example is shown in Figure 7.5. In Figure 7.5(a), the secondary user senses two channels (i.e., $i - 1$ and i) and then stops. In this case, channel $i - 1$ is unoccupied while i is occupied. As a result, the achievable data rate is $kWT_t/(T_t + 2T_s)$, where W is the channel bandwidth and k is the spectral efficiency. However, if the secondary user does not stop and continues sensing channel $i + 1$ (Figure 7.5), which is occupied by the primary user,

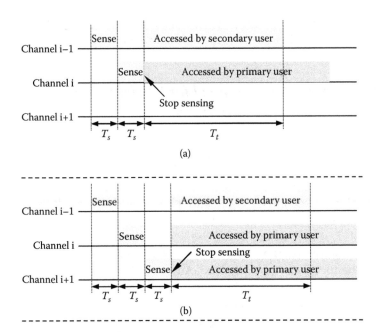

Figure 7.5 Overhead of channel sensing.

the data rate of the secondary user becomes $kWT_t/(T_t + 3T_s)$. Here, the data rate increases, and therefore a better performance can be achieved if the secondary user stops sensing at channel i. By using the theory of optimal stopping, an optimal number of sensing channels can be obtained for which the data rate is maximized.

The maximum number of idle channels within n adjacent channels (where channel $1, \ldots, n$ are sensed) can be obtained from

$$\mathcal{M}_n(x_1, \ldots, x_n) = \max_{\substack{1 \le i \le j \le n \\ j - i + 1 \le N_a \\ \mathcal{F}(i, j) \le N_f}} \sum_{k=i}^{j}(1 - x_k) \tag{7.1}$$

where $\{i, i+1, \ldots, j\}$ denotes a set of adjacent channels, $\mathcal{F}(i, j)$ is a function indicating the number of fragments of idle channels, N_a denotes the maximum number of adjacent channels that a secondary user can simultaneously access, and N_f is the maximum number of spectrum fragments that can be aggregated and accessed by the secondary user. The reward function of sensing n channels can be defined as follows:

$$\mathcal{R}_n = \frac{T_t}{T_t + nT_s}\mathcal{M}_n(x_1, \ldots, x_n). \tag{7.2}$$

To obtain the optimal value of n, this problem can be solved using backward induction starting from the maximum number of channels a secondary user can sense.

7.3.3 Cognitive MAC Protocol for Efficient Discovery of Spectrum Opportunities

When a channel is periodically sensed, the sensing period can be optimized to maximize the discovery of spectrum opportunities.[10] For example, if the sensing period is large (i.e., the channel is infrequently sensed), spectrum opportunities will not be fully discovered. However, if the sensing period is small, since a secondary user cannot sense and access all the channels simultaneously, the sensing overhead becomes large. Therefore, the proportion of $T_{s,i}/T_{p,i}$, where $T_{s,i}$ is the sensing time interval and $T_{p,i}$ is the length of time between two consecutive channel sensings (i.e., sensing period) of channel i, can be optimized. An example is shown in Figure 7.6. In Figure 7.6(a), the sensing period is too small, and therefore channel access by a secondary user is not maximized due to large overhead. On the other hand, in Figure 7.6(c), the sensing period is too large, and the amount of spectrum opportunities found is small. In Figure 7.6(b), the sensing period is optimally chosen so that the maximum amount of spectrum opportunities are found.

Let u_i denote the average time that channel i is occupied by the primary users. An optimization problem can be formulated as follows:

$$\text{Maximize:} \quad \sum_{i=1}^{N}(1 - u_i) - SSOH_i(\mathbf{T}_p) - UOPP_i(T_{p,i}) \qquad (7.3)$$

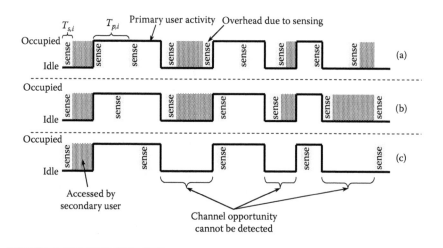

Figure 7.6   Sensing time and sensing period.

where N is the total number of channels, $\mathbf{T}_p$ is a vector of sensing periods for all channels, the function $SSOH_i(\mathbf{T}_p)$ is defined as *sensing overhead* and the function $UOPP_i(T_{p,i})$ is defined as *unexplored opportunity*. While sensing overhead can be simply expressed as the ratio of sensing time and sensing period, unexplored opportunity is obtained by considering the probability distribution of the *off* period for a primary user. From this optimization problem, the optimal vector of sensing periods can be obtained by using a numerical method.

In addition, channel-switching latency incurred due to searching idle channels can be optimized through a channel sensing sequence. In this case, a secondary user senses the channels one by one, and stops if an idle channel is found. To obtain the optimal channel sensing sequence, the probability for a channel to become idle is calculated based on the historical channel sensing data.

7.3.4 Cognitive MAC with Learning-Based Optimal Channel Selection

In Ref. [9] a channel selection based on stochastic control theory was proposed. A system model with a total number of N channels was considered. The activity of primary users on each channel was modeled as an on-off process where the durations of *on* and *off* periods are geometrically distributed random variables.

To estimate the probability that a time slot is idle, a Bayesian learning algorithm was applied. The posterior distribution f_t^{idle} of the random variable $x \in [0, 1]$ indicating idle channel was updated by the standard Bayes rule as follows:

$$f_t^{idle} = \frac{Y_i(t) + 1}{Y_b(t) + Y_i(t) + 2} \tag{7.4}$$

where $Y_b(t)$ and $Y_i(t)$ are the number of times that the channel is busy and idle, respectively. Similarly, the successful channel access probability was updated as follows:

$$\hat{p}_t^{succ} = \frac{Y_s(t) + 1}{Y_s(t) + Y_f(t) + 2} \tag{7.5}$$

where $Y_s(t)$ and $Y_f(t)$ are the number of times that transmission of a packet is successful and unsuccessful, respectively.

The optimal channel allocation was formulated as a multi-armed-bandit problem.[21] There are two steps in this channel allocation algorithm. The first step is to learn (i.e., estimate) the probability of successful transmission in each channel. The second step is to choose the channel with the highest probability of successful transmission. This problem can be solved by using a dynamic allocation index for each channel which a function of the state $x_i(t)$ of channel i. This index can be defined as follows:

$$v_i(x_i) = \max_\tau \frac{\mathcal{E}\left[\sum_{\tau=1}^{t_1} \beta^\tau \mathcal{R}_i(t) | x_i(1) = x_i\right]}{\mathcal{E}\left[\sum_{\tau=1}^{t_1} \beta^\tau x_i(1) = x_i\right]} \tag{7.6}$$

where β is the discount factor, τ is the stopping time, and $\mathcal{R}_i(t)$ is the reward function defined as follows:

$$\mathcal{R}_i(t) = \begin{cases} 0, & \text{channel is busy} \\ L, & \text{successful transmission} \\ -\alpha L, & \text{transmission has failed} \end{cases} \tag{7.7}$$

where L is the packet size and α is the energy cost factor indicating the energy consumption of transmitting a packet of size L bits.

In this case, the expected reward in period t can be obtained from

$$\mathcal{E}[\mathcal{R}(t)] = \hat{p}_t^{idle} \hat{p}_t^{succ} L - \hat{p}_t^{idle}(1 - \hat{p}_t^{succ})\alpha L. \tag{7.8}$$

7.3.5 Ad Hoc Secondary System MAC (AS-MAC) for Spectrum Sharing

In Ref. [12], an ad hoc secondary system MAC (AS-MAC) was proposed for an ad hoc network overlaid on a GSM cellular network (Figure 7.7). This AS-MAC protocol observes user activity on GSM channels. Then, based on this information, AS-MAC provides functionality for ad hoc nodes to access the available channels without interfering with the GSM users.

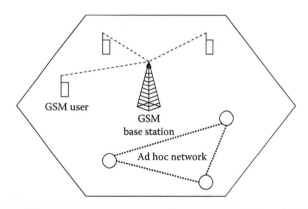

Figure 7.7   An ad hoc network of secondary users overlaid on a GSM network.

With AS-MAC, a cognitive radio detects the boundary of the time slot of GSM transmission by decoding the Frequency Correction Channel (FCCH) from the GSM base station. Then, Cell Global Identity (CGI) and Cell Channel Description (CCD) are decoded from the Synchronization Channel (SCH) and the Broadcast Control Channel (BCCH). The frequency used in the current GSM cell can be determined from this information. After the information is obtained, the cognitive radio node observes the channel usage by sensing user activity at the beginning of a time slot. With time slot size of 577 μs, the sensing duration of a cognitive radio node is less than $T_s = 15$ μs, and the sensing starts after the GSM guard band.

If a cognitive radio node in the ad hoc network has data to transmit, it senses the channel to detect an available time slot. Then, to avoid collision, this sender waits for a time period of T_{rts} and transmits an RTS message. This T_{rts} is uniformly distributed between 40–140 μs. The RTS message contains the number of time slots for data transmission, which is referred to as NAV similar to that in the IEEE 802.11 CSMA/CA protocol. After the receiver receives the RTS message, it senses the channel for an available time slot and transmits the CTS message. This CTS message contains the ID of the receiver, the NAV, and the time slot for communication. This time slot is selected from among the free slots in a transmission frame. After the sender receives the CTS message, the data is fragmented into multiple packets and transmitted over designated time slots. To enhance reliability of transmission, a selective ARQ is used in this AS-MAC. After all the packets are transmitted, the sender waits for an ACK message. This ACK message contains a list of successfully received packets. Based on this information, the sender retransmits only the lost packet. The AS-MAC protocol is shown in Figure 7.8.

Note that in designing this AS-MAC, the propagation delay was taken into account. This propagation delay from the transmission could result in 200 μs of overlapping duration of transmission from the GSM base station to mobile and transmission between ad hoc nodes. To avoid this problem, the guard band can be lengthened, but it results in 30% performance reduction.

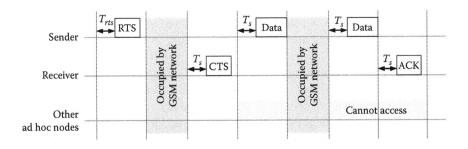

Figure 7.8 AS-MAC protocol.

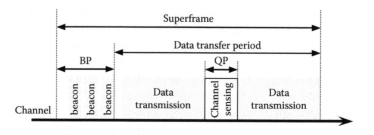

Figure 7.9   Superframe structure of C-MAC.

Therefore, in AS-MAC, overlapping duration is allowed if interference to GSM users can be maintained below the target level.

7.4   MAC Protocols with Synchronization Between Cognitive Radios

7.4.1   Slotted Beaconing Period and Rendezvous Channel

One of the challenges in designing MAC protocols for multihop cognitive radio networks is the synchronization among secondary nodes. Each secondary node must inform other nodes about the available channels and the channel to be accessed. To support this synchronization, a cognitive MAC (C-MAC) protocol was proposed in Ref. [22]. The major component of this C-MAC protocol is the rendezvous channel (RC), which is used mainly for information exchange among secondary nodes.

The major functions of the rendezvous channel are (1) to synchronize among available channel accesses, (2) to discover neighboring nodes, and (3) to balance channel load. In the C-MAC protocol, each channel has a superframe structure (Figure 7.9), and one channel is identified as a rendezvous channel. A superframe is composed of a beacon period (BP) and a data transfer period (DTP). The beacon period is divided into multiple slots. The first two slots are used by the new secondary nodes to join the network, while the rest are used by the existing nodes. Here, every node operating in different channels periodically visits the rendezvous channel to resynchronize with the network. As a result, after a new secondary node scans all channels, the rendezvous channel will be found and the new node can obtain information on the channel usage and activity of other nodes for the purposes of neighbor node discovery and load balancing.

Based on the beacon signal during the beacon period, a moving mobile node can observe channel allocation in the new location. This information (i.e., new set of neighboring nodes and their transmission schedule) can mitigate the hidden terminal problem. With the rendezvous channel, this M-MAC protocol also supports group communication (i.e., multicast or broadcast). A node with a group message visits the rendezvous channel and finds a transmission schedule. Then, the group message is transmitted for a few superframes. Since all other nodes will visit the rendezvous channel periodically, the group message will be guaranteed to be received by all nodes in the network. Finally, C-MAC can detect primary users by using both in-band and out-of-band channel sensing. If a primary user is found, the secondary node informs its neighbors by using the beacon signal, and the transmissions of all nodes in the occupied channel are rescheduled.

7.4.2 Dynamic Open Spectrum Sharing (DOSS) Protocol

Instead of limiting the secondary users to operate in only one spectrum, dynamic open spectrum sharing (DOSS)[15] was proposed to allow secondary users to adaptively select an arbitrary spectrum. DOSS was designed for an ad hoc cognitive radio environment in which a busy tone is used to mitigate the hidden and exposed terminal problems. This busy tone is transmitted in a narrow-band channel which can be mapped from a wideband data channel. Let the band corresponding to the busy tone channel and the data channel be denoted by $[f_l, f_u]$ and $[F_l, F_u]$, respectively. The mapping can be expressed as follows:

$$\mathscr{G}(z) = \frac{1}{F_u - F_l}((f_u - f_l)z + f_l F_u - f_u F_l) \tag{7.9}$$

where $\mathscr{G}(z)$ is a mapping function and z is a design parameter. An illustration of this mapping is shown in Figure 7.10.

When secondary nodes have data to transmit, the DOSS protocol works as follows. First, both the transmitter and the receiver locate idle spectrum by listening to the busy tone channel. Then, the transmitter sends a REQ message over the control channel to indicate the center frequency and the bandwidth. From the REQ message, the receiver identifies the spectrum to be used by the transmitter. A REQ_ACK message is then transmitted by the receiver over the control channel. After receiving the REQ_ACK message, the transmitter starts transmitting data. While the receiver receives data,

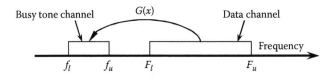

Figure 7.10   Channel mapping in DOSS.

it transmits a busy tone on the mapped narrow-band channel to inform the other nodes of the transmission. If the transmitted data is successfully received, the receiver replies with DATA_ACK.

7.5 Cognitive MAC Based on Channel Allocation and Scheduling

7.5.1 Multichannel Cognitive Radio MAC with Rate and Power Adaptation

The major challenges in a multichannel cognitive radio network include (1) a secondary user must choose an optimal set of channels to access, (2) a secondary user must choose the transmission parameters (i.e., rate and power) to satisfy the transmission requirements (e.g., interference to the primary user), and (3) collision due to the transmission of multiple secondary users must be mitigated. In Ref. [13] these challenges were considered jointly.

First, the channel allocation was solved by enumerating and searching for an optimal channel selection. The algorithm starts from a minimum number of channels. Then, a rate and power allocation algorithm is performed. If there is a feasible optimal solution for rate and power, the algorithm terminates and the secondary users utilize these channels. Otherwise, the algorithm continues searching until the constraint on the maximum number of channels is reached. In this case, if the maximum number of channels accessed by the secondary users is M and the total number of available channels is N, where $N > M$, the total number of combinations to be searched is $\sum_{m=1}^{M} \frac{N!}{m!(N-m)!}$.

For rate and power allocation, an optimization problem can be formulated as follows:

$$\text{Minimize:} \quad \sum_{i=1}^{m} (e^{r_i} - 1)\frac{I_0}{g} \tag{7.10}$$

$$\text{Subject to:} \quad \sum_{i=1}^{m} r_i W_i = R_{req} \tag{7.11}$$

$$0 \leq r_i \leq \ln\left(1 + \frac{P_{max}g}{I_0}\right) \tag{7.12}$$

where r_i is the transmission rate on channel i, m is the number of channels to be accessed by secondary users, I_0 is noise power, g is channel gain, W_i is the spectrum size corresponding to channel i, R_{req} is the transmission rate requirement of secondary users, and P_{max} is the maximum power. The objective of this optimization problem is to minimize the transmit power, which is a function of transmission rate. The constraint in (7.11) indicates that the total transmission rate must meet the requirement of secondary users. The constraint in (7.12) limits the transmit power. With these constraints, an algorithm was proposed[13] to obtain the solution. Note that the rate and power allocation algorithm for cognitive radio has been studied extensively in the literature.[23–28]

For the channel contention among multiple secondary users, an extension of the CSMA/CA protocol in the IEEE 802.11 standard was proposed. In this case, a free-channel table (FCT) is used to maintain a list of unoccupied channels by other secondary users, and a dedicated control channel is used to exchange information among secondary users. The detailed description of the protocol is as follows:

1. A secondary user listens to the control channel and updates the FCT if a transmission is detected.
2. If a secondary user wants to transmit data, it transmits an RTS message on the control channel. This RTS message contains the FCT and the duration of the flow (DOF). However, if the FCT at the source is empty (none of the channels is available), the secondary user backs off and retransmits the RTS message.
3. Upon receiving the RTS message, the destination secondary user obtains a set of available channels by comparing the FCT in the RTS message with the local FCT. Then, the rate and power allocation algorithm is performed on a candidate set of available channels. If there is a solution of this algorithm, the set of channels and the transmission parameters are included in the CTS message. Otherwise, an empty flag is used to indicate to the source secondary user that there is no feasible solution.
4. After the source secondary user receives the CTS message, it transmits an Echo-CTS (ECTS) message over the control channel to inform other secondary users of upcoming transmissions. Then, the source secondary user starts transmitting data over the set of available channels by using the optimal rate and power control strategy.

7.5.2 Statistical Channel Allocation MAC (SCA-MAC) for Cognitive Radio

A statistical channel allocation MAC (SCA-MAC) protocol for secondary nodes with a channel aggregation feature was proposed in Ref. [29]. In this protocol, a control channel is used for exchanging CRTS and CCTS messages (Figure 7.11). For the radio transceiver at the sender and the receiver, it is assumed that m consecutive channels can be aggregated for

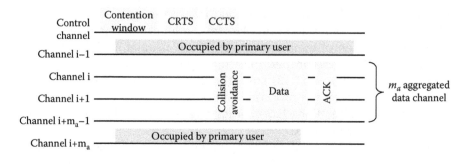

Figure 7.11 The SCA-MAC protocol.

transmission. The selection of the range of channels is formulated as an optimization problem as follows:

$$\{i^*, m_a^*\} = \arg \max_{i, m_a} \alpha([i, i + m_a - 1], L) \tag{7.13}$$

where $\alpha(., .)$ is the successful rate of transmitting a packet of length L on aggregated channel $[i, i + m_a - 1]$. This successful rate is computed from

$$\alpha([i, i + m_a - 1], L) = \alpha_c \alpha_L \tag{7.14}$$

where α_c is the probability of successful channel allocation for a given range of channels q. This probability can be obtained from

$$\alpha_c = \left(1 - \frac{\overline{\tau}_c n \overline{m}_c}{(1 - \tau) q}\right)^m \tag{7.15}$$

where τ is the channel utilization of primary users, $\overline{\tau}_c$ is the average channel utilization of neighboring secondary nodes, n is the number of neighboring nodes, and $\overline{m}_c$ is the average number of aggregated channels of neighboring nodes. Also, α_L denotes the probability that a packet with length L can be transmitted in an idle aggregated channel $[i, i + m_a - 1]$ for a duration T, and is given by

$$\alpha_L = \prod_{j=i}^{i+m-1} \Pr\left(C_j : T \geq t_{0,j} + L/M | T \geq t_{0,j}\right) \tag{7.16}$$

where $\Pr\left(C_j : T \geq t_{0,j} + L/M | T \geq t_{0,j}\right)$ is the probability that channel j is idle during $[t_{0,j}, T]$. Based on this successful rate estimation, a numerical solution can be obtained for i and m_a for the best performance.

7.5.3 MAC-Layer Scheduling in Cognitive Radio Networks

A scheduling algorithm for multihop cognitive radio networks was proposed in Ref. [17]. This scheduling algorithm determines the time slot and channel for transmissions by the nodes in the network. However, each node in the network has a different set of available channels. The scheduling problem was formulated as an integer linear program. Also, a heuristic-based distributed algorithm was proposed.

Let the network be presented as a graph containing nodes and links. Let $x_{l,j,k}$ be a binary variable defined as follows:

$$x_{l,j,k} = \begin{cases} 1, & \text{if link } l \text{ accesses time slot } k \text{ in channel } j \\ 0, & \text{otherwise.} \end{cases} \tag{7.17}$$

Let $\mathbf{L}_i$ denote the set of links used at node i, $\mathbf{C}_l^{(l)}$ denote the set of available channels for link l, and $\mathbf{C}_i^{(c)}$ denote the set of available channels at node i. The first constraint for the scheduling policy is defined as follows:

$$\sum_{l \in \mathbf{L}_i} \sum_{j \in \mathbf{C}_l^{(l)}} x_{l,j,k} \leq 1, \quad \forall i, k. \tag{7.18}$$

This constraint indicates that the same time slot cannot be allocated to more than one link at the same node. The second constraint is

$$\sum_{\forall k} \sum_{j \in \mathbf{C}_l^{(l)}} x_{l,j,k} = 1 \tag{7.19}$$

which ensures that each link in the network is allocated with a time slot and channel for transmission. The third constraint is

$$x_{l,j,k} + x_{l',j,k} \leq 1 \tag{7.20}$$

for all links l' at nodes that are neighbors of a node with link l. This constraint is used to ensure that none of the neighboring nodes transmits in the same time slot in the same channel. Based on these constraints, an integer linear program can be formulated with the objective of minimizing the number of time slots allocated for transmission in all links given a set of available channels in the network.

Since solving the above integer linear programming formulation incurs a large computational complexity, a heuristic-based distributed algorithm was proposed. This algorithm consists of two phases. In the first phase, each node chooses the time slot and the channel to access for all the outgoing links. In the second phase, this information is propagated throughout the network and each node adjusts the time slot and the channel accordingly.

7.5.4 Rental Protocol for Dynamic Channel Allocation

When the radio spectrum is not being used, primary users can sell the spectrum opportunities to the secondary users. In such a case, a spectrum leasing/rental protocol is required to support negotiation (e.g., on price and spectrum share) between primary and secondary users. A credit-token-based rental protocol for dynamic spectrum access was proposed in Ref. [30]. In the system model, a primary user sells spectrum share to multiple secondary users. Secondary users bid for the spectrum offered by the primary user (or primary service provider). A credit token is used as the medium for negotiation between primary and secondary users.

The proposed protocol works as follows. First, the primary user advertises information on the abailable spectrum to be leased (i.e., channel and the corresponding time period). Then, the interested secondary users send spectrum request messages to the primary user, and the primary user replies with the auction period (i.e., time period during which the spectrum will be allocated to secondary users). Secondary users provide information on the amount of credit token bid (or price), size of spectrum, and starting time and ending time to access the spectrum. The primary user (or primary network controller) chooses the secondary users that are going to share the spectrum, the clearing price to charge for credit tokens, and the time period allocated for each secondary user.

7.5.5 Centralized Cognitive MAC Protocol for Coordinated Spectrum Access

Spectrum assignment to secondary users can be performed by a centralized controller. A protocol to support this spectrum leasing, namely, the dynamic spectrum access protocol (DSAP), was developed.[31] The system model for spectrum leasing considered in DSAP is shown in Figure 7.12. A *DSAP client* is a secondary node which leases spectrum from a *DSAP server*, which is a primary user (or primary service provider). A *DSAP relay* is used to delegate spectrum leasing for DSAP clients that are not in the coverage of a DSAP server. At the DSAP server, a *RadioMap*, which is a spectrum database, is used to maintain information on spectrum allocation and leasing.

With DSAP, spectrum leasing works as follows. The DSAP client sends a request for spectrum usage (i.e., *ChannelDiscover* message). The DSAP server searches for available spectrum in the RadioMap and responds by sending a *ChannelOffer* message to the DSAP client. This message contains information on available spectrum that can be leased to the DSAP client. Based on this information, the DSAP client chooses the spectrum access parameters (e.g., frequency and transmit power). Then, a *ChannelRequest* message is sent back to the DSAP server. If the DSAP server agrees to lease the spectrum, the RadioMap is updated and a *ChannelACK* message is sent

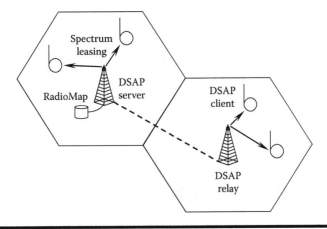

Figure 7.12 System model of spectrum leasing of DSAP.

to the DSAP client. In addition, a *ChannelReclaim* message is used by the DSAP server to reallocate or terminate lease of the spectrum to the DSAP client.

7.6 Adaptive Cognitive MAC Protocols and Performance Analysis Models

7.6.1 Adaptive MAC Protocol for Cognitive Radio

Different MAC protocols can be used to achieve different objectives under different circumstances. For example, a fixed-assignment TDMA MAC protocol can achieve better performance under heavy load conditions, while a CSMA/CA provides better performance under light load conditions. Therefore, based on the different network conditions and characteristics, a cognitive radio can select an appropriate MAC protocol dynamically. A framework for MAC protocol adaptation, namely, MultiMAC, was proposed in Ref. [38]. The structure of MultiMAC is shown in Figure 7.13. In MultiMAC two simple policies for MAC protocol selection were considered. The first policy was based on the load level in the network. The second policy was based on the type of traffic. For best-effort traffic, a low-complexity protocol such as Aloha can be used. However, for QoS-sensitive traffic, a fixed-assignment TDMA scheme would be more appropriate.

7.6.2 Queueing Performance Model for Cognitive MAC Protocols

To investigate the QoS performance of cognitive radio transmission, a queueing model was proposed in Ref. [32] that considered both physical

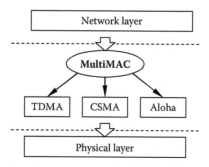

Figure 7.13   Structure of MultiMAC for MAC protocol adaptation.

and MAC parameters. The system model considered a dynamic cognitive radio scenario for which available spectrum opportunities are aggregated and the cognitive users use this aggregated spectral bandwidth as a single channel. This dynamic cognitive radio design can be achieved by OFDM in the physical layer. In this queueing model, it was assumed that there are M secondary users sharing the spectrum with N primary users. The packets arriving at all of the secondary users are assumed to be aggregated in a single queue, and the secondary users share the same channel. This queueing model can be shown as in Figure 7.14.

Packet arrivals at each secondary user were assumed to follow a Poisson process. For primary user i, the spectrum is occupied for period $T_{oc}^{(i)}$ and available for period $T_{av}^{(i)}$. Therefore, the probability that k channels are available can be obtained from

$$r_k = \sum_{c=1}^{\frac{N!}{k!(N-k)!}} \left(\prod_{i \in \mathbf{S}_c^{(k)}} (1 - \tau_i) \prod_{j \in \{1,2,\dots,N\}/\mathbf{S}_c^{(k)}} \tau_j \right) \qquad (7.21)$$

where $\tau_i = T_{oc}^{(i)}/(T_{oc}^{(i)} + T_{av}^{(i)})$ is the probability that primary user i occupies the channel, and $\mathbf{S}_c^{(k)}$ is a set of primary users for which k channels are

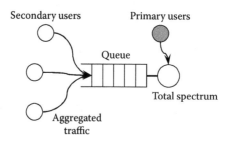

Figure 7.14   Queueing model for cognitive radio.

occupied. For secondary users, the capacity (i.e., transmission rate) can be expressed as

$$C = \frac{T - T_s}{T} \times r_k \times W \times \log_2 \left(1 + \frac{P}{N_0 \times r_k \times W} \right) \qquad (7.22)$$

where T is the length of a time slot, T_s is the time used to sense the channel, W is the total spectrum, P is the transmit power, and N_0 is the noise power. An M/G/1 queueing model was used to analyze the transmission performance for cognitive radio, where the service rate was assumed to be equal to the capacity of the secondary users.

This queueing model can be extended by considering the correlation of spectrum access by primary users. In this case, a multidimensional Markov chain can be established to obtain the performance measures. In this context, the issue of admission control to limit the number of secondary users will be important to provide QoS performance guarantees to the cognitive radio users.

7.7 Open Research Issues in Cognitive MAC Protocol Design

Several open research issues in designing MAC protocols for dynamic spectrum access in cognitive radio networks are as follows:

■ *MAC protocol for underlay spectrum access:* While most of the above MAC protocols are proposed for overlay dynamic spectrum access, cognitive MAC protocols need to be designed for underlay spectrum sharing. For spectrum underlay, power control is a very important issue. Transmit power must be chosen such that the receiver can receive data correctly and the primary users are not interfered with. Approaches used for designing MAC protocols for ultra-wideband (UWB) radio can be applied for underlay spectrum sharing.[33–36]

■ *Congestion in control channel:* In a cognative radio network, synchronization among secondary nodes is performed through a control channel. However, when the number of secondary nodes is large or traffic load is heavy, this control channel may become the bottleneck of the system. To avoid congestion in the control channel, dynamic control messaging in multiple channels may need to be developed.

■ *Performance analysis models:* Traditionally, MAC protocol performance for dynamic spectrum access is evaluated through simulations. However, analytical models (e.g., queueing models) can be developed to obtain the QoS performance results with less computational complexity. Also, analytical models can be used along with an optimization method to obtain the optimal MAC parameter settings.

7.8 Cognitive MAC Protocols for the IEEE 802.22 Standard-Based WRANs

IEEE 802.22 is the new wireless air-interface standard based on cognitive radio that is being designed for wireless regional area networks (WRANs).[37] The frequency spectrum allocated to this standard will be the same as that currently allocated to television (TV) service. The TV bands are unoccupied in many parts of the US and many other regions of the world. Therefore, in the US, the spectrum allocation policy for TV bands is being revised to allow coexistence of unlicensed services (i.e., based on IEEE 802.22 networks). In North America, the range of operational frequency for IEEE 802.22 networks is 54–862 MHz, while the standard is being extended to operate in the 41–910 MHz band to meet international regulatory requirements.

IEEE 802.22 networks would support fixed point-to-multipoint connections. The base station (BS) in a cell controls all the connections from consumer premise equipment devices (CPEs) in the cell. Transmissions in both uplink and downlink will be based on orthogonal frequency division multiple access (OFDMA) using a channel of 6 MHz bandwidth. Various modulation and coding schemes with spectral efficiency ranging from 0.5 bit/symbol/Hz to 5 bits/symbol/Hz will be supported in the standard. A higher throughput can be supported by a channel bonding technique for which multiple 6 MHz channels can be used simultaneously for transmission. Since IEEE 802.22 networks must operate without interfering with the incumbent TV service, and also multiple IEEE 802.22 networks may operate in the same or overlapping areas, dynamic spectrum sensing and spectrum access will be important components in this standard.

7.8.1 802.22 MAC Frame Structure

Figure 7.15 shows the superframe structure of IEEE 802.22 MAC. At the beginning of a superframe, a preamble and a superframe control header (SCH) information are transmitted by the BS over all available TV channels. The preamble is used to protect incumbent service, while the SCH is used by the CPE to synchronize with the BS—it contains all the information for a CPE to initiate a connection. Note that, due to the requirements of the FCC, two channels (e.g., channels $i - 2$ and $i + 2$ in Figure 7.15) remain unused to prevent interference with the TV service.

A superframe consists of multiple transmission frames, and the structure of a transmission frame is shown in Figure 7.16. Transmissions in each

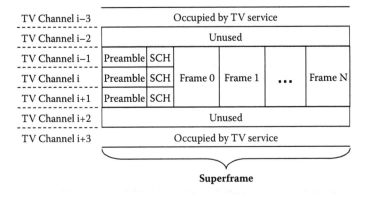

Figure 7.15   Superframe structure of IEEE 802.22 standard.

frame are time-slotted and are based on OFDM subcarriers. In one frame, there are upstream and downstream subframes. A downstream subframe consists of a preamble to indicate the beginning of the frame. Then, US-MAP and DS-MAP are used to indicate the structure of the upstream and downstream subframes. In the downstream subframe, there is a contention interval for CPE initialization (e.g., ranging), urgent coexistence situation (UCS) notification, and self-coexistence to detect other IEEE 802.22 networks. In both subframes, a transmission burst is allocated for each CPE, both upstream and downstream. Note that the structure of the IEEE 802.22 superframe, frame, subframe, and burst are controlled by the BS.

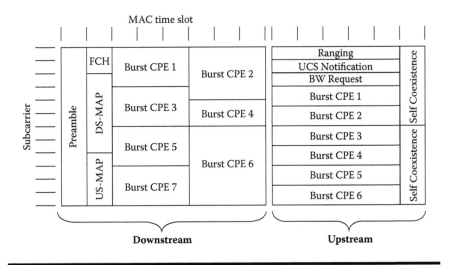

Figure 7.16   Frame structure of IEEE 802.22 standard.

For a CPE to join the network, it scans and identifies available TV channels that are not occupied by the TV service. Based on these available channels, the CPE then scans for SCH transmission from the IEEE 802.22 BS. Since the transmission is based on the superframe structure, the smallest amount of time that a CPE senses the channel is equal to the superframe duration. After an SCH is received, the CPE can initiate a connection by sending the request to the BS. The BS assigns the burst and allows the CPE to transmit based on the channel availability.

7.8.2 Spectrum Management and Sensing in 802.22 MAC

In 802.22, both BS and CPE perform channel sensing periodically. Since channel sensing can be either in-band or out-of-band, two interfaces are required at the CPE. One directional antenna is used for communication with the BS, and another omnidirectional antenna is used for channel sensing. The BS instructs the CPE to sense a TV channel, and the sensing result is sent back to the BS to construct a spectrum occupancy/availability map for a cell. This map is used for spectrum management. The MAC protocol should be designed to support the required spectrum management functions including channel switching, channel suspension/resuming, and inclusion/removal of channels from a channel access list.

To detect incumbent TV service, IEEE 802.22 supports fast and fine in-band sensing mechanisms (Figure 7.17) to improve the efficiency of the system. Fast sensing (e.g., energy detection) is performed in a small time interval to reduce interruption in data transmission. Results of this sensing are used to analyze the occupancy of channels. If high transmission energy is detected, fine sensing is performed in the target channel. This fine sensing algorithm takes a longer period of time (e.g., up to 25 ms) to find a particular signal signature from the incumbent service. These fast and fine sensing mechanisms must ensure the requirements of the IEEE 802.22 standard. Specifically, digital TV signals with power above threshold −116 dBm should be detected with a probability of at least 0.9 (i.e., the maximum

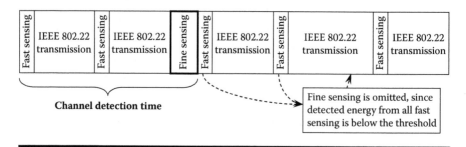

Figure 7.17   Fast and finesensing mechanisms.

value of the misdetection probability is 0.1).[39] For an entire period of channel sensing, the incumbent service should be detected within no more than two seconds after the incumbent service starts utilizing the channel.

7.8.3 Challenges in Designing IEEE 802.22 MAC

The major challenges in designing cognitive MAC protocols for IEEE 802.22 networks can be summarized as follows:

- *Coexistence with incumbent service:* To avoid interference with incumbent service, in-band channel sensing would be required. In this case, a CPE must perform in-band channel sensing and report the sensing result back to the BS. The BS can then construct a spectrum occupancy/availability map, and subsequently perform an optimal channel allocation for communications with all CPEs.
- *Self-coexistence:* Since multiple IEEE 802.22 networks can operate in the same or overlapping areas and the number of unused TV channels is limited, IEEE 802.22 networks would require a careful channel access scheme to avoid interference with each other. This issue is important especially in terms of competition and cooperation among IEEE 802.22 network providers. For self-coexistence, the 802.22 networks may compete or cooperate with each other.
- *QoS support:* Development of a QoS framework to support different traffic types in IEEE 802.22 networks is an open research issue. Several types of traffic will need to be supported, including constant-bitrate, real-time, and best-effort traffic. In this case, traffic scheduling, resource allocation, and admission control in the QoS framework must be designed to support the dynamic and opportunistic channel access mechanism in the standard.

7.8.4 Enhancements to the IEEE 802.22 Air Interface

The problems of self coexistence and hidden incumbent service were considered in Ref. [18]. In particular, a channel allocation algorithm based on a graph coloring model was presented for self-coexistence. Also, an enhanced MAC protocol was proposed for detection of hidden incumbent service. The utility of an IEEE 802.22 network was defined in terms of the bandwidth of the operating channel. Transmissions are constrained by the fact that neighboring BSs cannot use the same or overlapping channels. Based on the utility definition and the constraint, an undirected graph $G = \{V, \mathcal{E}, \mathcal{B}\}$ was defined, where V is the set of vertices (i.e., BSs), $\mathcal{E}$ is the set of all undirected links (i.e., a link between two vertices indicates that transmissions from the corresponding BSs interfere with each other), and $\mathcal{B}$

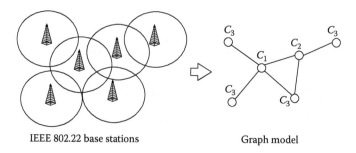

IEEE 802.22 base stations   Graph model

Figure 7.18   Graph coloring model.

is the set of available channels unused by incumbent service. An example of this graph model is shown in Figure 7.18.

To obtain the solution of this graph coloring model, it was assumed that the topology information is available to all BSs. Three different objectives were defined for this graph coloring model. The first objective was to maximize the total utility of all BSs. The second objective was to provide proportional fair utility. That is, the BS which interferes with the least number of other BSs is prioritized to maximize the utility. The third objective was to provide complete fairness utility regardless of the interference.

To obtain the optimal solution, a two-phase algorithm, namely, the utility graph coloring (UGC) algorithm, was proposed. In the first phase, the graph was colored by a standard graph coloring algorithm to obtain the minimum number of colors to meet the constraint. In the second phase, a progressive algorithm was used to collect the number of occurrences of a color in the graph. This information was used together with the number of edges corresponding to the vertices (i.e., the number of other BSs that one BS interferes with) to assign channels to the BSs. The assignment of channels was performed to achieve the above three objectives. For example, with an objective of maximizing utility, the color with the highest number of occurrences was assigned the largest bandwidth channel. For the other colors, the minimum bandwidth channel was allocated.

To mitigate interference with incumbent service, the MAC protocol was enhanced by using dynamic multiple outband broadcasting. In the standard IEEE 802.22 MAC protocol, messages from a BS are transmitted in only one channel. On the other hand, in the scheme proposed in Ref. [18] messages from a BS are periodically broadcast in different channels and can be changed dynamically depending on feedback from the CPEs. As a result, if a CPE detects incumbent service through in-band channel sensing, the CPE will still be able to report this information to the BS. Another enhancement was the spectrum usage report, which is broadcast by the BS to the CPEs. Using this report, the CPEs can change the uplink channels CPEs use to transmit to the BS.

7.8.5 Dynamic Frequency Hopping MAC for IEEE 802.22

In an IEEE 802.22 network, in-band channel sensing for a single frequency band is inefficient, since the BS and CPEs must stop transmission during the sensing period. To improve the performance of channel sensing, dynamic frequency hopping (DFH) for IEEE 802.22 networks was proposed.[40] According to the standard, an IEEE 802.22 network is designed to operate in a single channel (i.e., non-hopping mode), and the channel sensing mechanism must be performed every two seconds. In contrast, with DFH, the network can hop over a set of channels. While utilizing one channel, the 802.22 network may perform out-of-band sensing for other channels. The decision on hopping can be based on the result of channel sensing. In this way, transmissions in the 802.22 network will not be interrupted due to in-band channel sensing. Also, if multiple BSs perform collaborative channel sensing, the problem of self-coexistence of BSs in the same dynamic frequency-hopping community can be mitigated. An example of frequency hopping is shown in Figure 7.19. When WRAN2 accesses channel i, it can perform out-of-band sensing of channel $i - 1$ simultaneously. If an incumbent signal is not detected on channel $i - 1$, WRAN2 hops to channel $i - 1$.

A community of IEEE 802.22 networks supporting DFH operation was defined. In this community there is a leader to manage the member networks, collect channel occupancy/availability information from the networks in the same community, calculate the hopping pattern, and broadcast this information back to the other member networks. Based on this mechanism, all IEEE 802.22 networks must be able to communicate with the leader through the coexistence beacon protocol (CBP), which was designed for wireless internetwork communication. To form a community, first the leader is selected among the BSs with the smallest MAC address. Then, the leader chooses the hopping channels and broadcasts the information to other networks. The BSs that receive this message may join the community.

After a community is formed, the leader decides the hopping pattern for a certain period of time. All member networks perform channel sensing and

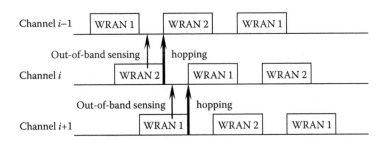

Figure 7.19   Dynamic frequency hopping (DFH).

send the sensing results to the leader. A new hopping pattern is decided and broadcast back to the community if incumbent service is detected. The protocol also supports community merging and splitting.

7.9 Summary

We have provided a survey on the medium access control (MAC) protocol design for dynamic spectrum access in cognitive radio networks. The basic cognitive MAC functions and the design challenges have been stated. A survey of the related works has been provided. To this end, we have described the MAC features in IEEE 802.22-based wireless regional area networks (WRANs). Several enhancements to the basic MAC protocol defined in the IEEE 802.22 have been reviewed.

References

[1] FCC, Spectrum Policy Task Force Report, No. 02-155, Nov. 2002.
[2] J. Mitola, "Cognitive Radio for Flexible Multimedia Communications," in *Proc. MoMuC'99*, pp. 3–10, 1999.
[3] S. Haykin, "Cognitive Radio: Brain-Empowered Wireless Communications," *IEEE Journal on Selected Areas in Communications*, vol. 23, no. 2, pp. 201–220, Feb. 2005.
[4] L. Luu and B. Daneshrad, "An Adaptive Weaver Architecture Radio with Spectrum Sensing Capabilities to Relax RF Component Requirements," *IEEE Journal on Selected Areas in Communications*, vol. 25, no. 3, pp. 538–545, April 2007.
[5] S. Srikanteswara and C. Maciocco, "Interference Mitigation Using Spectrum Sensing," in *Proceedings of International Conference on Computer Communications and Networks (ICCCN'07)*, pp. 39–44, Aug. 2007.
[6] Y. Youn, H. Jeon, H. Jung, and H. Lee, "Discrete Wavelet Packet Transform based Energy Detector for Cognitive Radios," in *Proceedings of IEEE Vehicular Technology Conference (VTC), Spring 2007*, pp. 2641–2645, April 2007.
[7] T. Erpek, A. Leu, and B. L. Mark, "Spectrum Sensing Performance in TV Bands using the Multitaper Method," *IEEE Signal Processing and Communications Applications (SIU'07)*, pp. 1–4, June 2007.
[8] M. Gandetto and C. Regazzoni, "Spectrum Sensing: A Distributed Approach for Cognitive Terminals," *IEEE Journal on Selected Areas in Communications*, vol. 25, no. 3, pp. 546–557, April 2007.
[9] A. Motamedi and A. Bahai, "MAC Protocol Design for Spectrum-Agile Wireless Networks: Stochastic Control Approach," in *Proceedings of IEEE International Symposium on New Frontiers in Dynamic Spectrum Access Networks (DySPAN'07)*, pp. 448–451, April 2007.
[10] H. Kim and K.G. Shin, "Efficient Discovery of Spectrum Opportunities with MAC-Layer Sensing in Cognitive Radio Networks," *IEEE Transactions on Mobile Computing*, vol. 7, no. 5, pp. 533–545, May 2008.

[11] Q. Zhao, L. Tong, A. Swami, and Y. Chen, "Decentralized Cognitive MAC for Opportunistic Spectrum Access in Ad Hoc Networks: A POMDP Framework," *IEEE Journal on Selected Areas in Communications*, vol. 25, no. 3, pp. 589–600, April 2007.

[12] P. Papadimitratos, S. Sankaranarayanan, and A. Mishra, "A Bandwidth Sharing Approach to Improve Licensed Spectrum Utilization," *IEEE Communications Magazine*, vol. 43, no. 12, pp. 10–14, Dec. 2005.

[13] T. Shu, S. Cui, and M. Krunz, "Medium Access Control for Multi-Channel Parallel Transmission in Cognitive Radio Networks," in *Proceedings IEEE Global Telecommunications Conference (GLOBECOM'06)*, Nov. 2006.

[14] L.-C. Wang and A. Chen, "On the Coexistence of Infrastructure-Based and Ad Hoc Connections for a Cognitive Radio System," in *Proceedings of International Conference on Cognitive Radio Oriented Wireless Networks and Communications 2006*, June 2006.

[15] L. Ma, X. Han, and C.-C. Shen, "Dynamic Open Spectrum Sharing MAC Protocol for Wireless Ad Hoc Networks," in *Proceedings of IEEE International Symposium on New Frontiers in Dynamic Spectrum Access Networks (DySPAN'05) 2005*, pp. 203–213, Nov. 2005.

[16] G. Auer, H. Haas, and P. Omiyi, "Interference Aware Medium Access for Dynamic Spectrum Sharing," in *Proceedings of IEEE International Symposium on New Frontiers in Dynamic Spectrum Access Networks (DySPAN'07)*, pp. 399–402, April 2007.

[17] M. Thoppian, S. Venkatesan, R. Prakash, and R. Chandrasekaran, "MAC-Layer Scheduling in Cognitive Radio Based Multihop Wireless Networks," in *Proceedings of IEEE International Symposium on a World of Wireless, Mobile and Multimedia Networks (WoWMoM'06)*, June 2006.

[18] S. Sengupta, S. Brahma, M. Chatterjee, and S. Shankar N, "Enhancements to Cognitive Radio Based IEEE 802.22 Air-Interface," in *Proceedings of IEEE International Conference on Communications (ICC'07)*, pp. 5155–5160, June 2007.

[19] Q. Zhang, J. Jia, and X. Shen, "Optimal Spectrum Sensing Decision for Hardware-Constrained Cognitive Network," *Cognitive Wireless Communication Networks*, Eds. E. Hossain and V.K. Bhargava, Springer 2007.

[20] Y. S. Chow, H. Robbins, and D. Siegmund, *Great Expectations: The Theory of Optimal Stopping*, Houghton Mifflin Company, 1971.

[21] J. C. Gittins, *Multi-Armed Bandit Allocation Indices*, Wiley, 1989.

[22] C. Cordeiro and K. Chllapali, "C-MAC: A Cognitive MAC Protocol for Multi-Channel Wireless Networks," in *Proceedings of IEEE International Symposium on New Frontiers in Dynamic Spectrum Access Networks (DySPAN'07)*, pp. 147–157, April 2007.

[23] A. T. Hoang and Y.-C. Liang, "Maximizing Spectrum Utilization of Cognitive Radio Networks Using Channel Allocation and Power Control," in *Proceedings of IEEE Vehicular Technology Conference (VTC), Fall 2006*, Sept. 2006.

[24] H. Islam, Y.-C. Liang, and A. T. Hoang, "Joint Beamforming and Power Control in the Downlink of Cognitive Radio Networks," in *Proceedings of IEEE Wireless Communications and Networking Conference (WCNC'07)*, pp. 21–26, March 2007.

[25] W. Wang, T. Peng, and W. Wang, "Optimal Power Control Under Interference Temperature Constraints in Cognitive Radio Network," in *Proceedings of IEEE Wireless Communications and Networking Conference (WCNC'07)*, pp. 116–120, March 2007.

[26] H.-S. T. Le, and Q. Liang, "An Efficient Power Control Scheme for Cognitive Radios," in *Proceedings of IEEE Wireless Communications and Networking Conference (WCNC'07)*, pp. 2559–2563, March 2007.

[27] K. Hamdi, W. Zhang, and K.B. Letaief, "Power Control in Cognitive Radio Systems Based on Spectrum Sensing Side Information," in *Proceedings of IEEE International Conference on Communications (ICC'07)*, pp. 5161–5165, June 2008.

[28] D. Zhang and Z. Tian, "Adaptive Games for Agile Spectrum Access Based on Extended Kalman Filtering," *IEEE Journal of Selected Topics in Signal Processing*, vol. 1, no. 1, pp. 79–90, June 2007.

[29] A. C.-C. Hsu, D. S. L. Wei, and C.-C. J. Kuo, "A Cognitive MAC Protocol Using Statistical Channel Allocation for Wireless Ad hoc Networks," in *Proceedings of IEEE Wireless Communications and Networking Conference (WCNC'07)*, pp. 105–110, March 2007.

[30] D. Grandblaise, K. Moessner, G. Vivier, and R. Tafazolli, "Credit Token Based Rental Protocol for Dynamic Channel Allocation," in *Proceedings of International Conference on Cognitive Radio Oriented Wireless Networks and Communications 2006*, June 2006.

[31] V. Brik, E. Rozner, S. Banerjee, and P. Bahl, "DSAP: A Protocol for Co-ordinated Spectrum Access," in *Proceedings of IEEE International Symposium on New Frontiers in Dynamic Spectrum Access Networks (DySPAN'05)*, pp. 611–614, Nov. 2005.

[32] S. N. Shankar, "Squeezing the Most Out of Cognitive Radio: A Joint MAC/PHY Perspective," in *Proceedings of IEEE International Conference on Acoustics, Speech and Signal Processing (ICASSP'07)*, vol. 4, pp. IV-1361–IV-1364, April 2007.

[33] K. Lu, D. Wu, Y. Qian, Y. Fang, and R.C. Qiu, "Performance of an Aggregation-Based MAC Protocol for High-Data-Rate Ultrawideband Ad Hoc Networks," *IEEE Transactions on Vehicular Technology*, vol. 56, no. 1, pp. 312–321, Jan. 2007.

[34] N. J. August, H.-J. Lee, and D.S. Ha, "Enabling Distributed Medium Access Control for Impulse-Based Ultrawideband Radios," *IEEE Transactions on Vehicular Technology*, vol. 56, no. 3, pp. 1064–1075, May 2007.

[35] I. Broustis, S.V. Krishnamurthy, M. Faloutsos, M. Molle, and J.R. Foerster, "Multiband Media Access Control in Impulse-Based UWB Ad Hoc Networks," *IEEE Transactions on Mobile Computing*, vol. 6, no. 4, pp. 351–366, April 2007.

[36] H. Jiang and W. Zhuang, "Effective Packet Scheduling with Fairness Adaptation in Ultra-Wideband Wireless Networks," *IEEE Transactions on Wireless Communications*, vol. 6, no. 2, pp. 680–690, Feb. 2007.

[37] C. Cordeiro, K. Challapali, D. Birru, and S. Shankar, "IEEE 802.22: The First Worldwide Wireless Standard Based on Cognitive Radios," in *Proceedings of IEEE International Symposium on New Frontiers in Dynamic Spectrum Access Networks (DySPAN'05)*, pp. 328–337, Nov. 2005.

[38] C. Doerr, M. Neufeld, J. Fifield, T. Weingart, D.C. Sicker, and D. Grunwald, "MultiMAC: An Adaptive MAC Framework for Dynamic Radio Networking," in *Proceedings of IEEE International Symposium on New Frontiers in Dynamic Spectrum Access Networks (DySPAN'05)*, pp. 548–555, Nov. 2005.

[39] "Functional Requirements for the 802.22 WRAN Standard r47," IEEE 802.22-05/0007r45, September 2005.

[40] W. Hu, D. Willkomm, M. Abusubaih, J. Gross, G. Vlantis, M. Gerla, and A. Wolisz, "Dynamic Frequency Hopping Communities for Efficient IEEE 802.22 Operation," *IEEE Communications Magazine*, vol. 45, no. 5, pp. 80–87, May 2007.

Chapter 8

Cognitive Radio Adaptive Medium Access Control (MAC) Design

Fei Hu, Rahul Patibandla, and Yang Xiao

Contents

8.1 Introduction

The cognitive radio (CR) concept has been defined as "the point at which wireless personal digital assistants (PDAs) and the related networks are sufficiently computationally intelligent about radio resources and related computer-to-computer communications to detect user communications needs as a function of use context and to provide radio resources and wireless services most appropriate to those needs" [1]. Many regulatory bodies like the Federal Communications Commission (FCC) in the United States have observed that the radio spectrum is inefficiently utilized. According to the FCC, a cognitive radio is defined as one that can sense its environment and then alter its operating frequency, power, and modulation techniques in order to use the spectrum efficiently. A congnitive radio has the following features: (a) it should be able to identify and detect the channel in the available band and to tune itself to that particular channel; (b) after identifying the channel, it should then establish the network connection and operate in that particular channel; (c) to obtain the best throughput, the primary aspect for any type of system, CR implements better bandwidth for efficient data transmission and also error control and correction schemes to obtain the best throughput; and (d) in order to optimize the received signal strength it can adjust the direction of its antenna.

With the help of the CR techniques, the entire wireless spectrum is used for communication by cognitive radio networks. In the case of cognitive radio network access, the unused portion of the spectrum uses a different spectrum scheme in order to reduce the interference among the users.

CRs can be classified into various types based on different parameters [2]. (1) *Full cognitive radio* takes every parameter into account that is being observed by the wireless node or network, and then bases the decisions on the changes in the transmission or reception. (2) *Spectrum sensing cognitive radio* senses the entire spectrum, detects the part of the spectrum that is left unused, and then shares this part of the spectrum with the other users without causing any interference with the primary users (licensed users). (3) *Licensed band cognitive radio* uses the spectrum that is particularly meant for licensed user access. It initially checks for primary user activity on the particular channel of the spectrum; if the primary user is active, then it switches to the other channel, and if the primary user is not active, then it gives access to the unlicensed user (secondary user) and monitors the entire channel for the primary user. An example of licensed band cognitive radio is

IEEE 802.22. (4) *Unlicensed band cognitive radio* uses the unlicensed parts of the spectrum that are available for secondary users only. Therefore, there is no need for the cognitive radio to sense the entire spectrum before the secondary users use the channel. An example of unlicensed band cognitive radio is IEEE 802.19.

The performance of wireless networks is improved by introducing various medium access control (MAC) protocols. The slotted seeded channel hopping algorithm is one of these protocols [3]. In this protocol the users share pseudorandom codes for accessing the medium in a time-slotted manner. A dynamic channel assignment algorithm is also proposed for the MAC layer. In this algorithm control messages like request-to-send (RTS) and clear-to-send (CTS) are adopted. A control channel is mainly used for exchanging these messages. The data transmissions take place along the data channel. A disadvantage of using these protocols is the control channel saturation problem. Hence in order to overcome this disadvantage, the dynamic open spectrum sharing MAC protocol is proposed [3].

The MAC layer of a CR network is responsible for spectrum sensing (through neighboring message exchange), spectrum sharing, and spectrum access scheduling issues. In the next section, we will first briefly discuss several MAC schemes in the literature. Then we will analyze some special MAC design issues such as spectrum sharing and scheduling.

8.2   Existing Cognitive Radio MAC Layer Protocols

8.2.1   Dynamic Open Spectrum Sharing MAC Protocol (DOSS)

The dynamic open spectrum sharing (DOSS) protocol [3] classifies the MAC protocol of a cognitive radio network (CRN) into five basic steps. These are: (1) detecting the presence of the primary users, (2) setting up three frequency bands (busy tone band, control channel band, and data band), (3) mapping the spectrum, (4) negotiating the spectrum allocation, and (5) transmitting the data.

8.2.1.1   Detecting the Presence of Primary Users

The secondary users make use of the spectrum only when the primary users are not accessing it. This involves frequent message exchange among neighbors to reach a global view of channel availability information.

8.2.1.2   Setting up Three Operational Frequency Bands

The control channel is mainly used to help the radio receiver to identify the particular channel in which it can operate. After identifying the particular channel in which the radio can operate, it can then use the data band to start transmitting the data. The busy tone band is mainly used to alleviate the hidden and exposed terminal problems.

8.2.1.3 Spectrum Mapping

The spectrum mapping is used to establish a one-to-one mapping between the narrow bands and the wide bands. The authors in [3] also explain the mapping between the busy tone bands and the data channels. With the spectrum mapping in place, the receiver can set the busy tone on the spectrum in which it is receiving a transmission, and thereby inform its neighbors about the spectrum that it is currently using. On obtaining the spectrum information in this way, the neighbors will stop sending their data.

8.2.1.4 Spectrum Negotiation

In this step the sender and the receiver negotiate on a particular channel for the data transmission. In the process of negotiation, initially the nodes sense the spectrum and identify the channel availability. After identifying the channel, the sender sends the REQ packet to the receiver. This REQ packet has the information about the channel parameters. After receiving the REQ packet, the receiver checks with its own available channels, and picks the common channel that is available to both the sender and the receiver. The receiver then sends back a REQ_ACK packet. After receiving the REQ_ACK packet from the receiver, the sender realizes that the channel is available for transmission and starts its transmitter for the data transmission.

8.2.1.5 Data Transmission

The sender sends the DATA packet to the receiver. Receiver acknowledges the sender with the DATA_ACK packet only after it receives the DATA packet. The sender realizes that the transmission is successful only after it receives the DATA_ACK packet. If the sender does not receive the DATA_ACK packet from the receiver, it realizes that the transmission is unsuccessful and sends the DATA packet again.

8.2.2 Common Spectrum Coordination Channel Protocol (CSCC)

Another CR MAC scheme is called the common spectrum coordination channel protocol (CSCC) [6]. In the CSCC, all the users share a common control channel for spectrum coordination purposes. Control information is exchanged by the user using the narrow-band radio. All the users should periodically broadcast requests over the spectrum if they intend to use the spectrum. CSCC provides access to those users who have transmitted requests for accessing the spectrum. Other users will remain idle and not transmit any information.

8.2.3 *Distributed Channel Assignment Protocol (DCA)*

The distributed channel assignment protocol (DCA) is proposed in [7]. According to this protocol, all the users on the network share a one-way handshake signal for data transmission. The transmitting and receiving secondary users stop their transmission and reception upon identifying the primary user. In this protocol, all the users use two antennas [7]. One antenna is used for the traffic channel, and the other is used for the spectrum sensing and data transmission/reception on the channel.

8.2.4 *Slotted Seeded Channel Hopping Algorithm (SSCH)*

In the slotted seeded channel hopping (SSCH) algorithm [8], each user switches among the multiple channels, so that there is a significant increase in the network capacity. SSCH is a distributed protocol. Synchronization is not required among the users in the network. A slot is defined as the time that a user spends on a particular channel. A longer slot time will decrease the channel switching overhead but further increase the delay affecting the packets. A user maintains a channel schedule in which the channels that the user plans to switch to in the particular time slot are listed.

Table 8.1 summarizes several MAC protocols.

8.3 MAC Layer Spectrum Sharing Issues

In the MAC layer of CR networks, a primary concern is how to efficiently share the sensed spectrum among secondary users. The spectrum sharing process normally consists of five important steps, as follows [2]. (1) *Spectrum sensing*: The cognitive radio senses the entire spectrum for a vacant channel and checks whether that channel is not being accessed by the primary user. (2) *Spectrum allocation*: After identifying the channel, the next process is to assign the channel for a particular user. (3) *Spectrum access*: Effectively sharing the spectrum by preventing collisions among the users. (4) *Transmitter-receiver handshake*: This is mainly used by the transmitter to inform the receiver about the portion of the spectrum that is to be used. After identifying a portion of the spectrum for usage, the transmitter sends a handshake signal to the receiver in order to notify the receiver that it is going to receive the packets through the same portion of the spectrum. (5) *Spectrum mobility*: When the primary user starts to access that particular portion of the spectrum, the secondary user should vacate that portion and

Table 8.1   Existing MAC Layer Protocols

Protocols			
DOSS Protocol [3]	*CSCC Protocol [6]*	*DCA Protocol [7]*	*SSCH Protocol [8]*
Main feature			
The users select the spectrum arbitrarily for communication purposes.	The users share a common control channel for spectrum coordination purposes.	This protocol takes care of data exchange and signaling on the network among the users.	Each user switches among the multiple channels, yielding significant increase in the network capacity.
Advantages			
1. It does not use fixed spectrum allocation. 2. It prevents hidden and exposed terminal problems. 3. It supports unicast and multicast issues. 4. It is scalable. A real time and efficient spectrum allocation protocol.	1. It allows flexibility among the spectrum sharing procedures. 2. Advanced power control and multihop routing procedures can be implemented. 3. Terminal startup procedures with the new network can be avoided when the user enters a new physical area. 4. Collaborative spectrum usage can be used in this protocol. 5. This protocol has multihop routing capabilities.	1. It is a peer-to-peer cognitive radio network protocol. 2. Negotiation takes place among the two nodes only. 3. All nodes use one-way handshake for the information exchange.	1. The capacity of ad hoc wireless multihop networks is significantly improved. 2. It uses a single radio and does not use a dedicated control channel. 3. Synchronization takes place among the users that intend to communicate. 4. It works in both single-hop and multihop environment.

(Continued)

Table 8.1   Existing MAC Layer Protocols (Continued)

Disadvantages			
1. It requires multiple radio transmitters and receivers.	1. It causes the control channel saturation problem.	1. Restrictions on the channel assignment due to hardware configurations.	1. It introduces long transmission delays.
2. Increases the device cost.	2. It cannot handle hidden and exposed terminal problems.	2. Potential inefficiency in the channel assignment.	2. The communication overhead is significant for short flows.
	3. It cannot limit the traffic going through the control channel.		
	4. It doesn't support multicast communications.		

Applications			
Used in wireless ad hoc networks operating over open spectrum.	Used for coordinating radio devices in unlicensed bands.	Used in wireless emergency communication networks.	For increasing the capacity of IEEE 802.11 networks by using frequency diversity.

should move to another vacant portion of the spectrum in order to maintain the communication without any interruption.

In the dynamic open spectrum sharing protocol [3], the use of three different frequency bands has been proposed. They are the data band, the control channel, and the busy tone band [3]. The *data band* is mainly used for transferring the data from one point to the other. The *control channel* can be regarded as the logic channel, which is mainly used to carry the network information instead of the actual voice or data messages that are transmitted over the network. As to the *busy tone* band, initially the sender sends a request packet to the particular receiver over the control channel. The request packet contains the information about the channels of the sender. After the receiver receives the request packet from the sender, it then compares the received channel information with its own channel information. After comparing and identifying the channel information, the receiver picks a channel that is common to both the sender and the receiver and then sends an acknowledgment packet to the sender.

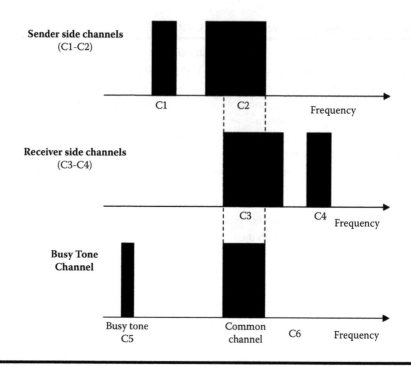

Figure 8.1   Use of busy tone in CR networks.

This acknowledgment packet has the information about the channel that is common to both. After sending the acknowledgment packet, the receiver turns on the busy tone, which informs the other neighbors not to send data over this channel, as this channel is allotted to a particular user (sender). See Figure 8.1 on the use of the busy tone [3].

One classification of spectrum sharing is *spectrum overlay and underlay* [2].

■ *Spectrum overlay*: This technique is characterized by the mode of spectrum sharing between the primary and the secondary users [9]. Spectrum overlay does not place transmission power constraints on the secondary users. The secondary users can identify the spectrum available for their data transmission without causing any intrusion on the primary users. Spectrum overlay makes use of the time domain or space domain [9]. In the time domain, the secondary users utilize the time slots based upon the primary users' traffic load. In the space domain, the secondary users utilize the frequency bands that are not being used by the primary users.

■ *Spectrum underlay*: This technique is also used to identify the spectrum sharing between the primary and the secondary users. Spectrum

Table 8.2   Spectrum Overlay and Spectrum Underlay

Spectrum Overlay	Spectrum Underlay
Spectrum overlay does not impose transmission power constraints on the secondary users.	Spectrum underlay imposes transmission power constraints on the secondary users.
Spectrum overlay makes use of time domain and space domain.	Spectrum underlay does not make use of time domain and space domain.
Sensing the rapidly varying spectrum characteristics is the major problem in the spectrum overlay. Errors occur due to the real-time spectrum sensing.	The unresolved problem in the spectrum underlay aggregate interference [9]. Hardware implementation of the underlay radios is still an issue.
Characteristics of the spectrum sensor should be taken into account while making spectrum access decisions [9].	Due to high power consumption, designing the high rate ADC is still a challenging issue.
Typical application is the reuse of certain TV bands which are not used for TV broadcast.	Underlay radios (UR) use fast frequency hopping and large bandwidth.

underlay normally imposes transmission power constraints on the secondary users. The secondary users operate in a lower noise band compared with the primary users. The shorter the distance between any two secondary users, the higher the data rate they can achieve. Due to power constraints (low power), underlay radios are used to spread the spectrum over a large bandwidth. The underlay radios use the spread spectrum technique or the wideband OFDM technique to spread the spectrum over a large bandwidth. Table 8.2 compares spectrum overlay and underlay.

The other classification of spectrum sensing techniques is the division into internetwork and intranetwork sharing techniques [2], as follows.

■ *Internetwork spectrum sharing*: In distributed internetwork spectrum sharing, the distributed QoS-based dynamic channel reservation (D-QDCR) scheme is used [2]. In the D-QDCR scheme, the number of reserved channels is dynamically adjusted with respect to the traffic situation [10]. This algorithm prevents a base station from managing a group of channels. A mobile switching center is not used, since the base stations are able to exchange the channel information among themselves.

The mobile cellular network is basically divided into clusters of small cells. Each cluster incorporates seven cells and each cluster has a base station. Assume that there are M channels divided into three groups and distributed among the cluster [10]. The particular cluster's base station can acquire the channel group as long as the channel group is not being controlled by a neighboring cluster's base station. Different coloring schemes are used for each group. Even when a group is acquired by the base station, it can look for the particular coloring scheme in order to make sure that it has acquired the particular group of channels.

The channels are normally classified into control channels and communication channels. The control channels are further classified into forward setup and reverse setup channels. The forward setup channels are the set of channels from the mobile hosts to the base stations, and the reverse setup channels are the set of channels from the base stations to the mobile hosts.

In order to establish communication with the channels, a wireless channel is first required to start the connection between the mobile host and the corresponding base station. Second, the wired links are provided between the source and the destination base stations. Third, a wireless channel is reserved between the destination mobile host and the destination base station. The communication session is dependent on the wireless channel. If the wireless channel is not obtained, then the communication session cannot be established [10]. The communication session is established after obtaining the wireless channel. At the end of the session, these wireless channels are used by the other mobile hosts.

■ *Intranetwork sharing:* These techniques can be further classified into two types.

(a) Cooperative spectrum sharing: In cooperative spectrum sharing, all the nodes work together and communicate using the common protocol [11]. With the cooperative spectrum sharing technique, the primary user creates the open spectrum. The primary users can create the unlicensed bands in the spectrum. The coexistence of the secondary users with the primary users can be obtained by using this approach. Using this approach, the secondary users are allowed to utilize the licensed spectrum without causing any interference with the primary users. The secondary users are controlled by the open spectrum, and the network is made transparent to the primary users.

The main problem in the open spectrum approach is designing an efficient way of managing the spectrum for the secondary users [13]. This approach also requires frequent coordination and information exchange among the users.

Table 8.3   Cooperative Spectrum Sharing and Noncooperative Spectrum Sharing

Cooperative Spectrum Sharing	Noncooperative Spectrum Sharing
Pros	
1. All nodes work together and communicate with each other by means of a common protocol.	1. All nodes work independently based on the local observations of the network.
2. Efficient usage of the spectrum is possible and is mainly used for simple deployment of various applications.	2. Nodes work based on the interference of the neighboring nodes.
3. No common channel control problem exists.	3. This technique reduces the control traffic and simplifies the allocation of the communication resources [13].
4. This improves scalability and reduces the deployment costs.	4. Spectrum management for the secondary users is possible in this technique.
Cons	
1. Efficient spectrum management for the secondary users is the main problem in this technique.	1. Even though fair service is guaranteed for each user, the communication overhead cannot be reduced.
2. The management scheme maximizes interference and does not provide fairness among the users.	2. Neighbors cannot exchange coordination information frequently.
3. This technique normally increases the stress on the communication resources of a particular network [13].	3. Common channel control problem exists.
4. Security is a major issue in this technique.	4. This increases the deployment costs.

(b) Noncooperative spectrum sharing: This is a device-centric spectrum management scheme in which the users do not coordinate with each other for exchanging information. The users behave independently on the local observations. These local observations can be from the neighbors. This noncooperative spectrum sharing approach reduces the control traffic and simplifies the allocation behavior. In the noncooperative spectrum sharing technique, neighbors cannot exchange coordination information frequently. Table 8.3 shows the advantages and disadvantages of cooperative spectrum sharing and noncooperative spectrum sharing.

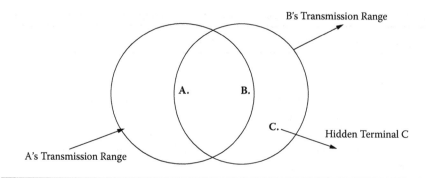

Figure 8.2   Hidden terminal problem.

8.4   Handling of CR Hidden and Exposed Terminal Problem

The processing of hidden terminal problems in a CR network is somewhat different from general wireless LAN (WLAN) cases. Let's first review the problem scenario, shown in Figure 8.2. Assume that A and B are two CR nodes with certain transmission ranges. The transmission range of A is intersecting with the transmission range of B. Node C is present in the transmission range of B but it is out of range of node A's transmission. Node A cannot recognize node C. Node A starts transmitting packets to node B, and node C is also transmitting packets to node B, and as a result there will be a collision of the packets at node B. This is known as the hidden terminal problem.

The hidden terminal problem in a CR network can be reduced by using the *busy tone* approach. In this method, during A's transmission of packets to B, A sets a busy tone during the entire period of transmission, and when C wants to transmit to B, it hears the busy tone by the means provided in another band, and will hold itself from sending the packets to B as long as the busy tone is present. Node A stops the busy tone after the end of its transmission, and then node C can start sending its packet to B [5].

The exposed terminal problem is shown in Figure 8.3. The RTS/CTS mechanism is used to explain the exposed terminal problem. Let us consider that node A has sent an RTS packet to node B. Node D doesn't lie in either of the transmission ranges of A or B. Consider that node D has transmitted an RTS packet to node C just before A's transmission. Now node C sends the CTS packet to node D. Since node C is in the transmission range of node B, node B will refrain from sending the packet CTS to A. As a result, node A will not receive a CTS packet from node B. Hence A stops transmitting the data to node B. This is known as the exposed terminal problem.

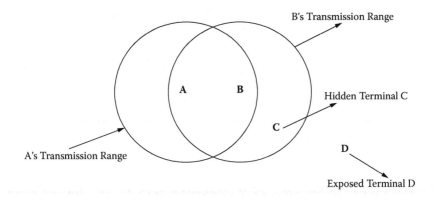

Figure 8.3   Exposed terminal problem.

The same busy tone approach can also be used to reduce the exposed terminal problem. Node A sends the RTS packet and sets the *busy tone* on the particular band. Before A's transmission, node D transmits the packet to node C. Since node C is in the transmission range of node B, it can sense the busy tone set by node A. After hearing the busy tone, node C stops sending the RTS packet to D. Now node B can send the RTS packet to A.

Some problems still persist with the busy tone solution.

First, there is a need for new channels. One channel is used for transmitting and receiving the data, and the other is used for setting up the busy tone for preventing the hidden and exposed terminal problems. This results in the expansion of the spectrum, i.e., more spectrum is used, and more hardware is required for maintaining the additional channels.

Second, collisions still persist when the transmission range of the data channel is greater than that of the busy tone band.

Third, when the busy tone band has a larger transmission range than the data channel, some of the data transmission will be suppressed.

8.5   The Use of MAC Layer Configuration

In order to determine the various sets of channels for communicating between the various secondary users in a cognitive radio network, the idea of MAC layer configuration is used in [17]. Using MAC layer configuration, it is easy to discover the global network topology and to identify the location of a node in the network. It is also easy to identify the common set of channels that are available to all the nodes in the network [17].

Consider the following packet transmission diagram [17] (Figure 8.4).

A_k^M consists of an array of N elements. The kth element corresponds to the availability of the channel set of node k as known to node M.

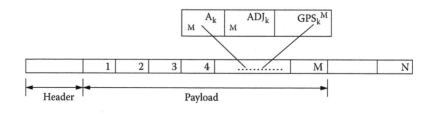

Figure 8.4 MAC layer configuration.

ADJ_k^M refers to the array of N elements where the k^{tb} element corresponds to the adjacent node of node k as known to node M.

GPS_k^M refers to the array of N elements in which the k^{tb} element corresponds to the GPS location of k and is known to node M.

In the diagram it is shown that each node k transmits a packet during the k^{tb} time slot. The payload is divided into N segments. Node k can know about the status of node M by knowing the configuration of the corresponding segment. The segment has all the bits zero if node k does not know about M. The node k is in receiving mode during the remaining time slots. The node k will not be receiving or transmitting when the corresponding segment has all zero bits. The node k has to update its segment content after receiving each frame. Each node k knows about the available sets only after completing the N frames. They also know about the GPS location and the adjacency information of all the nodes in the CR network. After the end of the N^{tb} frame, the node k will set the global channel set and construct the CR network topology.

8.6 Control Channel Saturation Problem

The control channel saturation problem occurs when there are too many data channels to manage [4]. A congested control channel cannot coordinate the use of different data channels very well, and this results in conflict among the channels, and hence causes inefficient use of the channels.

Normally nodes on the data channel exchange RTS/CTS packets one at a time. When the nodes try to exchange more than one RTS/CTS packet through the control channel, the control channel saturation problem ensues. Hence it is better to use multiple channels for exchanging multiple RTS/CTS packets simultaneously. This scheme has been proposed in the multi-control-channel medium access control (MCMAC) protocol [4]. In the process of choosing among the multiple channels for exchanging their packets, the neighbors of the receiver and the transmitter are informed about the corresponding channel. According to this protocol, there are two control channels available. The first one is called the default control channel (DCC) and the other is called the alternate control channel (ACC). All the users on the channel initially have the same DCC, and at the same time all the users on the channel know about all the other users' DCC. Normally the station checks for the free data channel that is available and corresponds to the particular DCC. The transmission occurs only when the station is not acting in the NAV (native allocation vector) mode. When the station is in the NAV mode, it then switches to the ACC. This is the case where the station acts in multiple channels for exchanging multiple packets. The two cases where the multiple channels are used are (1) when the station is acting in NAV, and (2) when the station is not acting in NAV.

The control channel saturation problem can be mitigated by using three techniques adopted in [3]. These techniques are as follows.

(1) Limiting the traffic going through the control channel.
(2) Adjusting the bandwidth ratio of the control channel and the data channel. To transmit the data packet successfully, this ratio consideration is important. The ratio of the control traffic to the data traffic is set below a predetermined threshold.
(3) Allowing the common control channel to migrate onto the current control channel on the basis of the traffic load. The initial control channel is used for the initial communications among the nodes. This control channel then migrates to the other channel, which has a better central frequency and a different bandwidth than the previous channel.

8.7 MAC Layer Security

Spectrum sensing helps the secondary users to identify the vacant portion of the spectrum not being used by the primary users and to share the spectrum with the primary users without causing any interference. There are various security issues associated with the spectrum sensing property of cognitive radios, such as robust identification of the primary users and trustworthy distributed spectrum sensing [14, 15].

The secondary users periodically sense the spectrum in order to identify the spectrum holes. These spectrum holes are used for transmitting and receiving data by the secondary users. When a secondary user senses that the primary user is accessing the spectrum or determines that the primary user is going to access the spectrum, it then vacates the spectrum and moves to the another spectrum or lowers its transmitting power. If there is another secondary user instead of the primary user, then the first secondary user shares the spectrum with the new secondary user. Security problems will occur when the secondary user cannot detect the presence of the other user. Another problem is that the secondary user can misidentify the other secondary user as the primary user, and as a result interference occurs, preventing the other secondary user from accessing the spectrum.

In order to overcome this problem, the primary user signals should use different network characteristics for identification purposes [15]. The primary users are left unmodified. The information flow is only from the primary user to the secondary user, but not bi-directional. Identification of primary users can be done by considering information about their location. Received signal strength is mainly used to measure the distance from the primary user. The gap between the two different users can be obtained by measuring the arrival time of the signal [15]. The weighted sequence probability ratio test (WSPRT) is one of the solutions for trustworthy distributed spectrum sensing [15].

8.8   Optimization MAC Layer for Cross-Layer Design

Cross-layer design is used to obtain effective wireless internetworking. Cross-layer design [16] means that the information from one layer is exchanged among the other layers. The quality of service (QoS) is the main factor to be taken into account in the implementation of cross-layer design. In order to maintain the end-to-end QoS, it is mandatory to consider the hop-to-hop behavior across the wireless links. The MAC-layer parameters can be modified to improve routing-layer techniques. For instance, timely spectrum sensing information from the MAC layer can be used for proactive ad hoc network routing algorithm updates. Thus packet loss can be avoided even as new channels are used in links.

A spectrum handoff in the MAC layer normally causes a certain delay [2]. The delay has an adverse effect on such network performance components as routing QoS support. Besides affecting network performance, the spectrum handoff can also change such channel parameters as path loss, interference, link error rate, etc. [2]. In order to determine the delay caused by the spectrum handoff, information about the MAC layer is required. The transport layer and the application layer should also be aware of the delay. The TCP timeout setup should be updated in light of the new estimation of spectrum handoff delay.

Spectrum sharing is based on the spectrum sensing capability of the CRN. The main challenge in spectrum sensing is that the spectrum is sensed opportunistically, which means that the whole spectrum cannot be sensed every time. The spectrum is sensed at particular time intervals, and the higher layers should be aware of this switching schedule to avoid link outage.

8.9 Conclusions

In this chapter, we have covered MAC design issues in CRNs. Spectrum sharing is one of the top MAC design concerns. The common channel saturation problem and hidden/exposed terminal problem have been explained in detail, and the solutions to overcome these problems have also been discussed. MAC layer scheduling involves considering both how to determine and how to facilitate the common set of channels for effective communication with various nodes. Various spectrum sharing mechanisms are also explained. Some of the important spectrum sharing techniques explained are the intra network spectrum sharing and the internetwork spectrum sharing techniques. Some of the main important security problems associated with the cognitive radio MAC layer are explained briefly. Optimization of the MAC layer with the upper layers of the wireless architecture model is also important to reduce protocol overhead.

References

[1] J. Mitola III, "Cognitive Radio: An integrated agent architecture for software defined radio," May 2000. Available at http://citeseer.ist.psu.edu/568482.html.

[2] I.F. Akyildiz, W.-Y. Lee, M.C. Vuran, and S. Mohanty, "NeXt generation/ dynamic spectrum access/cognitive radio wireless network: A survey," *Elsevier Computer Networks*, pp. 1–30, January 2006.

[3] L. Ma, X. Han, and C.-C. Shen, "Dynamic open spectrum sharing MAC protocol for wireless ad hoc networks," *Proc. of DySPAN 2005*.

[4] H. Koubaa, "Fairness-enhanced multiple control channels MAC for ad hoc networks," *Proc. of IEEE VTC-spring 2005*.

[5] A. Jayasuriya, S. Perreau, A. Dadej, and S. Gordon, "Hidden vs. exposed terminal problem in ad hoc networks," http://www.itr.unisa.edu.au/~sgordon/doc/jayasuriya2004-hidden.pdf.

[6] D. Raychaudhuri and X. Jing, "A spectrum etiquette protocol for efficient coordination of radio devices in unlicensed bands," In *Proceedings of the 14th International Symposium on Personal, Indoor and Mobile Radio Communication*, IEEE 2003.

[7] P. Pawelczak, R.V. Prasad, L. Xia, and I.M Niemegeers, "Cognitive radio emergency networks—requirements and design," *Proc. of DySPAN 2005*.

[8] P. Bahl, R. Chandra, and J. Dunagan, "SSCH: slotted seeded channel hopping for capacity improvement in IEEE 802.11 ad hoc wireless networks," *Proc. of MobiCom'04*. 2004.

[9] Q. Zhao and A. Swami, "A survey of dynamic spectrum access: signal processing and networking perspectives," *Proc. of IEEE International Conference on Acoustics, Speech, and Signal Processing (ICASSP): special session on Signal Processing and Networking for Dynamic Spectrum Access*, April, 2007.

[10] A. Boukerche, T. Huang, and K. Abrougui, "Design and performance evaluation of a QoS-based dynamic channel allocation protocol for wireless and mobile networks," *In Proceedings of 13th IEEE International Symposium on Modeling, Analysis, and Simulation of Computer and Telecommunication Systems (MASCOTS'05)*, 2005.

[11] J.M. Peha, "Approaches to spectrum sharing" *In IEEE Communications Magazine*, February 2005, vol. 43, issue 2, pp. 10–12.

[12] L. Cao and H. Zheng, "Distributed spectrum allocation via local bargaining," *In Sensor and Ad Hoc Communications and Networks Conference*, September 2005.

[13] H. Zheng and L. Cao, "Device-centric spectrum management," *Proc. of DySPAN 2005*.

[14] R. Chen and J.-M. Park, "Ensuring trustworthy spectrum sensing in cognitive radio networks," *IEEE Workshop on Networking Technologies for Software Defined Radio Networks* (held in conjunction with IEEE SECON 2006), September 2006.

[15] R. Chen, J.-M. Park, and K. Bian, "Robust distributed spectrum sensing in cognitive radio networks," Report TR-ECE-06-07, Dept. of Electrical and Computer Engineering, Virginia Tech, July 2006. See http://www.arias.ece.vt.edu/pubsall.html.

[16] J.L. Burbank and W.T. Kasch, "Cross-layer design for military networks," *IEEE Military Communications Conference*, October 2005, pp. 1912–1918, vol. 3.

[17] S. Krishnamurthy, M. Thoppian, S. Venkatesan, and R. Prakash, "Control channel based MAC-layer configuration routing and situation awareness for cognitive radio networks," *IEEE Military Communications Conference*, October 2005, pp. 455–460, vol. 1.

ROUTING LAYER OF COGNITIVE RADIO NETWORKS

Chapter 9

Cognitive Routing Models*

Luca De Nardis and Maria-Gabriella Di Benedetto

Contents

---

*An earlier version of this work was presented in the invited paper Cognitive routing in UWB networks, by L. De Nardis and M. G. Di Benedetto, published in the *Proceedings of the IEEE International Conference on UWB 2006*, pp. 381–386 (September, 2006).

9.1   Introduction

The introduction of the cognitive principle into the logic of a wireless network requires extending the cognitive concept to rules of operation that take into account the presence of several nodes in the network as well as their instantaneous configuration. In this perspective, the design goal moves from the definition of a single smart device to a network of smart devices that must be capable of efficiently coexisting in a given geographical area by using cooperation. This goal requires the integration of cognitive principles into the rules of interaction between nodes in the network: the set of wireless nodes should form a social network that must be modeled and analyzed as one entity in order to optimize the design of network functions such as resource management and routing.

In this investigation we focus on the introduction of the cognitive principle into the logic of a wireless network as regards routing. To this aim, we first review existing investigations on the application of the cognitive principle to the routing problem. Next, we describe our approach to cognitive routing for wireless networks, originally proposed in [1]. We assume that the routing function incorporates measurements of the instantaneous behavior of the external world, as represented for example by current network status in terms of interference suffered by an overlaid network. The framework that we consider for our research refers to low data rate and low cost networks for mixed indoor/outdoor communications investigated within the IEEE 802.15.4a Task Group [2,3]. Within this group, an impulse radio ultra-wideband (IR-UWB) physical layer, capable of providing the accurate ranging information required for accurate positioning, was adopted. The IEEE 802.15.4a Task Group concluded its activity in March 2007, when the new standard was released [3].

The chapter is organized as follows. In Section 9.2 we review previous work on the cognitive routing problem, and provide a description of the main contributions on this topic. In Section 9.3 we introduce our proposed approach, starting from the model for the routing module, and describe strategies for route selection that take into account UWB features (power limitation, synchronization, battery limitation, interference, etc.) and coexistence issues. In Section 9.4 we define a routing cost function that incorporates the model of Section 9.3. The approach is analyzed and investigated by simulation as described in Section 9.5. Section 9.6 concludes the chapter.

9.2   Previous Work

Research activities related to the introduction of cognition into the routing process have been carried out in the last fifteen years, with particular interest devoted to the introduction of learning capabilities into the routing algorithm. In the following we will start our review from earlier works on cognitive routing, which mainly addressed the case of fixed and wired networks and focused on the optimization of internal network behavior without considering the problem of interaction with external systems [4–7]. We will then analyze more recent works, where the growing interest in cognitive radios led to the proposal of routing protocols capable of coping with the frequent topology changes due to the channel switching of the cognitive radios forming a network [8–11].

In [4] the authors propose the application of computing intelligence to the routing problem by introducing a set of agents inspired by the behavior of ants in an ant colony. The agents, which can be implemented in the form of probe packets, explore the network in order to collect information on average end-to-end delay, and then propagate backward in order to update the intermediate routers according to the collected information.

The authors move from previous work on artificial colony-based routing and introduce learning capabilities by means of a reinforced learning mechanism based on artificial neural networks. The proposed solution can be summarized as follows:

- An artificial neural network is implemented in each router. The neural network receives as input the probability of selecting each possible next hop towards a given destination, and the average trip time towards that destination using each possible next hop, and provides as output the new values of probabilities and estimated trip times to the same destination for each possible next hop.
- At each hop, a forward ant traveling to a given destination selects the next hop by using the artificial neural network.
- When an ant propagates backward from the destination to a previously visited node, it updates the weights of the neural network and the routing table according to the measured trip time to the destination, thus modifying the behavior of the neural network and the choices of the following ants.

Simulation results reported in [4] show that the introduction of learning capabilities can improve routing performance, leading to a slight increase in throughput and a significant reduction in end-to-end delay.

The approach proposed in [4] for the behavior of a single node can actually be mapped onto the cognitive cycle as defined by Mitola. Each node in the network observes the system status by receiving the measurements provided by the ants, and takes decisions according to the observations. Furthermore, both the system status and the impact of previous decisions are taken into account in the learning process, impacting future decisions. Overall, network behavior is thus the result of independent cognitive cycles taking place in each network node.

The concept of cognitive routing is addressed more thoroughly in [5]. In this work the authors move the learning capability from the node to the packet by introducing the concept of cognitive packets. A cognitive packet (CP) is divided into four parts: the ID field (for identifying the packet and its class of service), the DATA field (carrying user data), and two special fields related to the cognitive routing algorithm—the *cognitive map* field and the *executable code* field. The cognitive map (CM) contains a network map, that is, an estimation of the state of the network based on previous information collected by the packet. The executable code implements a decision-taking algorithm that operates using the CM field as an input, and a learning algorithm for the update of the CM. Furthermore, the decision-taking and learning algorithms take into account a predefined goal set for the packet, that is, a performance metric to be optimized, such as minimum delay or maximum throughput.

Nodes in the network play essentially two roles: (1) they provide storage capability in the form of *mailboxes* that can be read or written by cognitive packets, and (2) they execute the executable code contained in each received packet.

Whenever a CP is received by a node, the node executes the code stored in the executable code field of the packet; the input to the code consists of the cognitive map stored in the node itself, and the content of the mailbox in the node. As a result of the code execution, any of the following actions can be performed.

- The cognitive map in the packet is updated.
- The mailbox in the node is written.
- The packet is sent on an output link.
- The packet is kept in a buffer waiting for a given condition to be met.

The authors compare the performance of their cognitive packet network with a straightforward shortest path algorithm, and show that even in the case of very simple learning and decision-taking algorithms, their approach can improve performance in terms of packet loss and delay. Even larger improvements in network performance can be obtained when more complex learning algorithms, such as neural networks, are implemented in the executable code field.

The approach proposed in [5] poses, however, several implementation challenges, in particular in terms of routing overhead due to the code to be stored in each packet. Later evolutions of the approach moved back to a more traditional approach, where the learning and decision-taking code is stored in the nodes, and its execution is triggered by the arrival of cognitive packets [6]. Furthermore, cognitive packets only constitute a small fraction of overall packets and do not carry any user data information, leading to a solution similar to the one later proposed in [4] and described above.

In the original formulation of the CPN approach, the cognitive map field poses an overhead issue as well, since the number of observations grows with path length, and thus with the size of the network. In order to solve this issue, a modified version of the protocol was proposed in [7] in order to improve scalability and reduce overhead, making the protocol potentially suitable for wireless networks as well.

In [8,9] the authors propose a routing metric that models the end-to-end delay by taking into account both the average delay introduced by collisions on a single frequency band, and the delay introduced by each channel switch required along the path.

The work presented in [10] addresses the same problem by proposing a solution for spreading the information on the positions of the nodes and the channels available to each node in order to enable efficient routing. The proposed information exchange protocol, based on a broadcast packet exchange, is however only tested in a very favorable scenario, characterized by an error-free channel and collision-free medium access.

An additional characteristic of cognitive radio networks that may impact routing is the fact that the network can be formed by devices complying with different wireless standards. Furthermore, a network node can potentially support more than one wireless network interface. The routing protocol proposed in [11] deals with this aspect by introducing a routing metric that models the different characteristics of each radio link available between network nodes. The metric is used to build a routing tree between a base station and wireless nodes in the network.

Channel switching is only one of the possible solutions to allow coexistence between cognitive secondary users and primary users. Ultra-wideband radio offers an alternative solution. Thanks to the huge bandwidth used by the UWB signal and the low power levels allowed by regulation, a UWB signal is in most cases invisible to the primary user. The main problem in routing within a UWB network is thus how to cope with the interference caused by primary users. This goal can be achieved by including the interference generated by such users among the routing criteria. A cognitive routing model that addresses this problem, originally proposed in [1], is illustrated in the following sections.

9.3 Routing Strategy

As indicated in Section 9.1, our research is framed within the area of UWB ad hoc and self-organizing networks. As a consequence we assume that the MAC strategy adopted in the network is based on our previously investigated $(UWB)^2$ protocol [12,13]. The basic hypothesis of $(UWB)^2$ is uncoordinated access in an Aloha-like fashion. The Aloha approach that forms the basis of $(UWB)^2$ was actually voted with a large majority as the medium access strategy for the IEEE 802.15.4a standard, although a CSMA approach is also available for optional operational modes.

As regards routing strategy, key issues that must be taken into account in the selection of a multihop route can be listed as follows.

- *Synchronization*: The assumption of an uncoordinated MAC protocol leads to a significant synchronization overhead. In particular, control routing packets, such as route request and route reconstruct packets, introduce heavier overhead, since synchronization must be acquired between terminals that, in the worst case, are not aware of each other. On the other hand, transmission of data packets over active connections may require lower overhead, since transmitter and receiver preserve at least coarse synchronization between two consecutive packets.

- *Power*: Smart management of available power in order to optimize network performance while meeting the emission limits for UWB devices is required. As a consequence, power issues should be paramount in route selection in order to efficiently make use of available power. The concept of power-aware routing for ad hoc networks was widely analyzed in past investigations [14,15]. Figures 9.1 and 9.2 show the impact of using hops vs. power as the routing metric.

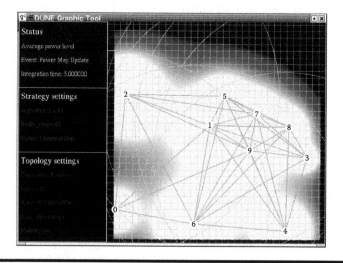

Figure 9.1 Spatial power distribution in the case of minimum number of hops as the route selection strategy. Brighter spots on the map correspond to higher average spatial power density levels.

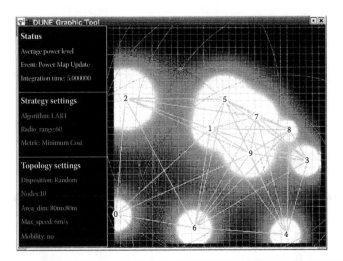

Figure 9.2   Spatial power distribution in the case of minimum energy route selection strategy. Brighter spots on the map correspond to higher average spatial power density levels. Note that compared to Figure 9.1 this route selection strategy reduces the average spatial power density.

- *Multiuser interference (MUI)*: Selecting power-optimized routes, by itself, is not sufficient for guaranteeing the efficient use of power at the network level. The selection of a route in a high density region, in fact, may provoke increased required power to achieve an acceptable packet error rate (PER) on all active links of such a region due to increased interference, leading thus to inefficient power use. MUI should therefore be taken into account in route selection. This can be achieved by considering network topology, as shown in Figure 9.3 vs. Figure 9.4. Figure 9.3 shows the minimum-energy route, which is likely to cause high interference due to high network density (see, for example, node 9). Contrariwise, Figure 9.4 shows an alternative route that takes into account network topology, and therefore avoids the high-density region.
- *Link reliability*: Node mobility and variable network conditions (due to link setup and releases, nodes switching on and off) may cause high instability in selected routes, leading to frequent route reconstruction procedures, and thus high overhead. Poor reliability can easily lead to poor quality of service (QoS). In order to reduce instability, link reliability should be incorporated into the route selection procedure.
- *Traffic load*: The above criteria may potentially cause a terminal that particularly fits one or more criteria to be more frequently selected than others. For example, a nonmobile terminal may guarantee

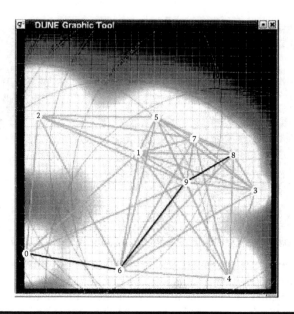

Figure 9.3 Example of minimum-energy route subject to high interference. The highlighted route between nodes 0 and 8 is potentially subject to high interference, in particular at intermediate node 9.

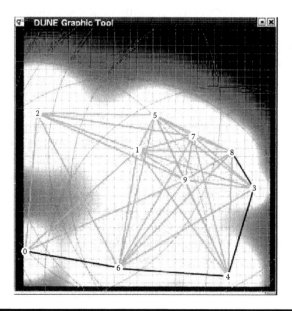

Figure 9.4 Example of a topology-sensitive minimum-energy route between nodes 0 and 8. Note that compared to Figure 9.3 this route avoids highly interfered nodes such as node 9.

greater reliability, and may therefore experience heavier traffic, with the consequence of reduced battery autonomy. This negative effect can be avoided if traffic load of each terminal is taken into account in route selection.

■ *End-to-end delay*: As observed above, link reliability is crucial for QoS when required, such as in ftp and http transfers. On the other hand, in the case of voice and multimedia traffic, having a low end-to-end delay is far more important than correctly delivering all packets. Delay should therefore also be taken into account in route selection in order to assure acceptable delays for time-sensitive traffic classes.

■ *Battery autonomy*: Transmission power is not the only source of power consumption in a node, and route selection should also take into account power consumption due to processing in the node, as for example during the receiving action or the execution of code implementing MAC and routing algorithms. Energy efficiency in the selection of the end-to-end path should consider the residual energy in each node, and attribute higher costs to nodes that are running low on energy.

■ *Coexistence*: The above criteria refer to an isolated UWB network and ignore the environment in which the UWB network operates. Due to coexistence, however, in particular with narrow-band systems, route selection must be able to adapt to external interference. This is where we introduce a cognitive mechanism into the operating principle of the routing module.

Note that according to the above criteria, the route selection process must in some cases integrate tradeoffs between opposite requirements. The power minimization component, for example, leads to routes composed of several hops. On the other hand, the end-to-end delay favors routes with few hops.

9.4   Cognitive Routing Cost Function

In this section, we introduce a cognitive routing cost function that is defined as the sum of different subcosts that in turn take into account each of the routing criteria defined in the previous section. The total cost corresponds, therefore, to a linear combination of subcosts, where each additive component is weighted by a specific subcost coefficient.

According to the criteria defined in the previous section, the cost function over a generic link between nodes x and y should account for the following subcosts: synchronization, transmission power, multiuser interference, reliability, traffic load, delay, autonomy, and coexistence. A general expression for the routing cost function can be thus written as follows:

$$Cost(x, y) = c_{Sync}(t) \cdot Sync(x, y) + c_{Power}(t) \cdot Power(x, y)$$

$$+ c_{MUI}(t) \cdot MUI(x, y) + c_{Reliability}(t) \cdot Reliability(x, y)$$

$$+ c_{Traffic}(t) \cdot Traffic(y) + c_{Delay}(t) \cdot Delay(x, y)$$

$$+ c_{Autonomy}(t) \cdot Autonomy(y)$$

$$+ c_{Coexistence}(t) \cdot Coexistence(y). \quad (9.1)$$

Note that some terms in Equation (9.1) depend on the status of both transmitter x and receiver y, while others such as the traffic, autonomy, and coexistence terms only take into account the status of receiver y. Subcost coefficients are assumed to be dependent upon time t. This assumption is desirable in order to account for time-varying properties of the network, such as variable topology, traffic features, and degree of cognition in the nodes.

In the following we analyze and propose a possible way for defining each term of the cost function separately.

9.4.1 Synchronization Term

This term can be defined as follows:

$$Sync(x, y) = \delta(x, y), \quad (9.2)$$

where $\delta(x, y)$ is 0 if nodes x and y already share an active connection, and 1 otherwise.

Given the $(UWB)^2$ access protocol, synchronization between transmitter and receiver must be acquired from scratch for all random packets involved in setting up a link.

9.4.2 Power Term

We define the power term as follows:

$$Power(x, y) = \left(\frac{d(x, y)}{d_{max}} \right)^{\alpha}, \quad (9.3)$$

where $d(x, y)$ is the distance between x and y, d_{max} is the maximum transmission distance from x as estimated by x that still guarantees a target signal-to-noise ratio (SNR), and α is the path loss exponent. This term takes into account the power required to transmit over the link between x and y for a given SNR, normalized by the maximum transmit power. The SNR characterizing link (x, y) is in fact

$$SNR = \frac{P_T(x, y)/A(d(x, y))}{P_N} = \frac{P_T(x, y)/(A_0 \cdot d^\alpha(x, y))}{P_N}$$

$$\Rightarrow P_T(x, y) = SNR \cdot P_N \cdot (A_0 \cdot d^\alpha(x, y)), \tag{9.4}$$

where $P_T(x, y)$ is transmission power, $A(d)$ is attenuation over link (x, y), and P_N is noise power. For a target SNR and given bit rate, the transmitted power corresponding to d_{max} is thus

$$P_{max} = SNR \cdot P_N \cdot (A_0 \cdot d^\alpha_{max}). \tag{9.5}$$

One has thus

$$\frac{P_T(x, y)}{P_{max}} = \frac{SNR \cdot P_N \cdot A_0 \cdot d^\alpha(x, y)}{SNR \cdot P_N \cdot A_0 \cdot d^\alpha_{max}} = \left(\frac{d(x, y)}{d_{max}}\right)^\alpha. \tag{9.6}$$

In order to compute the power term, the receiver node y must have an estimate of distance $d(x, y)$; this information is expected to be provided by the UWB ranging module. An estimate of P_{max} at node x may also be required, except in the case where all terminals have the same P_{max}, when an explicit computation of this quantity is not necessary.

9.4.3 MUI Term

This term takes into account the potential impact of a transmission from x to y on the neighboring nodes of x.

With regard to MUI, a node x should be avoided if either of the following conditions is met:

(1) x has a large number of neighbors that could be adversely affected by its transmission; or
(2) x has a neighbor at a very short distance which would be subject to strong interference during transmission by x.

Given the ranging capability provided by the UWB physical layer, we propose to use distance information in order to model the impact of x as determined by the two above conditions. A possible way to achieve this goal is to define the MUI term as follows:

$$MUI(x, y) = \frac{1}{N_{Neigh}(x) - 1} \cdot \sum_{n=1, n \neq y}^{N_{Neigh}(x)} \left(1 - \frac{d_{min/y}}{d(x, n)}\right)^2, \tag{9.7}$$

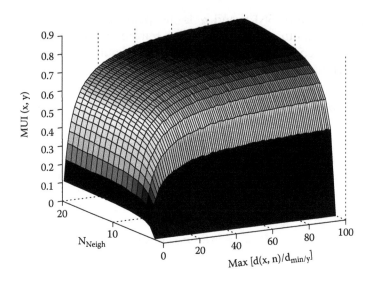

Figure 9.5 Value assumed by the MUI term for the (x, y) link as a function of the number of neighbors of x, N_{Neigh}, and of the maximum value of the ratio $d(x, n)/d_{min/y}$ between the distance from x to a generic neighbor n, and the distance to the closest neighbor excluding y.

where

- N_{Neigh} is the number of neighbors known to x;
- n is the generic neighbor, excluding y; and
- $d_{min/y}$ is the distance between x and its closest neighbor, excluding y.

The value assumed by the term defined in Equation (9.7) as a function of the number of neighbors and the maximum value of the ratio $d(x, n)/d_{min/y}$ is shown in Figure 9.5. Note that when both the conditions previously defined are satisfied, the MUI term assumes high values, thus discouraging the inclusion of the (x, y) link in the selected route.

9.4.4 Reliability Term

We measure the reliability of a link (x, y) as the combination of two factors:

1. the number of packets exchanged between x and y within a predefined observation interval (the higher the number, the higher the expected stability of the link), and
2. the MUI potentially affecting the intended receiver y.

According to this approach, the reliability term can be defined as follows:

$$Reliability\,(x, y) = \frac{1}{2} \left[\frac{1}{N_{packets}\,(x, y)} \right.$$

$$\left. + \frac{1}{N_{Neigh}\,(y) - 1} \cdot \sum_{n=1,\, n \neq x}^{N_{Neigh}\,(y)} \left(1 - \frac{d_{min/x}}{d\,(y, n)}\right)^2 \right], \qquad (9.8)$$

where

- $N_{packets}\,(x, y)$ is the number of packets y received from x in the last observation interval;
- $N_{Neigh}\,(y)$ is the number of neighbors known to y;
- n is the generic neighbor, excluding x; and
- $d_{min/x}$ is the distance between y and its closest neighbor, excluding x.

The stability of the link, expressed by the number of packets that y has received from x at a given time, implicitly takes into account node mobility. Expected MUI also affects reliability and is evaluated as proposed for the MUI term, but with reference to receiver y. As an alternative, y could provide an estimation of future interference based on the interference observed in the past.

9.4.5 Traffic Term

The analytical expression for this term may be written as:

$$Traffic\,(y) = \frac{1}{B_{max}\,(y)} \sum_{i=0}^{N_{active}\,(y)-1} B_i, \qquad (9.9)$$

where:

- $B_{max}\,(y)$ is the maximum overall rate that can be guaranteed by node y,
- B_i is the rate of the ith active connection involving y, and
- $N_{active}\,(y)$ is the total number of active connections at y.

As anticipated in Section 9.3, this term avoids unfair selection of routes by increasing the cost of routes including nodes already involved in many active connections.

9.4.6 Delay Term

This term is defined as follows:

$$Delay(x, y) = 1. \tag{9.10}$$

As a first approximation, the end-to-end delay can be considered to be proportional to the number of hops; in this case, this term is constant.

9.4.7 Autonomy Term

We give the following expression to the autonomy term:

$$Autonomy(y) = 1 - \frac{ResidualEnergy(y)}{FullEnergy(y)}, \tag{9.11}$$

where *FullEnergy(y)* is the energy available in y when the node is first turned on, and *ResidualEnergy(y)* is the energy that is left at time of evaluation of the term.

9.4.8 Coexistence Term

The coexistence term can be defined as follows:

$$Coexistence(y) = \frac{MeasuredExternalInterference(y)}{MaximumInterference(y)}. \tag{9.12}$$

Note that the introduction of this term requires that the UWB receiver is able to measure the level of narrowband interference.

9.5 Simulations

The cognitive routing strategy described in the previous sections was tested by simulation. The routing model was implemented in the framework of the OMNeT++ simulation tool by combining the routing cost function with the Dijkstra shortest path algorithm. During simulations, the computation of the shortest path was carried out by a central node that communicated the path to each node starting a new data connection. The overhead generated by the central cognitive node for the collection of the cost values and the transmission of path information to interested nodes was neglected in the analysis for the sake of simplicity.

The simulation analysis focused on the effect of three terms: end-to-end delay, autonomy, and coexistence. The effect of other terms was analyzed in previous investigations, as described in [16].

Table 9.1   Simulation Setting

Parameter	Setting
Number of nodes	50
Area	150 m × 150 m
Network physical topology	Random node positions, averaged on 10 topologies
Channel model	802.15.4a (see [18])
User bit rate R	64 kb/s
Transmission rate	1 Mb/s
Available transmission power	74 μW (FCC limit for 1 GHz bandwidth)
Traffic model	Constant bit rate connections with average duration 15 s
DATA packet length	576 bits (+ 64 bits for Sync trailer)
UWB Interference Model	Pulse Collision (see [19])
Transmission settings	$N_s = 10$, $T_s = 100$ ns, $T_m = 1$ ns

9.5.1   Simulation Scenario

We considered a network of UWB devices basically following IEEE 802.15.4a Task Group specifications, and adopting thus a time-hopping impulse radio transmission technique [17]. Furthermore, all devices adopted the $(UWB)^2$ MAC protocol [12,13].

The main simulation settings are presented in Table 9.1.

9.5.2   External Interference

In order to analyze the impact of a cognitive cost function on system performance in the presence of external interferers, we introduced interference sources modeled as wideband interferers.

Each interferer was characterized by an emitted power P_{Tx}, an activity factor a, a transmission bandwidth B_{INT}, and a carrier frequency f_c. An interferer was randomly added or removed from the system every T_{Switch} seconds, in order to take into account variable interference conditions. The interference characteristics in terms of bandwidth and carrier frequency were chosen in order to model a WiMax [20] transmitter at 3.5 GHz, which constitutes at the present day one of the most relevant coexistence scenarios for UWB systems [21].

The settings used for generating the interferers are presented in Table 9.2.

9.5.3   Cost Function Settings

In the simulation we compared three different coefficient sets in the scenario defined in Sections 9.5.1 and 9.5.2. The coefficient sets are presented

Table 9.2 External Interferer Settings

Parameter	Setting
P_{Tx}	10 mW
Position	Randomly selected
Activity factor a	Uniform random variable in (0,1)
Carrier frequency f_c	3.5 GHz
Transmission bandwidth B_{INT}	10 MHz
Update time period T_{Switch}	100 s
Initial number of interferers	2

in Table 9.3. Note that the coefficients of the other terms are set to zero in the investigation presented in this work (see [16] for the analysis of other terms).

Set 1 only takes into account delay in the determination of the best path. Given the definition of the delay cost term in Section 9.4.6, set 1 leads to the selection of the path characterized by the minimum number of hops.

Set 2 favors the selection of paths minimizing the autonomy cost (see Section 9.4.7), and aims at maximizing network lifetime.

Set 3 leads to the selection of paths involving nodes suffering external interference in a minor way, thus aiming at the best possible coexistence between UWB and external interferers.

9.5.4 Simulation Results

Two runs of simulations were carried out.

The first run focused on network performance in the presence of external interference. Performance was expressed by throughput and end-to-end delay.

The second run analyzed network lifetime both in the presence and in the absence of external interference. Network lifetime was expressed by the time at which the first node ran out of battery from network startup.

In the first simulation run, network performance was analyzed for the three coefficient sets defined in Section 9.5.3 in the presence of external interference.

Table 9.3 Cost Function Coefficient Sets

Coefficient	Set 1	Set 2	Set 3
C_{Delay}	1	0.0001	0.0001
$C_{Autonomy}$	0	1	0
$C_{Coexistence}$	0	0	1

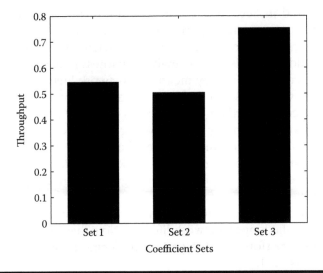

Figure 9.6   Throughput for the coefficient sets defined in Table 9.3.

Throughput and end-to-end delay in the three cases are shown in Figures 9.6 and 9.7, respectively. The results highlight that the adoption of a routing cost function that takes into account measured external interference (set 3) significantly improves both throughput and delay compared to the case where only UWB network internal status is considered in the route selection (sets 1 and 2).

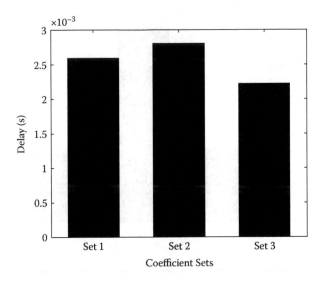

Figure 9.7   End-to-end delay for the coefficient sets defined in Table 9.3.

As discussed in Section 9.3, however, a cost function that takes into account only one specific aspect (e.g., power, interference, or delay) in route selection may lead to unfair energy consumption among terminals. In order to address this issue, we analyzed fairness in energy consumption for the three coefficient sets by measuring network lifetime.

Two cases were considered: absence of external interference, and presence of interferers according to the settings of Table 9.2.

Previous work on energy-aware routing suggested that a routing cost function that takes into account the residual autonomy of the nodes leads to high fairness and thus to long lifetime [15]. Results obtained in the absence of external interference, as presented in Figure 9.8, are in agreement with the above statement.

Set 2, which takes into account battery autonomy in route selection, leads in fact to the longest network lifetime. Note that set 3, in the absence of interference, performs end-to-end delay minimization, and leads to the same results as set 1.

The introduction of external interference according to the settings in Table 9.2 significantly affected the behavior of the three coefficient sets. Results in the presence of interference are presented in Figure 9.9, showing that set 2 is no longer the optimal choice in terms of network lifetime. The selection of nodes close to external interference sources causes, in fact, high power consumption in such nodes due to retransmission attempts, and reduces network lifetime. Set 3 is in this case the best choice, since it guarantees similar network lifetime while providing better network performance.

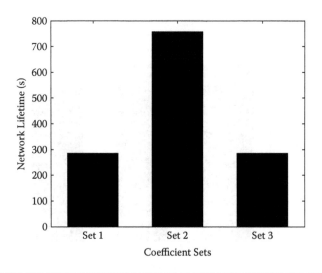

Figure 9.8 Time of first node death as a function of the coefficient set without external interference.

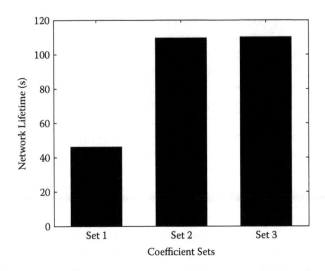

Figure 9.9 **Time of first node death as a function of the coefficient set in the presence of external interference.**

9.6 Conclusions

In this chapter we analyzed the problem of introducing a cognitive approach into the routing problem. Existing contributions on this topic have been reviewed, identifying the main solutions proposed to this problem for both wired and wireless networks. Next we focused on the ultra-wideband case, analyzing the problem of optimal choice of a multihop route in a network of low data rate UWB terminals of the IEEE 802.15.4a type. Based on this analysis we proposed a cognitive routing cost function that takes into account the status of both the UWB network and the external environment by means of additive cost terms weighted by a set of coefficients.

The adoption of different sets of coefficients allows for a straightforward tuning of the cost function. Different sets can be adopted to support traffic with different characteristics. Noninteractive data traffic, such as ftp transfers (for example), requires a high degree of data integrity but can tolerate high end-to-end delays. The cost function can be customized for this traffic class by increasing the relative weight of the reliability cost term, while reducing the weight of the delay term. Contrariwise, voice-like traffic can tolerate a relatively high PER, but poses strong constraints on the end-to-end delay. In this case the roles of the reliability and delay terms are inverted, with the latter term characterized by a much higher relative weight than the former one.

In the results shown in Section 9.5, we focused on a single traffic scenario, characterized by low bit rate connections at constant bit rate, and we investigated the impact of a subset of the cost function terms on network performance and lifetime by means of computer simulations.

The results show that with the introduction of information related to external interference, the routing strategy acquires the capability of adapting network behavior to the external environment, leading to a significant increase in network performance. Furthermore, the reduction of PER and retransmission attempts obtained by taking into account external interference sources in route selection contributes to achieving a fair power consumption among nodes, and thus a long network lifetime.

The proposed cognitive routing approach focuses on a subset of the actions defined in the cognitive cycle; in particular, the algorithm observes the network status, decides by selecting the best route given the observation data, and acts by modifying the routing tables of the nodes involved in the path. We foresee that the introduction of a learning capability based on the result of previous decisions can further improve the performance of the algorithm. A possible solution to introduce such learning capability is to allow network nodes to modify the cost function coefficients on the basis of the impact of previous routing decisions on network performance. In order to do so, however, several challenges must be addressed, including the definition of the feedback mechanisms and of the algorithm for coefficient tuning. Addressing such challenges will be the main subject of our future research activities on this topic.

References

[1] L. De Nardis and M. G. Di Benedetto. Cognitive routing in UWB networks. In *IEEE International Conference on UWB 2006*, pp. 381–386 (September 2006).

[2] IEEE 802.15.TG4a official Web page. http://www.ieee802.org/15/pub/TG4a.html.

[3] L. De Nardis and G. M. Maggio. Low data rate UWB networks. In *Ultra Wideband Wireless Communications*, pp. 315–339. New York: John Wiley & Sons, (2006).

[4] X. Jing, C. Liu, and X. Sun. Artificial cognitive BP-CT ant routing algorithm. In *Proceedings of International Joint Conference on Neural Networks*, pp. 1098–1103 (July 2005).

[5] E. Gelenbe, Z. Xu, and E. Seref. Cognitive packet networks. In *Proceedings of the 11th IEEE International Conference on Tools with Artificial Intelligence*, pp. 47–54 (November 1999).

[6] E. Gelenbe and P. Liu. QoS and routing in the cognitive packet network. In *Sixth IEEE International Symposium on a World of Wireless Mobile and Multimedia Networks*, pp. 517–521 (June 2005).

[7] R. Lent. Linear QoS goals of additive and concave metrics in ad hoc cognitive packet routing. *IEEE Transactions on Systems, Man and Cyberrnetics, Part B.* **36**(6), 1255–1260 (December 2006).

[8] G. Cheng, W. Liu, Y. Li, and W. Cheng. Spectrum aware on-demand routing in cognitive radio networks. In *2nd IEEE International Symposium on New Frontiers in Dynamic Spectrum Access Networks, 2007,* pp. 571–574 (April 2007).

[9] G. Cheng, Y. Li, and W. Cheng. Joint on-demand routing and spectrum assignment in cognitive radio networks. In *IEEE International Conference on Communications,* pp. 6499–6503 (June 2007).

[10] S. Krishnamurthy, M. Thoppian, S. Venkatesan, and R. Prakash. Control channel based MAC-layer configuration, routing and situation awareness for cognitive radio networks. In *IEEE Military Communications Conference,* vol. 1, pp. 455–460 (March 2005).

[11] B. Zhang, Y. Takizawa, A. Hasagawa, A. Yamaguchi, and S. Obana. Tree-based routing protocol for cognitive wireless access networks. In *IEEE Wireless Communications and Networking Conference,* pp. 4204–4208 (March 2007).

[12] M. G. Di Benedetto, L. De Nardis, M. Junk, and G. Giancola. $(UWB)^2$: Uncoordinated, wireless, baseborn medium access control for UWB communication Networks. *Journal On Mobile Networks and Applications.* **10**(5), 663–674 (October 2005).

[13] M. G. Di Benedetto, L. De Nardis, G. Giancola, and D. Domenicali. The Aloha access $(UWB)^2$ protocol revisited for IEEE 802.15.4a. *ST Journal of Research.* **4**(1), 131–142 (May 2007).

[14] V. Rodoplu and T. H. Meng. Minimum energy mobile wireless networks. *IEEE Journal on Selected Areas in Communications.* **17**(8), 1333–1344 (August 1999).

[15] C. K. Toh. Maximum battery life routing to support ubiquitous mobile computing in wireless ad hoc networks. *IEEE Communications Magazine.* **39**(6), 138–147 (June 2001).

[16] L. De Nardis, M. G. Di Benedetto, and S. Falco. Higher layer issues, Ad Hoc and Sensor Networks, in *UWB Communications Systems—A Comprehensive Overview,* pp. 1–59. Hindawi Publishing Corporation, (2006).

[17] M. G. Di Benedetto and G. Giancola. *Understanding Ultra Wide Band Radio Fundamentals.* Prentice Hall, (2004).

[18] IEEE 802.15.4a channel model final report, rev.1 (november 2004). Available at ftp://ieee:wireless@ftp.802wirelessworld.com/15/04/15-04-0662-00-004a-channel-model-final-report-r1.pdf (November 2004).

[19] M. G. Di Benedetto and G. Giancola. A novel approach for estimating multiuser interference in impulse radio UWB networks: The pulse collision model, *Signal Processing.* **86**(9), 2185–2197 (September 2006).

[20] IEEE 802.16e-2005 standard. Available at http://www.ieee.org/.

[21] A. Rahim and S. Zeisberg. Evaluation of UWB interference on 3.5 GHz fixed WiMax terminal. In *16th IST Mobile and Wireless Communications Summit,* pp. 1–5 (July 2007).

Chapter 10

Routing in Cognitive Radio Networks

Fei Hu, Mark Lifson, and Yang Xiao

Contents

10.1 Introduction

As we discussed in previous chapters, a cognitive radio network (CRN) is a network that performs communication through dynamic channels. Each channel is composed of different characteristics (corner frequency, bandwidth, power, etc.). The channel borders are defined by their frequency widths, possibly changing with time as primary and secondary users access/release portions of the spectrum. When a frequency range is unused, there exists a possibility for channel assignment. In a typical CRN environment a spectrum-agile radio (SAR) is used to scan a large frequency range and operate on a number of small channels within a specified frequency band. When the SAR detects an unused frequency band, called a "hole," it can begin sharing the frequency with other CRN radios, as long as a path exists from source to destination. Given a selection between optimal bands, the ones chosen for operation are based on variables such as the number of secondary accesses at the source or destination and the activity of licensed users.

Before the discussion of routing issues, we first review some basic terms. An interface is considered to be a wireless device with at least a half-duplex radio transceiver. A channel is the smallest unit of frequency that can be operated on and may be a small subset of the entire radio spectrum, a defined frequency band, or a union of many small subsets. Each channel has a specific bandwidth and is used to transmit and receive, and as such is the smallest frequency a transceiver can operate on. A mapping is defined as the commitment of one interface to one channel so that nodal communication can occur within the channel. A frequency band is a larger portion of the spectrum that may contain many channels. In CRNs there is what is called a spectral opportunity (SOP), which is a frequency band characterized by parameters such as bandwidth, center frequency, etc., and which may be available only for a short while and within a specific location before it is reclaimed by a licensed user. If a wireless device is equipped with a number of interfaces, then many channels can be mapped, thus increasing the potential throughput of the device. There can be interference between numerous channels, especially when many devices attempt to operate on a certain frequency band. The solution to reducing interference is to either increase the frequency band and/or filter the channels so that they don't overlap. The overlap is a result of the transceivers' operational range being nonideal and propagating signals outside of the defined boundaries. A multichannel network is considered to be any network that uses more than one channel to access nodes. Each node (or wireless device) may know either all of the channels, as in static networks, or a small subset of the total channels, as in CRN. A multihop network is characterized by the type of transmissions that are occurring: if the source and destination are separated by at least one node it is considered a multihop route, and if most communication on a network occurs with this nodal separation then it is considered a multihop network.

The networks reviewed in this chapter are categorized into two architectures: either centralized or distributed. A centralized network contains a single entity with massive computing power and memory, which is used to execute spectrum allocation, access procedures, and resolve communication from the surrounding nodes in the cluster. A distributed network requires all nodes to share spectral information and coordinate channel access. As opposed to a single entity, all nodes must disseminate information to each other. A distributed network is used when a centralized infrastructure is not preferable or deployable.

Additionally, the spectrum allocation behavior of the network can be either cooperative or noncooperative. In a cooperative network all nodes share interference data, so that any single node, before beginning communication, can check to ensure that no interference exists. A non-cooperative network exists when the nodes act selfishly and transmit without regard for interference, and is used in situations that require low communication overhead (such as the dissemination of interference data).

Last, the access technologies adhered to by the nodes can either be overlay or underlay. An overlay spectrum sharing scheme is when nodes transmit only on channels that do not contain any licensed user access; this technology ensures that there will be no interference with licensed users. An underlay spectrum sharing scheme occurs when nodes attempt to transmit information below the noise floor of a licensed user. Since anything above the noise floor is used for licensed user access, the underlay scheme allows for more frequency bands to be exploited. However, the algorithms reviewed in this chapter will consist solely of overlay access schemes, to avoid the unnecessary complexity of underlay methods.

The future for CRN will lie with different combinations of architectures, allocation behaviors, and access technologies. For example, an optimal military CRN network would have a distributed, noncooperative, underlay protocol, which would enable military nodes to transmit on any frequency while simultaneously managing the network without a susceptible central entity. A commercial application would likely be a centralized, cooperative, overlay protocol, to ensure fairness through the cooperative behavior and minimize the component cost associated with an underlay and distributed system.

In almost all routing algorithms, the best route between nodes is considered to be the one with the best overall score. A score can be calculated a number of ways and includes totals from individual nodes along a path. A combination of metrics can be used, such as link load, number of available channels, number of free interfaces, etc. Most frequently, the algorithms reviewed in this chapter attribute the best path to the one with the shortest delay. The two most common nonzero delays in a CRN are backoff delay and switching delay. When nodes transmit on the same channel, there are normally collisions. These collisions incur a backoff delay at the nodes that were attempting to transmit before collision; therefore the packets have to be retransmitted at a different specified time. The time it takes to retransmit a collided packet depends on the algorithm that is used, such as choosing a random time or an exponential backoff scheme. The other type of delay is called a switching delay. When a node has more than one interface that changes frequency bands, or when the interface changes between channels, the time it takes to perform either operation has a specific latency or delay. In CRNs, there is a tradeoff between these two delays, since transmitting on multiple channels incurs a switching delay while transmitting on a single channel incurs a backoff delay.

In this chapter, we will focus on the design of a routing layer protocol for a CRN, covering topics such as path discovery from source to destination, choosing the optimal path in a multihop network, servicing flows (scheduling), and finding the optimal network capacity.

The problem of locating paths is complicated due to the variance in frequency bands along with their respective parameters such as center frequency and bandwidth. Additionally, in a multihop network the channels available at the source node may not be the same as the channels available at the destination node. Therefore, a single path may have to be formed across multiple channels operating within different frequency bands and at different bandwidths. However, when a path is formed, any deviation in neighborhood interference or primary user access will require the path to be recalculated and possibly dropped. In this instance it would be necessary for a new path to be discovered and communication reestablished. Due to these particulars, routing protocols tend to use on-demand routing schemes.

Once a number of paths are formed from source to destination, the node must intelligently calculate the best route. This problem becomes apparent in CRNs due to the large number of variables affecting the optimal route. Many protocols fixate on a few variables and compute an algorithmic result to differentiate between multiple paths. Therefore it is necessary to represent the problem with an accurate model. It should be noted that using "hello" packets to find the shortest path may not necessarily be the best approach, as will be proven in the upcoming sections. In Sections 10.2 and 10.3 we summarize the protocols that utilize different metrics to determine the optimal route.

After communication has been established between source and destination, a node will receive data from multiple sources. Since the flows may enter on multiple channels, the node must serve the requests in an intelligent fashion to reduce switching delays. A scheduling algorithm is summarized in Section 10.3.

Section 10.3 will also introduce research that has been performed on multichannel multihop algorithms, and their influence on current routing protocols. We will present slight modifications to these algorithms that are necessary in order to solve the CRN routing problem. Additionally, in Sections 10.2 and 10.3, we will discuss research results that are specific to CRN routing design. Finally, Section 10.4 will compare the similarities of the routing algorithms, and outline the characteristics necessary for a routing layer.

10.2 Routing in Multichannel Multihop Non-CRN Networks

In this section, we first introduce general routing strategies in multichannel non-CRN wireless networks. The next section investigates specific routing requirements for multichannel CRNs.

In [2] the authors propose a method to exploit multiple channels that includes a strategy for interface assignment in a multiple-channel environment, and an algorithm to select high-throughput routes. Similar research was done in [3] that operated on general wireless mesh networks, producing results with IEEE 802.11 interfaces.

The structure of 802.11 is similar to CRN in that there are many available channels. Thus, there is a similar opportunity to improve network throughput as was introduced for CRN. However, the channels in the 802.11 standard are fixed, and the characteristics well known, i.e., center frequency, bandwidth, etc., unlike in the case of CRN. Additionally, in [3] the assumption is made that the nodes are not mobile and the traffic characteristics do not change very frequently (on the order of hours or days). The ideas presented in [2] and [3] improve network capacity by employing interface switching. In a wireless network, each node has at least one interface. Two nodes can communicate if they each assign an interface to the same channel. It would require multiple interfaces and constant interface reassignment for nodes to communicate in parallel. Research [3] attempts have been made to employ a switching protocol that is based on link load and link capacity. For example, if a highly intensive but short-duration request is made (e.g., HTTP), then the interfaces will be switched frequently and often, causing the throughput to be lower. This will not perform as well as [2], and in particular will not work for CRN's.

Instead of using link load and link capacity for channel assignment, the authors in [2] suggest a switching protocol where nodes have at least two interfaces under their control; at least one of those interfaces is "fixed," and the remaining one can be switched to as necessary. Additionally, the protocol requires any neighbors around a node to be aware of the node's assigned interfaces in order to communicate in an on-demand fashion.

Fixed interfaces are assigned to a fixed channel so that a particular node is permanently configured to send and receive. It is preferable if the neighboring nodes are assigned to differing channels in order to maximize network coverage. Each node contains two lists, *NeighborTable* and *ChannelUsageList*. The *NeighborTable* has entries for every neighbor and their fixed channels, while the *ChannelUsageList* tracks the channels and the number of interfaces affixed to that channel. When the network is first formed, each node chooses a random channel to assign an interface to and transmits hello packets (containing information on the assigned channel)

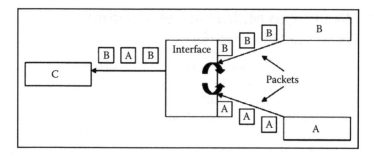

Figure 10.1   Channel switching delay.

on every single available channel. The nodes thereafter update their lists periodically and scan the *ChannelUsageList* for the number of occurrences related to their own fixed channel. If there are a large number of nodes utilizing the same channel, then a particular node will reassign its fixed interface to a less used channel. This reassignment does not occur as frequently as for the switched interfaces. The switching protocol in [2] suggests that switching interfaces be assigned based on the number of nodes using a channel. This simplifies the protocol, and at the same time avoids the high switching frequency associated with channel load and the low switching frequency associated with assigning based on network topology. Once the channels are assigned, each node must have a scheduling algorithm to avoid excessive switching delays. For example, if a node C needs to receive packets from A and B in alternating order, as shown in Figure 10.1, and there is only one switched interface available at C, then the node will be incurring a switch delay after every packet. To alleviate this delay, a channel can be assigned to a third state we call "semi-fixed" where the interface is active until either a timer expires (*MaxSwitchTime)* or at least a certain number of packets have been sent (*MaxBurstLength)*. Therefore, the node is bound to a channel for at most *MaxSwitchTime*. In Section 10.3 we will see a scheduling algorithm for CRN that is closely related to this method.

The routing protocol proposed by [2] operates on the previously mentioned switching protocol and attempts to find the highest-throughput route using the metrics of channel diversity and interface switching cost. The channel diversity is determined by the number of routes that utilize multiple channels and the interference of each link in the route. A route that incorporates many channels will receive a lower cost than a route with fewer channels. However, if multiple *links* in a route utilize the same channel, only one of those links can transmit at a time and will reduce the overall throughput. The diversity metric for a link is determined by the number of times a path utilizes the channel dedicated to that link. The total channel diversity is calculated as the summation of all the link diversity costs in the path, with a suitable algorithm to prevent double counting. In essence, if a specific route uses a channel multiple times, it will receive a higher diversity cost than a route that better distributes its links across channels. This method is straightforward and efficient because the size of each channel is assumed to be the same and the throughput at interfering nodes is simple to calculate. As we will show in Section 10.3, the fixed channel size is not valid for CRNs.

When a node tunes an interface to its destination's fixed channel, a delay is incurred. To calculate the best route that uses the fewest interface switches, the authors define an *active channel* as a switchable interface that is used on a channel for a long period of time. A channel is determined to be active if its *ChannelUsageFraction* is above a threshold. For example, the threshold at a node A is 0.5 and it contains an interface of X and Y. X transmits 10 packets on channel 1, and interface Y transmits 5 packets on channel 2. Then X's *ChannelUsageFraction* would be 10/15 and be considered active, while Y's *ChannelUsageFraction* is 5/15 and inactive. Therefore, a channel is deemed inactive if an interface is tuned to a particular channel but shows little or no activity. The switching cost is calculated as zero if a source's interface is active or fixed to the same channel as the destination, which may be either active or fixed. Otherwise, the switching cost is calculated as

$$switching\ cost = \frac{switching\ delay}{estimated\ packet\ transmission\ time}, \quad (10.1)$$

where the switching delay is the time to switch, and the packet transmission time the amount of time to transmit an average packet. Once the metrics have been defined, the authors in [2] propose a combined weighted metric of number of hops, diversity cost, and switching cost. The authors also suggest an on-demand route discovery approach that is similar to DSR and that parallels the CRN scheme we will summarize in Section 10.3. The method first requires a node to request a route discovery by sending an RREQ (route request) packet. Attached to the packet is the channel index and the switching cost associated with that index. The intermediate nodes then attach their index and switching cost, compute the diversity cost (for themselves), and transmit the packet if they have not seen the request previously or if the distance is shorter than what is currently stored. Once the destination receives the RREQs, it will then transmit an RREP (route reply) back to the source every time it encounters a route containing a lower cost than was seen before. This packet structure will be reviewed again in Section 10.3, as one of the algorithms discussed there uses RREQs and RREPs in its route discovery.

10.3   Routing in Multichannel Multihop CRN Networks

There are few prescribed solutions for routing in a CRN network. The research presented in [4–7] make different assumptions on the layering method used, which we divide into two categories: coupled and decoupled. On this basis we will divide the algorithms detailed in [4–7] into their respective categories and in the remainder of this section explore the advantages and disadvantages of both.

10.3.1   Decoupled Designs

In a typical networking situation, the MAC layer and the routing layer perform separate tasks. The MAC layer is responsible for ensuring delivery between two adjacent nodes, and the routing layer is the "overseer" in which the entire route is calculated and passed to the MAC layer. In this situation, the routing and MAC layers perform their respective functions independently of one another, and this is known as a decoupled design.

A CRN must first sense the available spectrum opportunities before it can begin designating a route. In a decoupled design, this function is performed and evaluated in the MAC and physical layers. The MAC layer then performs interface assignment; however, the methods discussed in Section 10.2 fall short of solving the problem in CRNs. First, it is assumed in [2] that the channels are static and fixed in bandwidth and power, whereas in a CRN these assumptions are inherently non-existent. Second, an assumption is made that many or most of the channels are assigned interfaces, which is impossible due to the variability and number of channels in a CRN. Third, static and fixed channels have specific interference properties that are simple to calculate, whereas CRNs tend to have more conflicts. Last, fixed channel sizes allow for a static packet size, whereas in CRNs it may not be feasible to switch at the packet granularity. The work in [4] proposes a solution to the previously outlined problems by using a separate MAC function to determine interface assignment.

The layered graph model in [4] can be used to reduce interference between nodes and utilize the network space efficiently when assigning interfaces. The structure of the layered graph model is composed of a set of vertices connected by weighted edges. The orientations and weights of these nodes and edges are the key ingredients used in finding the best interface assignments for a particular route. In this design the information that is required to assemble this graph is assumed to be disseminated by a single central entity (centralized). The information can include items such as the current topology, link states, channel capacities, etc. Theoretically, the authors state that simple modifications could allow for distributed systems. Therefore the authors in [4] attempt to define a structure that finds

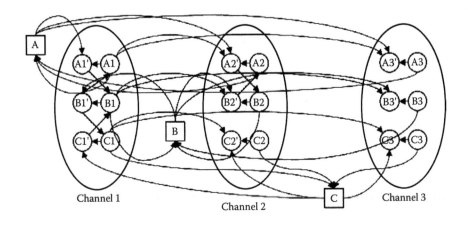

Figure 10.2   Network layer model [4].

an optimal path, and then assign interfaces along this path by using a dynamic graph that changes based on the spectral opportunities. In order to demonstrate the accomplishments in [4], it is necessary to understand how the graph appears before we begin discussing the technicalities. Figure 10.2 shows how a fully linked layered graph appears.

The graph contains layers, where the total number of layers is denoted as N. Each layer represents a channel available to the system. Each node has a subnode at every channel, and therefore has N subnodes, where the ith subnode is available at the ith layer. Every subnode has a directed edge ending at the node with which it is associated. Therefore every node has N inputs, one from each subnode. Every node then has N outputs, which are directed towards auxiliary subnodes denoted by a prime ($'$). An auxiliary subnode exists for every subnode and is used as a buffer between vertical edges (defined below). The auxiliary subnode has two inputs, one from the primary node and one from its associated subnode, which is specifically called an internal edge. The outputs of an auxiliary subnode are edges that connect to every subnode in the same layer, and these edges are known as horizontal edges. Finally, every subnode has N inputs, receiving data from every ith subnode, and it has N outputs, sending data to every ith subnode. These subnode edges are called vertical edges.

When a path wishes to change layers (change channels), the data is sent to the new layer's auxiliary node. This is designed so that no node can take more than one path when changing a signal. For example, if a node wished to forward its data onto a different channel, it would first forward the data to the auxiliary node of the new channel, and the auxiliary node would then push the data horizontally to the next node's subnode. The inverse is also true: if a subnode wishes to forward data to a new subnode on the same channel, then the data would be transmitted to the auxiliary node

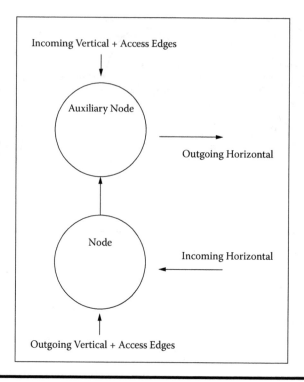

Figure 10.3 Node in layer structure.

on the same channel before being pushed horizontally. This encourages an auxiliary node to act as a multiplexer, selecting the data to push horizontally from nodes at other channels as well as the data from the nodes in the same channel, shown in Figure 10.3.

The weights are modified for the particular system being simulated/ implemented. It is wise to assign a negative value for vertical edges, as this will increase the throughput of the system. Internal edges are given a large positive value to prevent same-layer hopping (no horizontal movement). Access edges are given an extremely large weight to prevent the node from being used as a subnode. Finally, horizontal edges are given a dynamic variable that most likely will change with some metric such as data size. An example of weight assignments is provided in Table 10.1.

Once the structure of the graph is outlined, it is necessary to implement a method to construct the graph at a network level given the disseminated information from a node.

Construction of layered graph:

1. On every layer *i*, use a horizontal edge to connect every pair of subnodes on that layer.

Table 10.1 Weights for Algorithm

Edge	Cost
Access Edges	900
Internal Edges	30
Horizontal Edges	9
Vertical Edges	−9

2. If there is more than one free interface available at a node, then any two channels can communicate when the free interfaces are assigned. Therefore, connect each node's subnode to each other with a vertical edge.

3. If there is exactly one free interface available at a node, then every active subnode in the node should be mapped to every inactive subnode in that node with a vertical edge. This is because the free interface allows a mapping to activate any single inactive node if the need arises.

4. Lastly, all active subnodes in a node should be connected to each other with a vertical edge, since their interfaces are already assigned and communication can occur. The construction of the graph now includes all horizontal and vertical edges. Each inactive node can be assigned one free interface (or the node is disabled).

Figure 10.4 illustrates the four different combinations of free and assigned interfaces.

Now that the graph model is created, it is necessary to initiate a method

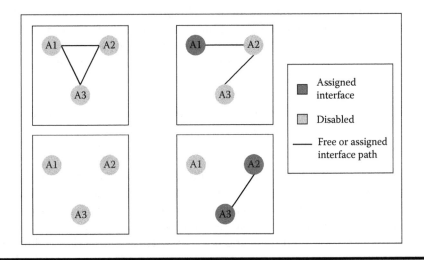

Figure 10.4 Interface patterns.

to assign the available free interfaces to channels. Note that the spectrum broker (the entity in a centralized system) induces a path-refresh every time a new SOP (spectral opportunity) is detected. The interface assignment is performed as follows.

A source node begins by determining routes using the most recent graph; it uses the shortest path method with the weights as outlined. As we will show with a different algorithm [5], this approach may not always yield the optimal solution. However, the simple shortest path route discovery is performed as follows:

Routing:

1. Sort all node pairs (source and destination) by some metric. This can be link capacity, traffic load, etc. Start with the heaviest or most intensive calculation.
2. While there exist node pairs that haven't had a calculated path, use the predetermined metric to calculate a proper path between source and destination. Assign the interfaces for the particular path.

The interface assignment in the graph layered model is performed completely independently of the routing, as required for a decoupled design.

Interface along a path:

1. For each inactive subnode along the calculated path, assign a free interface from its corresponding node and mark the subnode as active.
2. If after assignment there is only one interface left, then any inactive node can be assigned this interface and thus cannot communicate with the other inactive nodes. Therefore delete the vertical edges between any two inactive subnodes, as the interface cannot switch between the two channels the subnodes reside on.
3. If there are no more free interfaces, then there cannot be any connections with inactive subnodes and they must be disabled in both their vertical and horizontal edges. However, the horizontal edges should be stored so that when a free interface becomes available at a later point in time, the mappings can be restored.

Once the free interfaces have been assigned, the route will stay static until a new SOP is either detected or removed. If a node wishes to assign an interface in a distributed system, then the problem becomes more complex, and is only briefly mentioned in [4]. In a distributed system it is possible for a node A to calculate the interface assignments at the same time a node B begins transmission with the same interfaces. Thus there exists a situation in which B's transmission is invalidated as the interfaces are switched for A's purpose. Therefore it is the responsibility of the node to keep updating its tables, neighbors, and central entity until the "system" reaches a suitable steady state.

However, the solution in [4] requires a centralized environment containing massive amounts of memory and computing power, and as such can benefit from a distributed architecture.

As mentioned earlier, the shortest path solution when using weighted edges may not provide the optimal solution. In [5], another decoupled method is proposed which finds a Pareto optimal solution while also addressing the need for data-sensitive routing. The Pareto optimal solution performs data dissemination and route determination with the metrics of data quantity, channel capacities, spectrum availability, link propagation time, and secondary user link occupancy probabilities. However, the method in [5] differs from that in [4] because it focuses on choosing the best route, ignoring interface assignment and interference delays. Two suitable algorithms are presented for both single source to single destination and single source to multiple destinations (broadcast and multicast).

The objective of the algorithm in [5] is to find a path with the smallest mean capacity and shortest propagation delay, both of which add to the total delay, denoted by T_d:

$$T_d = propagationDelay + \frac{totalData}{dataCapacityofPath} \qquad (10.2)$$

Using this formula, the algorithm attempts to calculate a single dominant path, which uses a Pareto solution to find the dominant path.

Dominant path:
A path p is considered to be dominant over a path q:

1. if the propagation delay from source to sink on path p is less than or equal to that on path q for the same data, and the mean capacity is less for path p than path q, or
2. the delay from source to sink on path p is less than that on path q for the same data, and the mean capacity for path p is less than or equal to that for path q.

This means that the dominant path will be better in one aspect and at least equal to or better than another path in the other. Therefore it is not the optimal solution that is found, but rather the optimal Pareto solution. Although this simplifies the problem drastically, filling in the details for the mean path capacity requires some already estimated calculations for the channel use on every link. If this information is available prior to path calculation, then the use of a simple formula gives the probability that a secondary user can transmit on a free orthogonal channel. The following formula is defined as the admittance probability:

$$P_i = 1 - \prod_{k=1}^{\psi} \left(1 - P_k^i\right),\tag{10.3}$$

where P_k^i is defined as the probability a channel i is available at node k, and ψ is the number of orthogonal channels available to the node. From this, the mean capacity can be determined by multiplying the entire link capacity by the probability that any channel will be available. For example, if all the channels were available on a particular link, then P_i would be 1 and the channel would have the entire link capacity.

Therefore, the path chosen is a direct result of the data required to be sent from source to destination, which enables data-sensitive routing. This algorithm can also be extended to data-adaptive multicasting and broadcasting using spanning trees instead of link paths. The algorithm is computed the same way using a Pareto optimal solution. The authors' approach provides a new and interesting method for data routing for a CRN that has not been explored in the past. Both methods are heavy in discrete math and an eager reader is encouraged to explore the paper, as it solves many more complexities in routing. However, the algorithms presented in [5] do not model interference and as such cannot account for unforeseen licensed user access. Because of the randomness in primary user access, there is a fundamental problem with calculating the optimal path before transmission. If the path chosen was critically altered during transmission, then the information would be lost. Additionally this method requires a lot of information from the MAC layer and may benefit from a coupled design, introduced in the next part.

10.3.2   *Coupled Designs*

In order to address end-to-end optimizations, communication between the MAC and networking layers must occur. These are known as coupled designs. This hybrid layer assumes control over final route calculation and additionally schedules the single-hop transmissions between every node pair. When the layer provides such precise control, the quality of service improves dramatically [6]. Having the ability to determine single-hop transmissions allows for better prediction of link state and route selection. This improvement is due to the instantaneous route recalculation that occurs whenever the spectrum state changes. However, the trade off is increased overhead and algorithm complexity. Therefore, all the algorithms incorporating a coupled design attempt to minimize communication overhead and algorithm complexity.

The first simple coupled method is [6], which first invokes route discovery procedures for determination on current link quality. Once a number of routes have been established, the algorithm attempts to calculate all possible channel-routes and use a performance metric to determine the throughput performance for each full route (from source to destination). The source node then notifies each node in the route and all nodes share time schedules to predict link quality. However, the paper [6] restricts its algorithms to centralized architectures where a single entity has knowledge of the network topology. This main entity provides the computational power to determine the route and channel decisions. This design, however, breaks down the process into two complete algorithms: conflict-free scheduling and route-and-channel selection. The conflict-free scheduling algorithm removes the possible combinations (or vertices, to be shown later) where the channels conflict with one another once a route has been selected. The route selection is determined from the second algorithm. This algorithm finds the set of routes that achieve a data connection from source to destination—and invokes the first algorithm to remove conflicts. Both of these algorithms are outlined below.

Route and channel selection:
1. Find all possible routes to destination.
2. Find all possible combinations of channels on each route.
3. Execute conflict-free scheduling for each route-channel combination (full path).
4. Choose the best path with the outlined scheduling (determined from the conflict-free scheduling).

Conflict-free scheduling:
1. Make a graph where each single-hop link is a vertex. If two vertices have a conflict with regard to channel usage, an edge is drawn between them. This is known as a conflict graph. A channel usage is conflicted when two links are in close proximity and on the same channel, or if two users share a single link.
2. For each vertex (single hop), assign a weight that is equal to the number of edges it is associated with. Thus the weight is proportional to the number of conflicts a single-hop link has.
3. Find the weighted maximum independent sets of the conflict graph, where a maximal independent set is an independent set of vertices, i.e., a subset of the total vertices, such that no two vertices are joined by an edge. Delete the vertices and associated edges with the independent set.
4. Stop if the graph is empty, or move to the next node and repeat steps 1–3. A deleted independent set is a subset of vertices (or single-hop links) that can transmit without conflict. The number of independent sets is known as the *chromatic number* of the conflict graph.

The system throughput is inversely proportional to the number of edges available (*chromatic number*). The *chromatic number* should be calculated for all possible routes. However, this method increases exponentially in computation time as more nodes are added; therefore it is necessary to approximate the *chromatic number* using a maximum clique number. The method in [6] proposes to replace the method for conflict-free scheduling with an approximation function when the number of nodes is large.

Approximation of chromatic number:

1. Assign a weight to each vertex.
2. For each vertex in the conflict graph, delete all the vertices that are not connected to it.
3. Construct a set with the first element equal to the first vertex.
4. Add remaining vertices to the set, in order of largest weight.
5. Stop when all vertices are connected.
6. N_i = the number of elements in the set.
7. Repeat step 2 until all vertices have been iterated through.
8. The maximum clique number is now the largest number of elements in any set provided by a single vertex.

It should be noted that this algorithm attempts to calculate the shortest path based on channel capacity and link propagation time. The accuracy of this result has been proved to not always provide the optimal solution [5]. Nonetheless, the results of the experiment performed in [6] prove that their collaborative design outperforms a decoupled design by as much as 33% when three channels are assumed to exist. However, the collaborative design was assumed to utilize a central planner, which may degrade performance when more nodes and more channels (i.e., more overhead) are added to the system.

For a route to be calculated, it is necessary to constantly update nodes with disseminated information while satisfying requests for routing. The algorithm in [7] is a coupled design that uses on-demand routing, which means the routes are discovered at source request before transmission, and which is ideal for a CRN, as the topology changes quite often. The authors focus heavily on this type of routing and borrow from ad hoc on-demand distance vector (AODV) routing. The on-demand routing algorithm is an addition to AODV but with modified RREQ and RREP packets. There is also a focus on scheduling and its influence on route selection. A novel multiflow multifrequency scheduling algorithm is introduced to service flows entering at a node while quantifying backoff and switching delay at every node between a source and a destination. This information allows the selection of optimal frequency bands which is then attached to an RREQ message packet. The matching nodes attach their IDs and then forward

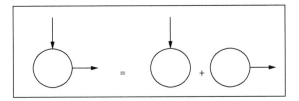

Figure 10.5   Flow decomposition.

the packet and its reply. The proposed algorithm is a coupled design because the on-demand routing is performed at the routing layer and this aids scheduling performed in the MAC layer.

The multiflow multifrequency scheduling algorithm first queues the flows by incoming and outgoing frequency bands. The node is always performing round-robin service based on the active frequency bands. Active frequency bands are a subset of the total bands that are available and are the ones used for data communication. If the servicing was performed by flow then there would be additional switching delays due to one flow possibly sharing the same frequency band as another. The total delays experienced from source to destination are node intersection delays and path delays. At each intersection of a node in a path there exist switching delays. A frequency switch incurs a nonzero switching delay. Additionally, when transmitting on the same frequency band, the node also experiences backoff delay. Therefore the node is constantly reducing both delays by modifying its band switching order accordingly. Figure 10.5 shows how flows are decomposed.

The two delays, switching and backoff, can be calculated using thoroughly tested procedures. The delay at a node is

$$D_N = \textit{switching delay} + \textit{backoff delay}, \tag{10.4}$$

where the switching delay is defined as

$$D_s = 2k(|Band_M - Band_I|). \tag{10.5}$$

The switching time is indicated within absolute value, and it references the bandwidth, as two pairs of bandwidths may have a different time. The k is a positive constant that indicates a factor of two because a switch from one channel will also require a switch back. The backoff delay is dependent on the method used to back off when a collision occurs; the definition, however, is not necessary to complete this discussion, so an eager reader is encouraged to examine [6] in its entirety.

It is obvious that when a new flow is added as a new frequency, the switching delay for the node increases. Conversely, if the flow was added to an existing frequency band, the backoff delay for the node would increase (as collisions increase). These two methods for delay add together to become the intersection delay for that particular node.

Additionally, there exists a path delay between nodes:

$$D_P = SwitchingDelay + BackoffDelay \qquad (10.6)$$

Similar to the case of the intersection delay, there exists a switching delay between every pair of nodes on a path. The formula looks very similar, but note that the band is for each node, not each frequency within a node:

$$D_s = \sum_{j=1}^{H} k(|Band_j - Band_{j+1}|) \qquad (10.7)$$

where H is the total number of nodes in that particular path. Similarly, the backoff delay can also be calculated for the route. Now, any node m can assign an appropriate frequency band that contains the smallest delay.

To begin transmission, a node transmits a route request packet (RREQ) similar to that found in AODV. However, in this case, the SOP information is attached to the packet and forwarded by the nodes that can identify with the route. When the destination node responds, it automatically knows the links that formed the route from the source. The destination then forms a packet that contains the bands to be used and leaves the intermediate nodes to cross-reference the RREQ with the RREP to assign bands for transmission.

Using on-demand routing alongside multifrequency multiflow scheduling can help disseminate information to path nodes in a distributed system. The delay modeling is also simple in that it uses information easily available to any node.

10.4   Discussion and Conclusions

In this section we will compare and contrast the methods based on their performance.

Many of the methods explored in the previous section were solutions to particular routing problems for CRNs. However, similarities existed across the different algorithms, and it is important to note their respective performances. Although the simulations performed in papers [4–7] were unlike each other, some conclusions can still be drawn.

The paper [4] made interesting claims as to how a CRN should be organized. Using a novel approach for communication between nodes, the method proposed finding the optimal path based on weighting the vertices and assigning interfaces as they become free. However, the scale of this global topology requires a central entity, or it would suffer from massive overhead in a distributed system. Additionally the computations necessary every time an SOP is detected may lag the system.

In [5], the authors took a different approach in which the best path was defined as being the one with the least total delay. The breakdown of the system also allows data to be routed according to data sensitivity. However, the assumptions made in the paper may limit the system. For example, having the number of orthogonal channels or the link load at particular nodes requires very high nodal communication, and as such exhibits high overhead. Also, interface mapping is not explicitly discussed, nor is licensed user interference. These reasons, coupled with heavy mathematical calculations, may make this algorithm impractical.

The authors in [6] make an attempt to integrate the MAC layer with the routing layer. The focus of [6] is on how best to service flows while minimizing the interference amongst nodes. Once the interference has been removed, the best path is chosen as the one with the highest channel capacity and the least link propagation time. This method is very abstract and requires more details to be implemented.

The last algorithm, [7], focuses on choosing a path solely on the basis of delay. In addition, the information detected at one node can be passed to others through packet attachment, an approach that is very close to AODV. Therefore, this method can be implemented with little work if the information for switching delay and backoff delay can be disseminated accordingly.

The breakdown for these methods is displayed in Table 10.2.

The algorithms are scored on a ranking of 1–4 (since there are four algorithms). According to the simulation results depicted in papers [4–7], the best methods are those that are coupled. Because the spectrum sensing must occur throughout the route calculation so as not to interfere with primary users, it is important that the node-to-node and source-to-destination functions be combined. An efficient algorithm should consider four factors:

1. *Performance metrics*: The quality of the metric is very important. The metric is what determines the system's reliability. Therefore, there is a tradeoff between the number of metrics and the size of

Table 10.2   Ranking of CRN Routing Methods

CRN	Metric	Implementation	Scalability	Overhead
[4]	2	2	3	4
[5]	3	4	4	3
[6]	4	3	2	2
[7]	1	1	1	1

the system. If an algorithm requires many metrics, the overhead increases. Also it should be noted that a greater number of metrics does not necessarily imply better system performance; the quality is what's important (or the relationship of the measurement to the CRN).

2. *Complexity*: The system should be simple to implement. In fact, if it can be piggybacked on an already well-known scheme (such as AODV), it will allow the design to come to fruition. In particular, basing algorithms on AODV and learning from protocols such as those discussed in [2] and [3] will lead to reliable and well-understood results.

3. *Scalability*: The system should be scalable to the point where an algorithm runs independently of the number of nodes.

4. *Routing overhead*: The lower the routing overhead, the greater the number of channels that can be used in a CRN, increasing system throughput performance.

It is important to note that many of the algorithms presented in [4–7] are in preliminary stages. The merger of each algorithm can be achieved to provide for a more powerful routing layer. It will soon be necessary to find a solution that solves the characteristics outlined in this chapter.

References

[1] I. F. Akyildiz, W. Y. Lee, M. C. Vuran, and S. Mohanty, "Next generation dynamic spectrum access/cognitive radio wireless networks: A survey," *Computer Networks*, 50:2127–2159, 2006.

[2] P. Kyasanur and N. Vaidya, "Routing in multichannel multi-interface ad hoc wireless networks," Tech. rep., UIUC, Dec. 2004.

[3] A. Raniwala, K. Gopalan, and T. cker Chiueh, "*Centralized channel assignment and routing algorithms for multichannel wireless mesh networks*," *SIGMOBILE Mob. Comp. Comm. Rev.*, **8**(10.2):50–65, 2004.

[4] C. Xin, "A novel-layered graph model for topology formation and routing in dynamic spectrum access networks," *Proc. of IEEE DySPAN 2005*.

[5] R. Pal., "Efficient Routing Algorithms for Multi Channel Dynamic Spectrum Access Networks," *Proc. of IEEE DySPAN 2007*, pp. 288–291.

[6] Q. Wang and H. Zheng. Route and spectrum selection in dynamic spectrum access networks. In *IEEE CNCC*, 2006.

[7] G. Cheng, W. Liu, Y. Li, and W. Cheng, "Spectrum aware on-demand routing in cognitive radio networks," *Proc. of IEEE DySPAN 2007*.

CROSS-LAYER CONSIDERATIONS IN COGNITIVE RADIO NETWORKS

IV

14

CROSS-LAYER
CONSIDERATIONS
IN COGNITIVE
RADIO NETWORKS

Chapter 11

Impacts of Cognitive Radio Links on Upper Layer Protocol Design

Qixiang Pang, Laxminarayana S. Pillutla, Victor C. M. Leung, and Vikram Krishnamurthy

Contents

11.1 Introduction

Radio frequency (RF) spectrum is a critical resource for many services that people across the world rely on for their safety, communications, employment, and entertainment.[1] The current static spectrum management policy allocates dedicated frequency bands to specific wireless services. The RF spectrum thus allocated can be vastly underutilized, a fact confirmed by the spectrum measurements in the major cities of America. The US Federal Communications Commission (FCC) found that spectrum access is an even more significant problem[2,3] than physical scarcity of RF spectrum, largely due to the current spectrum allocation policy that limits the ability of users with immediate needs to access available RF spectrum that has already been allocated to another user or service, even though the latter does not have an immediate need to use the spectrum.

These findings and the need for more efficient utilization of the RF spectrum lead to the proposal of *cognitive radio* (CR) technology[4-6] as a new mechanism for flexible usage of spectrum.

A cognitive radio is a radio that can change its transmitter parameters based on what it learns from the environment in which it operates.[5] CR is envisioned as a technology that allows unlicensed (secondary) users to operate in underutilized licensed frequency bands in an intelligent way without intruding on the privileges of licensed (primary) users.[5] A CR-enabled secondary user is capable of identifying and utilizing available channels in the RF spectrum. The ability of a user to change its frequency of operation is commonly referred to in the literature as *dynamic spectrum access* (DSA).[8-11]

Due to the specific characteristics of CR technologies, wireless CR networks need to address technical challenges that are different in many aspects from those of traditional infrastructure and ad hoc wireless networks.[3] So far, the majority of research on CR networks is concentrated at the physical and medium access control (MAC) layers of the protocol stack (such as spectrum sensing techniques and spectrum sharing solutions). However, the impacts of CR technologies on upper layers have not yet been widely explored. The spectrum-aware transmissions of the CR networks can affect the performance of higher layer routing and transport protocols.[12,13] Therefore, it is necessary to investigate the issues of upper layer protocol design caused by the new features of CR networks. A routing or transport protocol that is well matched to the CR link can substantially enhance the end-to-end performance (e.g., throughput and delay) and bring benefits to the end users of a CR network. Consequently in this chapter we study the impacts of CR links on the routing and transport layer protocols.

The rest of this chapter is organized as follows. In Section 11.2 we present the characteristics and representative architecture of CR networks. In Section 11.3, we discuss the technical challenges of the routing layer in CR networks and present a survey of existing routing solutions. Some open research issues on routing protocols are given. In Section 11.4, the technical challenges seen by the transport layer are discussed. The effects of DSA on the transmission control protocol (TCP) performance in CR networks are presented. Section 11.5 concludes the chapter.

11.2 Characteristics of Cognitive Radio Networks

Figure 11.1 shows a representative architecture of a CR network comprising CR-enabled nodes sharing the RF spectrum with licensed radios. The licensed radios belong to the primary users of the RF spectrum, and the CR-enabled nodes belong to the secondary users. To more efficiently utilize

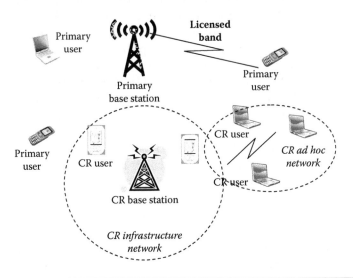

Figure 11.1 Cognitive radio network architecture. The CR users coexist with the primary users. The CR users can use both unlicensed bands and licensed bands that are free from primary users' activities.

the RF spectral resources, the CR-enabled nodes are specifically designed to incorporate two major capabilities:

- *Cognitive capability*: This refers to the ability of the CR to perceive RF spectrum usage (especially in the licensed bands) by the primary users and to adapt to the situation accordingly.[3,5]
- *Reconfigurability*: This refers to the ability of the CR to be configured dynamically according to the perceived environment, e.g., to utilize the spectrum holes found by its cognitive capability. The parameters that can be reconfigured may include operating frequency, modulation type, and transmission power.[3]

The capabilities of the CR node described above collectively enable DSA.[8] Among different DSA models proposed in the literature, the hierarchical access model is most compatible with the current spectrum management policy.[8] This model allows opportunistic access of the licensed spectrum by the unlicensed secondary users. The basic idea is to open the licensed spectrum to unlicensed secondary users, while limiting the interference perceived by the licensed primary users. Two approaches to spectrum sharing have been considered: spectrum underlay and spectrum overlay.[8]

In the spectrum underlay approach, severe constraints on the transmission power are imposed on the secondary users, which access the network by spreading their signals over a wide frequency band. By operating below the noise floor of the primary users, the interference from the secondary users as seen by the primary users is kept below a certain tolerable level. In the spectrum overlay approach, the secondary users access the portion of the spectrum that is not currently used by the primary users. As a result, there is virtually no interference to the primary users. There is no restriction on the power level of the secondary users. This method targets the spatial and temporal white spaces or spectrum holes created due to the primary users' inactivity. The CR networks that employ overlay or underlay access techniques are referred as overlay or underlay CR networks, respectively. Between the two spectrum sharing approaches, the overlay one has mainly been considered in the literature, and it also has the most significant impacts on the upper layer performance. Therefore, in this chapter, we consider only overlay CR networks.

Intermittent availability of RF spectrum, which results in intermittent link connectivity for data transmissions, is a prominent characteristic of overlay CR networks. Figure 11.2 illustrates the usage of RF spectrum in a CR network: the spectrum holes that are available to the CR nodes dynamically change in time and frequency. A generic CR link usage model can be described as follows. The number of available frequency channels varies

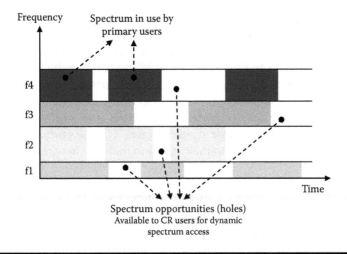

Figure 11.2   Illustration of dynamically changing spectrum holes. A cognitive radio can detect the availability of spectrum holes and utilize them for its transmissions.

in time between a maximum number and a minimum number of available frequency channels. Each CR user can use one or more of these channels depending on the traffic requirements and availability of the channels. The availability of a channel is subject to the existence and usage of primary users. The condition of a channel and the corresponding data rate sustainable over the channel depend on the interference level transmission power and distance, and receiver sensitivity and noise figure. When a channel is available, several CR nodes may contend to access the channel. If this channel becomes unavailable due to activities of primary users, the respective CR nodes will need to select alternative channels that are free of primary user activities. This is called a frequency spectrum handoff. While this is similar to a handoff in a cellular network, cellular networks can be engineered to minimize the handoff failure probability, whereas it is not possible to engineer a CR network to achieve some handoff failure probability objective, as the CR network has no control over the activities of the primary users. Therefore it can be seen that CR links exhibit a reduced level of reliability compared to conventional radio links.

The DSA and intermittent availability of spectrum bring new challenges to the design of upper layer protocols (i.e., those above the link layer), such as the routing and transport protocols. Conventional widely used routing and transport protocols may not be able to produce satisfactory performance.

For example, due to the intermittent availability and DSA, the operation channel of a CR may vary from time to time. When a CR node changes its operating frequency, a certain amount of delay is incurred for acquisition and synchronization before the data link becomes operational over the new channel. This is referred as the spectrum handoff latency. In addition, in a geographical area with heavy primary user activities, it may be impossible to find an alternative frequency channel immediately, which can lead to a new type of packet loss, namely, service interruption loss. As we know, the throughput of the widely used transport protocol TCP depends highly on delay and packet loss probability; therefore, it is worthwhile to investigate how the CR link usage model affects TCP performance and how to modify TCP to provide optimal performance in the presence of a CR link.

The CR link can have negative impact on the routing layer also. For example, in contrast to conventional multihop wireless networks, the use of broadcast messages for route discovery is not easy to implement in CR networks. It is hard to define a fixed common control channel to broadcast the messages. The DSA may incur rerouting due to spectrum handoffs, which should be taken into consideration when designing new routing solutions for ad hoc CR networks.

However, research on CR networks has so far focused primarily on spectrum sensing techniques and spectrum sharing solutions. The needs for new routing algorithms and transport protocols in a DSA environment lead to new directions in CR network research.

In the following sections, the challenges and requirements on the upper layer protocols are discussed. Existing proposals for CR-aware upper layer protocols are reviewed. Some performance evaluation results are presented. Research issues in seeking better protocols to better match the characteristics of CR networks and more efficiently utilize the radio resources are discussed.

11.3   Routing Protocols for Cognitive Radio Networks

11.3.1   Major Challenges

Routing techniques in traditional infrastructure and mobile ad hoc networks (MANETs) have been intensively studied. Many protocols, e.g., proactive, reactive, hierarchical, geographical, power aware, and multicast[14,15] are available to the MANET designers. The diverse routing protocols have their own advantages and disadvantages. Routing remains an important problem in CR networks, especially those with multihop communication requirements. The unique characteristics of DSA in CR networks necessitate the development of novel routing algorithms.

As opposed to traditional networks, a CR network may, by necessity, use multiple communication channels within the same network. Since transmissions on different channels do not interfere with each other, using multiple

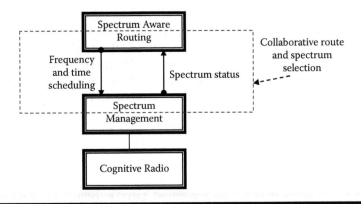

Figure 11.3   Joint route and spectrum selection using a cross-layer approach.

channels in parallel increases the network throughput. However, the CR in each node may only be able to transmit/receive over one channel at a time, which complicates routing and increases delay. In addition, in contrast to existing multichannel systems in which the number of channels and their frequencies are fixed, in a CR network, the number of available channels is not fixed and the available channels can be anywhere in the RF spectrum of interest. When the network has the ability to utilize multiple channels on a best-effort basis, traditional routing metrics such as number of hops, congestion, energy used, etc., may not be sufficient for the best routing decisions.

Next, we give an overview of the existing routing solutions[16–19] for multihop CR networks and discuss the open research issues in this subject. These existing routing solutions share a common property, i.e., *collaborations* between the routing and spectrum management layers, as shown in Figure 11.3. This is necessitated by the fact that the dynamic spectrum is intermittent in terms of both time and space.

11.3.2   Overview of Existing Routing Protocols for CR Networks

11.3.2.1   Collaborative Spectrum Selection and Routing Protocol

The interdependence between route selection and spectrum management has been investigated in Ref. 16.

First, a decoupled approach for route selection and spectrum management is studied. In this approach, the route selection is performed independently of the spectrum management. The decoupled approach can be regarded as a simple integration of existing routing and spectrum management algorithms. For route selection, the source node invokes path discovery to collect information from the nodes and selects the path using a specific performance metric (e.g., dynamic source routing with the shortest path metric[20]). For spectrum management, the nodes on the selected path can use an existing coordination protocol (e.g., HAMAC[21]) to schedule packet transmissions. The decoupled approach is a simple and modular solution. However, it has several drawbacks. Optimization in spectrum management focuses on single-hop instead of end-to-end performance. Also, it is difficult to predict link quality, since links switch between channels frequently.

Next, a cross-layer approach that considers joint route selection and spectrum management is proposed and studied. In this approach, some spectrum management tasks are integrated into the route selection function. In particular, each source node makes decisions on both route and channel (frequency) selection. This source-based routing technique is performed centrally using a global view of the network to show the upper bound in achievable performance. Each source node uses dynamic source routing to find candidate paths. It also schedules a time and channel usage for each hop. The nodes along the path follow the time and channel schedule to communicate with each other.

Simulations were performed for the cross-layer and decoupled approaches for routing and spectrum management. The results reveal that the cross-layer solution that jointly constructs routes and determines the operating spectrum for each hop outperforms the sequential approach where routes are selected independently of the spectrum allocation.[16]

11.3.2.2 Collaborative Routing Protocol Based on Layered Graph Model

Similarly to Ref. 16, a comparison of layered and cross-layer approaches is presented in Ref. 17, but the uniqueness of the latter work lies in the proposal of a novel graph modeling technique. In this model, each layer corresponds to a channel in the network. Each node is represented by subnodes forming the vertices in each layer. This model is exploited in Ref. 17 to construct routes. Using different cost functions for each edge, the required constraints for route selection, such as interference avoidance, can be achieved. This model provides an interesting solution for modeling CR networks with relatively static link properties.

The simulation results presented in Ref. 17 also reveal that the cross-layer approach is advantageous for routing in CR networks, since the availability of spectrum directly affects the end-to-end performance.

The two solutions for routing in CR networks given in Refs. 16 and 17 clearly show that cross-layer approaches that jointly consider route and channel selection are necessary for CR networks. Such approaches should also be used in designing other improved routing algorithms.

11.3.2.3 Spectrum-Aware Routing Protocol

In Ref. 18, the outline of a spectrum-aware routing protocol that is under development for the CR network is described. This protocol consists of the following functions.

1. *Dissemination of neighbor information:* In conventional wireless networks, neighbor discovery involves updating the identities and addresses of the adjacent nodes. In CR networks, due to the absence of a persistent common control channel between all the nodes, CR nodes need to periodically initialize their communication parameters in a coordinated manner to enable the nodes to exchange neighbor information.
2. *Joint route and spectrum discovery:* Once the network topology is configured, an optimal route is established from the source to the destination. Availability of the desired spectrum at a particular hop, transmitting activities of the primary users, interference from other secondary users, and the required capacity of the route all need to be taken into consideration to make the routing decision. CR networks are required to avoid generating interference to primary users. Therefore, an important routing metric is the level of primary user activity so that routes can be selected to avoid regions where primary user activities are frequent, as shown in Figure 11.4.

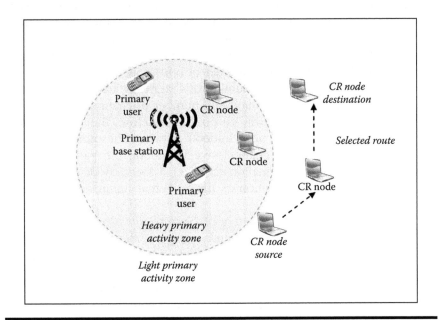

Figure 11.4 Primary spectrum usage-aware route discovery.

3. *Spectrum-adaptive route recovery:* In conventional wireless multi-hop networks, link failure is mainly caused by node mobility. However, in CR networks, the operating channel for the wireless link may abruptly change to avoid interference with primary users. In such cases, if the CR node needs to vacate a channel abruptly, it may lead to a link failure. To smooth the recovery from such link failures, a joint spectrum handoff and path rerouting method is adopted in Ref. 18.

Similarly to Refs. 16 and 17, the protocol presented in Ref. 18 is also a cross-layer approach; however, its performance is unknown because a performance evaluation was not provided.

11.3.2.4 Control-Channel-Based Routing Protocol

In Ref. 19, a control-channel-based routing protocol is presented. In this method, based on the global topology information collected through a common control channel, each node can compute the "best" routes to all nodes in the network. In addition to using the number of hops as a metric for route selection, the routing protocol proposed in Ref. 19 uses other routing metrics such as the number of frequency switches along a route and/or frequency of channel switches over a link. Two new routing strategies based on these metrics are outlined in Ref. 19:

1. *Routing based on minimum latency:* If the incoming and outgoing links of a node use different frequencies, then there is a channel switch at the node. A channel switch incurs some overhead and increases the node latency, and hence end-to-end delay. The shortest path in terms of the number of hops may not always yield the path with the minimum number of channel switches. On the other hand, a path with the minimum number of channel switches may be too long for practical purposes. Therefore, the route computation can be based on a tradeoff between the number of hops and the number of channel switches along the path. Using the global topology information collected, a source node can compute the route to a destination as the path that minimizes the overall latency.

2. *Routing based on frequency of channel switches over links:* Every node maintains a weighted average of the duration for which each channel is available along each outgoing link. Based on the weighted averages for the channels, each node computes the probability of a channel being available on a link. When a node has a packet to transmit, it selects the channel with the highest probability of being available. In this case, since every node makes a local decision about the channel to use for its transmission, this metric

might result in a path with a high number of channel switches. The appropriate strategy is to use this approach in conjunction with the minimum latency routing scheme described in (1).

Just as with the approach in Section 11.3.2.3, no numerical results are presented in Ref. 19 to examine the effectiveness of the proposed routing method. Also, a common control channel in a CR network is not easily established, as will be explained in Section 11.3.3, which limits the usability of this approach.

11.3.3 Open Research Issues

As discussed above, the efforts to tackle the routing problems in multihop CR networks have produced some initial results. However, the research in this field is still at an early stage[5, 16–19] and there are many open research issues that need to be addressed thoroughly. Some open problems for routing in multihop CR networks are listed below.

1. *Availability of a common control channel:* Conventional routing protocols require either local or global broadcasting messages for functionalities such as neighbor discovery, route discovery, and route establishment. In CR networks, when a primary user becomes active over a channel, this channel must be vacated by all CR nodes that are using the channel. If the control channel is established through DSA, then it is not immune from the above rule. In the presence of a network of primary user nodes dispersed over a wide area, the establishment of a common control channel for all secondary users through DSA may not be feasible, as the topology of the primary user network and the perceived primary user activities may vary over time and at different locations. On the other hand, if the transmissions of the primary users cover a wide area, e.g., television broadcasts, and the secondary users in the CR network are confined to a relatively small geographical area, then all the secondary users will perceive the same level of primary user activity. In this case, if a coordinated channel selection strategy is employed by all the CR nodes, the establishment of a common control channel may be feasible. Hence, the routing solutions for a CR network must consider the topologies and coverage of the primary and secondary users and the feasibility of establishing a common control channel.

2. *Intermittent link availability:* In CR networks, the reachable neighbors of a node may change dynamically. This is due to two reasons. First, the available channels may change due to the activities of the licensed primary users. Second, once a CR node has selected a channel for communications it may not be reachable through other channels from other nodes, unless more complex and expensive

multichannel radios are used. Therefore the connectivity assumption commonly used in the design of conventional mobile ad hoc networks is generally not applicable in CR networks due to the spectrum usage dynamics.

3. *Rerouting:* In CR networks, due to the intermittent link availability, a route established for a traffic flow can easily be disrupted. Fast discovery and recovery of link failure is needed in conjunction with an efficient rerouting mechanism. This is best accomplished through a cross-layer design approach.

4. *Analytical modeling of routing protocols:* Routing protocols for conventional multihop wireless networks are usually analyzed using graph models. However, in these networks, the communication channel is fixed and always available for each node to communicate with its one-hop neighbors. On the contrary, in CR networks due to DSA, modeling the time-varying topology and connectivity of a CR network becomes a challenging problem.

11.4 Transport Layer Protocols for Cognitive Radio Networks

11.4.1 Challenges and Issues

TCP has been used ubiquitously for reliable non-real-time data transmissions over the Internet. The transport layer issues in CR networks are relatively less well studied and understood,[5] although there has been some very recent work, namely Refs. 22 and 23, that studies TCP performance in CR networks from different angles.

TCP was originally developed for wired networks,[24] and later enhanced to cope with the channel errors in wireless networks. In general, the suggested solutions to improve the performance of TCP in wireless networks are classified into three basic groups based on their fundamental philosophies: end-to-end solutions, split-connection solutions, and link-layer solutions.[25] Among these three solution approaches, the link-layer solution is the most popular, since it does not require any modification of the network architecture or TCP operation.

The concept of DSA (see Section 11.2) in CR networks imposes unique challenges for transport layer protocols. In general, the performance of TCP depends on the round trip time (RTT) and the packet loss probability. The RTT of TCP in CR networks may be larger and more variable than in conventional networks due to the spectrum handoff phenomenon. The delay due to spectrum handoff, referred to as *observation time* in Ref. 22, can significantly degrade the performance of TCP in CR networks.[22]

Indeed, the observation time is the dominant component (compared to the primary user detection errors) responsible for TCP throughput reduction in CR networks. This is because an increase in the observation time can lead to increased retransmission timeouts, which subsequently cause the TCP sender to invoke the congestion avoidance mechanism.

In conventional wireless networks, packet losses are caused by congestion, channel contention, and channel errors. However, in CR networks, packet losses can also occur due to service interruptions over the wireless link caused by the primary users' activities. The behavior of TCP under this service interruption loss has been studied.[23] As these results[23] are very recent, we present them in more detail in this chapter.

11.4.2 Performance Study of TCP in CR Networks

11.4.2.1 Service Interruption Loss in Overlay CR Networks

The concept of DSA introduces a new type of loss, called service interruption loss, for the secondary users of an overlay CR network. Service interruption loss refers to the loss experienced by secondary users due to the intervention of primary users while transmitting the data.

Service interruption loss is different from the three major losses that occur in a conventional network: congestion loss caused usually by buffer overflow, channel error loss, and collision loss due to contentions between various users of the network. The service interruption loss due to DSA, on the other hand, does not depend on network conditions or channel characteristics, nor does it depend on the underlying MAC mechanism. Instead, it depends on the level of primary user activities in the spectrum, which in turn depends on extraneous factors such as the service used or provided by the primary user, geographical location, and time of day. In a given geographical region that has heavy primary user activities, it may not be possible for the secondary user to find an alternate channel immediately, thus leading to a longer duration service interruption loss.

TCP performance under the other three types of losses (congestion loss, channel error loss, and collision loss) has been studied extensively. Due to the advent of service interruption loss, however, TCP performance in an overlay CR network could be significantly different from that in a conventional wireless network.

11.4.2.2 Performance Results

The performance results of TCP subject to service interruption loss provided in Ref. 23 are summarized here. The numerical results were obtained through simulating a CR network model, which is shown in Figure 11.5. In the model, N_p primary users and N_s secondary users access their base stations via L shared radio channels. Each radio channel has a bandwidth

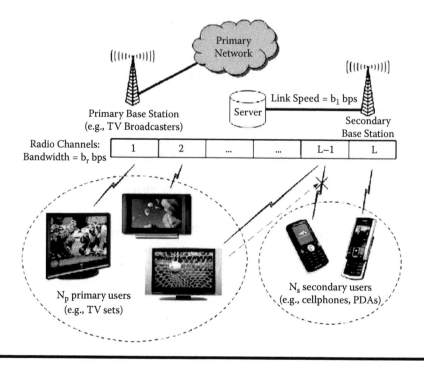

Figure 11.5   System model of an overlay CR network.[23]

of b_r bps. The primary base station connects to the primary network, while the secondary base station connects to a server via a wired link with a bandwidth of b_l bps. When needed, a primary user can force a secondary user to cease transmission, so that it can use the released channel. The details of the simulation settings can be found in Ref. 23. The following gives a summary of the simulation results and discussions.

Without any primary users (thus without the service interruption losses), the aggregate TCP throughput of the secondary users increases monotonically as the maximum number of channels (L_s) that are captured by the secondary users increases. Figure 11.6 shows that when there are 30 primary users, the throughput gradually increases as L_s is increased up to a certain point, beyond which the throughput decreases. Hence, there exists an optimal value of L_s that maximizes the aggregate TCP throughput.

This nonmonotonic behavior of TCP in an overlay CR network that implements DSA can be explained as follows. Intuitively, as L_s increases, we expect an increase in aggregate TCP throughput as the total bandwidth available becomes larger. However, an increase in L_s also increases the probability of service interruption due to the primary users, leading to increased packet losses. These packet losses (detected via timeouts or duplicated acknowledgements) lead to a decrease in TCP window size, and

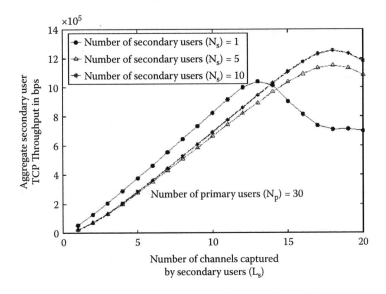

Figure 11.6   Aggregate secondary user TCP throughput versus the maximum channels for secondary users (L_s) in the presence of service interruption due to 30 primary users.[23]

hence a significant degradation of the aggregate TCP throughput of secondary users. Therefore, when adopting DSA in overlay CR networks, the secondary users need to be judicious in choosing the number of channels for data transmissions.

The study also reveals that the optimal value of L_s is related to the traffic load generated by the primary users (denoted by A, load factor). Figure 11.7 plots the optimal value (L_s^*) of L_s versus the load factor (A). The figure shows that for a given bandwidth, the optimal number of channels that the secondary users should capture is high when the load factor is small, and decreases as the load factor is increased up to a moderate value. At very large load factor values, the number of channels that the secondary users should capture to maximize throughput becomes insensitive as long as it is greater than some minimum value; i.e., L_s^* can take on any sufficiently large value.

The nonmonotone behavior of L_s^* with respect to the load factor A in Figure 11.7 can be explained as follows. When the load factor is small, the primary users rarely use the channels and the secondary users experience a small service interruption probability. Thus, the secondary users should take a chance and transmit in as many channels as possible to increase the TCP throughput. As the load factor increases, the service interruption probability also increases. In this case, L_s^* should be decreased to avoid

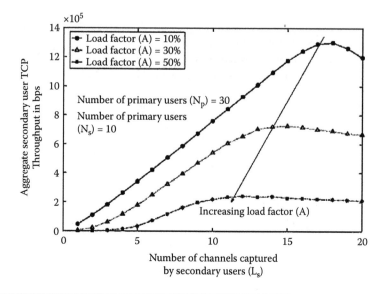

Figure 11.7 **Optimal number of channels (L_s^*) that the secondary users capture to maximize TCP throughput versus load factor (A) for different values of bandwidth.**[23]

service interruption by the primary users. The decrease of L_s^* (with respect to A) stops when the traffic load is extremely large. In this case, the primary users tend to occupy most of the channels most of the time, and the secondary users may not be able to acquire a channel. Therefore, the secondary users should take a chance and transmit packets on as many available channels as possible, even though they are prone to increased service interruption by the primary users. Due to the above nonmonotone behavior, the secondary users need to be judicious in setting the maximum number of channels L_s for data transmissions.

11.5 Conclusion

In this chapter, we have given an overview of the characteristics of the emerging cognitive radio technology, specifically the overlay dynamic spectrum access. We have discussed the new issues and challenges in upper layer protocol design resulting from the use of cognitive radio links. It has been shown that to perform well, new routing protocols tailored to cognitive radio networks must be able to collaborate with the dynamic spectrum access through cross-layer designs. It has also been shown that the performance of TCP, which is widely used over the Internet, is affected by a new source of packet losses, i.e., service interruption losses, in cognitive radio networks. To better utilize the efficiency of cognitive radio links in dynamic spectrum access while achieving an acceptable end-to-end data communications performance, many new research issues in routing and transport layer designs need to be addressed. Some of these open problems have been outlined in this chapter.

Acknowledgments

This work was supported by the Canadian Natural Sciences and Engineering Research Council through grant STPGP 322208-05.

References

[1]   B. Fette, *Cognitive Radio Technology*, Elsevier Inc., (2006).
[2]   Federal Communications Commission, *Spectrum Policy Task Force Report, ET Docket No. 03-222, Notice of Proposed Rule Making and Order*, (Dec. 2003).
[3]   S. Haykin, Cognitive Radio: Brain-Empowered Wireless Communications, *IEEE Journal on Selected Areas in Communications*, 23(2), 201–220 (Feb. 2005).
[4]   J. Mitola III and G. Q. Maguire Jr, Cognitive Radio: Making Software Radios More Personal, *IEEE Personal Communications*, 6(4), 13–18, (Aug. 1999).
[5]   I. F. Akyildiz, W.-Y. Lee, M. C. Vuran, and S. Mohanty, NeXt generation/dynamic spectrum access/cognitive radio wireless networks: A survey, *Elsevier Computer Networks*, 50(3), 2127–2159 (Sept., 2006).
[6]   G. Staple and K. Werbach, The End of Spectrum Scarcity, *IEEE Spectrum*, 41(3), 48–52 (Mar. 2004).
[7]   Federal Communications Commission. *Spectrum Policy Task Force Technical Report ET Docket No. 02-135, FCC*, (Nov. 2002).
[8]   Q. Zhao and B. M. Sadler, A Survey of Dynamic Spectrum Access: Signal Processing, Networking, and Regulatory Policy, *IEEE Signal Proc. Mag.* 24 (3), 79–89 (May 2007).
[9]   F. Akyildiz, W. Y. Lee, M.C. Vuran and S. Mohanty, A Survey on Spectrum Management in Cognitive Radio Networks, *IEEE Communications Magazine, Feature Topic on Enabling Technologies for Cognitive Radios and Dynamic Spectrum Access Networks* (Apr. 2008).
[10] J. A. Stine, Spectrum Management: the Killer Application of Ad Hoc and Mesh Networking, *Proc. IEEE DySPAN*, 184–193 (2005).
[11] Joseph Mitola III. *Cognitive Radio: An Integrated Architecture for Software Defined Radio*. PhD Dissertation, KTH Royal Institute of Technology, Stockholm, Sweden, (2000).
[12] P. Kolodzy, Application of Cognitive Radio Technology Across the Wireless Stack, *IEICE Trans. Commun.*, E88-B(11), 4158–4162 (Nov. 2005).
[13] D. Raychaudhuri et al., CogNet: An Architectural Foundation for Experimental Cognitive Radio Networks within the Future Internet, *Proc. First*

ACM/IEEE International Workshop on Mobility in the Evolving Internet Architecture, 11–16 (2006).

[14] I. Chlamtac, M. Conti, and J. J. -N. Liu, Mobile Ad Hoc Networking: Imperatives and Challenges, *Ad Hoc Networks*, 1(1), 13–64 (2003).

[15] *Routing Protocols for MANET*, http://en.wikipedia.org/wiki/List_of_ad hoc_routing_protocols (2007).

[16] Q. Wang and H. Zeng, Route and Spectrum Selection in Dynamic Spectrum Networks, *Proc. IEEE Consumer Communications and Networking Conference (CNCC)* (2006).

[17] C. Xin, A Novel Layered Graph Model for Topology Formation and Routing in Dynamic Spectrum Access Networks, *Proc. IEEE DySPAN*, 308–317 (2005).

[18] http://www.ece.gatech.edu/research/labs/bwn/CR/Projectdescription.html

[19] S. Krishnamurthy et al., Control Channel-based MAC-layer Configuration, Routing, and Situation Awareness for Cognitive Radio Networks, *Proc IEEE MILCOM* (2005).

[20] D. B. Johnson and D. A. Maltz, Dynamic Source Routing in Ad Hoc Wireless Networks, *Mobile Computing*, 153–181, Kluwer Academic Publishing, (1996).

[21] J. Zhao, H. Zheng, and G. Yang, Distributed Coordination in Dynamic Spectrum Allocation Networks, *Proc. IEEE DySPAN* (2005).

[22] A.M.R. Slingerland, P. R. Pawelczak, R. Venkatesha prasad, A. Lo, and R. Hekmat, Performance of Transport Control Protocol over Dynamic Spectrum Access Links, *Proc. 2nd IEEE Symp. New Frontiers Dynamic Spectrum Access Networks*, (Apr. 2007).

[23] T. Issariyakul, L. S. Pillutla, and V. Krishnamurthy, Tuning Radio Resource in an Overlay Cognitive Radio Network for TCP: Greed Isn't Good, *IEEE Comm. Mag* (in press).

[24] J. F. Kurose and K.W. Ross, *Computer Networking: A Top-Down Approach*, Addison-Wesley, (2008).

[25] H. Balakrishnan, V. N. Padmanabhan, S. Seshan, and R. H. Katz, A Comparison of Mechanisms for Improving TCP Performance over Wireless Links, *IEEE/ACM Trans. on Netw.* 5(6), 79–89 (Dec. 1997).

Chapter 12

On the Tradeoffs of Cross-Layer Protocols for Cognitive Radio Networks

Nie Nie, Katia Jaffrès-Runser, and Cristina Comaniciu

Contents

12.1 Introduction

Wireless networks are evolving toward networks of small, smart devices that opportunistically share the wireless spectrum with minimal infrastructure and coordination. The new generation of smart terminals can provide intelligent adaptive services by adjusting to the environment thanks to the software-defined radio (SDR) technology. This intelligence is incorporated into a cognitive cycle,[1] which allows the wireless device to gather information about its environment ("learn") and make decisions ("act") regarding its transmission parameters and possible access strategies. The development of this new technology is further motivated by the new paradigm of the FCC's spectrum management policy[2] that has adopted new rules to promote active spectrum sharing techniques in both licensed and unlicensed bands.

Distributed spectrum sharing techniques.　　The performance of radio networks is limited by interference, which reduces the nominal throughput of users in the network. Consequently, efficient interference management techniques, such as power control,[3] channel assignment,[4,5] or end-to-end interference-aware routing[6] are key elements in providing the quality of service (QoS) of such networks. A classic problem in cognitive networks is the distributed channel allocation scenario, where the users measure the available spectrum and dynamically decide which frequency they should use for transmission[4] based on their current measurements. Such channel measurements are usually related to their desired QoS metrics. For instance, when users require the highest possible throughput, they look for the best achievable signal-to-interference ratio (SIR), knowing their adaptive modulation and/or coding capabilities. Consequently, the users can cooperate to distributively assign the channels to the nodes and to mitigate interference in the network. Current distributed channel allocation schemes are based on graph coloring algorithms[4] and game theory.[7] In these algorithms, cooperation can be enforced at both the physical and medium access layers, where every node selects the channel that maximizes the SIR on its communication link, providing local performance improvements.

　　Based on the current allocation of the channels and powers, multihop routes can then be established. If costs related to QoS or energy expenditures are employed by the routing protocol, then the performance can be further improved at the network layer. However, as all these techniques (power control, channel allocation, and route allocation) are influenced by, and in turn influence, the distribution and the level of interference in the system, they are inherently cross-coupled and a joint optimization might lead to additional gains. This joint optimization is addressed via cross-layer design.

A cross-layer approach.　　Cross-layer design is a recent protocol definition technique that jointly optimizes the behavior of two layers of the protocol stack that have no common interface. According to Srivastava and Motani,[8] cross-layer design can be defined by "the violation of a reference layered communication architecture with respect to the particular layered architecture." Cross-layer design may build on the inherent flexibility of the software-defined radio architecture in the higher level decision protocols. Adjusting to the current environment by selecting transmission frequencies and waveforms needs communication and coordination between the physical layer and the upper layers. For example, if a set of nodes decides to adjust to a high or low multipath channel, they have to cooperate at the network level to concurrently change their transmission settings. For our particular case, to mitigate interference and reduce the energy spent for every data bit sent via cross-layer cooperation, additional information must be shared between the routing layer, the medium access control layer, and the physical layer, violating the reference OSI layered protocol model. An illustration of the cross-layer architecture is given in Section 12.2.2 and in Figure 12.1.

Designing cross-layer protocols has to be attempted carefully, as stated by Kawadia and Kumar,[9] as unintended cross-layer interactions can have undesirable consequences on the overall system performance. There is indeed a price to pay for designing a cross-layer implementation for our spectrum sharing problem in cognitive networks. Our proposed cross-layer implementation requires an iterative optimization of the transmission parameters and the source-destination routes, which results in increased energy consumption. This increase is due to the additional overhead triggered by the supplementary route updates created at each new routing iteration. There is also an overhead at the physical/medium access layers due to the packet exchange for channel and power updates required for enforcing cooperation at each new iteration.

The aim of this chapter is to show how and when the cross-layer implementation benefits more or less the overall network performance. When does the energy saved by the concurrent network and physical layer parameters' optimization outweigh the energy loss due to the cross-layer design overhead? How long does it take to get a final and stable minimum energy route configuration? For what kind of networks is a cross-layer approach beneficial?

Section 12.2 presents the cross-layer framework considered for the joint routing and spectrum sharing algorithm. A detailed implementation of this framework is provided in Section 12.3, presenting the solutions chosen for the routing protocol, channel allocation, and power control algorithms. Analysis of the overhead triggered by this cross-layered implementation is presented in Section 12.4. Section 12.5 presents the performance results of the proposed solution with respect to the overall energy consumption, the data delivery ratio, and the delay. These metrics are assessed for several network densities, providing a good insight into the tradeoffs that arise in cognitive networks with cross-layering.

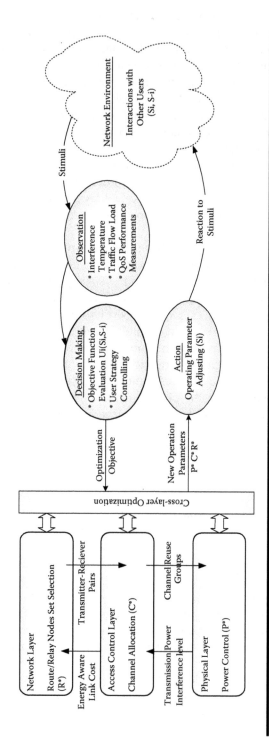

Figure 12.1   Cross-layer framework for cognitive radio networks.

12.2   A Cross-Layer Framework for Cognitive Radio Networks

12.2.1   System Model and Assumptions

We consider an example of a cognitive radio network consisting of a set of N nodes uniformly distributed in a square region of dimension $D \times D$. It is assumed that the nodes are fixed. In this network, a node generates continuous data traffic (worst-case scenario) that is transmitted towards a randomly chosen destination node. The traffic can be relayed through intermediate nodes, which also act as routers, forwarding packets to the destinations. To accomplish the transmissions, a node must determine the route of an outgoing packet according to a preset routing metric. The route with the minimal cost from the source to the destination is selected to forward the packets. If a node is selected to relay packets for multiple flows, the transmissions for different traffic flows at that node are time multiplexed.

There are various ways to define the routing metric. A simple hop count metric can be considered, as well as other performance-oriented metrics, such as congestion load or energy consumption, which are based on information originating from lower layer protocols. In this work, we define a link QoS measure for an arbitrary link (i, j) as the energy consumed for the correct transmission of a data bit[10,11] E_b^{ij}:

$$E_b^{ij} = \frac{Mp_i}{mRP_c(\gamma_{ij})},\tag{12.1}$$

where M is the packet length, m is the number of information bits in a packet, R denotes the transmission rate, and $P_c(\gamma_{ij})$ is the probability of the correct reception of a packet, which depends on γ_{ij}, the achieved link SIR.

In this work, the energy per bit over a link, E_b^{ij}, is considered as the routing metric to define the link cost. Accordingly, we define the energy per bit over a route, E_b^r, as the energy consumption for a data bit to travel along a route r (from its source to its destination):

$$E_b^r = \sum_{(i,j)\in r} E_b^{ij}.\tag{12.2}$$

The cognitive radio nodes in the network are assumed to be capable of measuring the spectrum availability and making a decision on the transmission channel. We assume that there are K frequency channels available for transmission, with $K < N$. Multiple users are allowed to transmit at the same time over a shared channel. By distributively selecting a transmitting frequency, the radios effectively construct a channel reuse distribution map with reduced co-channel interference.

The transmission link quality can be characterized by a target SIR, which is specific for the given application. The SIR measured at the receiver j associated with transmitter i can be expressed as:

$$\gamma_{ij} = \frac{p_i G_{ij}}{\sum_{k=1, k \neq i}^{N} p_k G_{kj} I(k, j) + \sigma^2},$$

(12.3)

where p_i is the transmission power at transmitter i, G_{ij} is the link gain between transmitter i and receiver j, and σ^2 denotes the received noise, which is assumed to be the same for all receiver nodes. $I(i, j)$ is the interference function characterizing the interference created by node i to node j and is defined as

$$I(i, j) = \begin{cases} 1 & \text{if transmitters } i \text{ and } j \text{ are transmitting} \\ & \text{over the same channel,} \\ 0 & \text{otherwise.} \end{cases}$$

(12.4)

For the nodes sharing the same frequency channel, their transmission powers affect their link quality and the interference temperature on that particular channel. It is assumed that the radio nodes are able to adjust the transmission power to improve the link quality and to enable the group of users who are transmitting over the same channel to meet a certain target SIR.

Analyzing Equation (12.3), we note that in order to maintain a certain SIR, the nodes can adjust at both the medium access control layer and the physical layer. At the medium access control level, the nodes can minimize the interference by appropriately selecting the transmission channel frequency, which leads to minimized values of the interference function in Equation (12.3). At the physical layer, for a feasible system, the nodes can adjust their transmission power level via distributed power control to reduce interference and ensure that all the nodes sharing the same channel meet the target SIR requirement at their intended receivers.

In the cognitive network considered, multihop routing is implemented at the network layer. Once the routes are set by the protocol, adjusting the power level and the channel values modifies the performance of the links of the network. In this case, the next hop receiver of a node may no longer be the optimal hop. Conversely, when the channels and the power values are defined, better routes may be chosen by the network protocol, optimizing the end-to-end performance of the communication. These interactions between individual layers are explored in the cross-layer solution described in the next section.

12.2.2 Cross-Layer Framework

This subsection defines a cross-layer framework designed to reduce the energy consumption of end-to-end data transmissions for the aforementioned cognitive radio network. The structure of this framework and the cross-layer interactions are illustrated in Figure 12.1. A two-way information exchange between the layers is necessary in our case.

In the downward direction, information from the upper layers determines the configuration of the lower layer protocols. In particular, at the network layer the routing protocol assigns the next hop identification to the medium access layer. Let R^* denote the route from a source node to its destination node. As a node may belong to several source-destination paths, there may be different next hop nodes for every relayed flow. As there is only one radio per node, a single next hop node has to be chosen by the protocol. The node that belongs to the link with the highest routing cost, i.e., the highest energy-per-bit metric (see Equation 12.1), is selected. In this case, we consider the link with the worst quality, and improve its performance through channel allocation and power control at the lower layers.

The medium access layer communicates the channel values to the physical layer. To reduce the co-channel interference and improve the quality of the links, the channel allocation mechanism at the medium access control layer creates a channel reuse map, denoted by C^*, based on the current transmitter-receiver pair configuration. Nodes are arranged into K groups with each group sharing the same transmission channel. For each group of nodes, the distributed power control algorithm adjusts the nodes' transmission power to a power assignment P^* at which all the nodes in the group meet the target link quality in terms of SIR at their intended receivers.

In the upward direction, lower layer information can affect the performances of the upper layer algorithms. Reconfigured transmission power assignment P^* leads to a reconstruction of the channel reuse map at the medium access control layer due to the changes of the interference temperature in the network. When channel reuse profile and transmission power distribution are updated, the quality of the links and therefore the link costs defined at the network layer change accordingly, which may trigger routing updates with new link costs.

Considering these interactions, an iterative algorithm is presented for this cross-layer framework to reduce the energy consumption of the end-to-end data transmission over the network. The macro-algorithm is given in Algorithm 12.1.

Algorithm 12.1

    (a) *Initial state.* Nodes operate at an initial transmission power and randomly select a transmitting channel. Routing link costs are computed and routes are built up by the routing protocol.

Iteration number $N_I = 0$.

(b) *Iteration begin.* Nodes determine their intended receiver with current route selection.

$N_I = N_I + 1$.

(c) Nodes select their transmitting frequency and adjust their transmission power level via channel allocation and power control algorithm.

(d) Routing link costs are computed.

(e) IF at least one route r can be found that reduces the total data transmission energy consumption E_{data}
THEN update the routes, GOTO (b)
ELSE algorithm terminated.

The measure E_{data} is defined as the sum of the energy per route over the network:

$$E_{data} = \sum_{r \in R^*} E_b^r. \tag{12.5}$$

It indicates the total energy requirement for every node in the network to successfully deliver one data bit to its destination. This cross-layer iterative algorithm converges towards a local minimum E_{data}. The convergence time of this algorithm is a function of the number of iterations N_I.

12.3 Implementation Aspects of the Cross-Layer Framework

Specific routing protocols, channel allocation mechanisms, and power control algorithms can be chosen for our above defined cross-layer framework for various application scenarios. The protocol and algorithms employed in this work are described in this section.

12.3.1 Routing Protocol

The choice of a routing protocol influences the way the routes are built in the network. In this implementation, an on-demand routing protocol, ad hoc on-demand distance vector routing (AODV),[12] has been chosen. The traditional hop count routing metric of the AODV protocol is replaced by the energy-per-bit metric E_b^{ij} of (12.1).

The main steps of the routing mechanism of AODV are described in the following. When a node S needs a route to some destination D, it broadcasts a *Route Request* to its one-hop neighbor nodes. Each intermediate node forwarding the Route Request packet records the reverse route back to node S. Once node D or a node having a route to D hears the Route Request, it generates a *Route Reply* packet including the information about the last known sequence number of D and the energy requirement to reach D (according to our energy-aware metric and given SIR measurements for each link on the path). This Route Reply packet is then sent back along the reverse route to node S. The source node S is now aware of the energy requirement of each hop from S to D along the path that conveyed this Route Reply.

Different replying nodes send back their Route Reply packets individually. Among these available routes, S selects the one that has the most recent sequence number or the lowest energy requirement given the same sequence numbers.

12.3.2 Channel Allocation and Power Control

In our previous work[13] we proposed a game-theoretic formulation of the channel assignment and power control problem. We have shown that an iterative algorithm for channel scheduling and power allocation can be implemented, which converges to a pure strategy Nash-equilibrium solution, i.e., a deterministic choice of channels and transmission powers for all the nodes.

In this game-theoretic formulation, the radio nodes are modeled as a collection of agents that distributively act to maximize their utilities in a cooperative fashion. The radios' decisions are based on their perceived utility associated with each possible action, which is related to the transmission power and to the channel selection. The players of the game are rational and aim to maximize their own utility. The utility function is defined by Equation (12.6), where s_i stands for the strategy chosen by node i, and s_{-i} the set of strategies chosen by all the other nodes in the game. The strategy of node s_i is set by the channel it has selected.

$$U_i(s_i, s_{-i}) = - \sum_{j \neq i, j=1}^{N} p_j(s_j) G_{ji} f(s_j, s_i) - \sum_{j \neq i, j=1}^{N} p_i(s_i) G_{ij} f(s_i, s_j) \quad (12.6)$$

$$\forall i = 1, 2, \ldots, N.$$

In Equation (12.6), we denote $P = [p_1, p_2, \ldots, p_N]$ as the set of discrete transmission powers for the N nodes and $S = [s_1, s_2, \ldots, s_N]$ as the nodes' strategy profile. $f(s_i, s_j)$ is an interference function defined as

$$f(s_i, s_j) = \begin{cases} 1 & \text{if } s_j = s_i, \text{ transmitter } j \text{ and } i \text{ choose} \\ & \text{the same strategy (same channel),} \\ 0 & \text{otherwise.} \end{cases}$$

This utility function characterizes the preference of a user for a particular channel, given the fact that the user knows that power control is employed by all the users sharing each given channel. It accounts for both the interference perceived by the current user, as well as for the interference that a particular user is creating for neighboring users sharing the same channel. Cooperation is imposed on the nodes to achieve a fair allocation of resources.

For the users sharing the same frequency channel, their transmission powers affect their link quality and the interference temperature on that particular channel. The goal of power control is to adjust the transmission powers of all users to improve the link quality and to enable the group of users who are transmitting over the same channel to meet a certain target SIR γ^*. For a feasible system with N users, a nonnegative power vector P^* can be obtained by solving the system of equations

$$P^* = (I - H)^{-1}\eta, \tag{12.7}$$

where $H = (h_{ij})_{i,j \in [1,...,N]}$ is the normalized link gain matrix such that $h_{ij} = \gamma^* \frac{G_{ij}}{G_{ii}}$ for $i \neq j$ and $h_{ij} = 0$ for $i = j$, $\eta = (\eta_i)_{i=1,...,N}$ is the normalized noise vector such that $\eta_i = \gamma^* \frac{\sigma}{G_{ii}}$, and σ^2 is the received noise power density.

For the K available frequency channels, each channel is shared by a group of users that transmits at the same frequency. Each group determines their transmission powers via power control. It is clear that by selecting different transmitting channels, a user will belong to different groups and will choose its operating power level with respect to the interference environment of that particular group. The population and the members of these user groups will change with respect to the channel strategy profile S.

Let $s_i = 1, 2, ..., K$ denote the choice of transmitting channel for user i, $i \in N$; then the power vector for the kth user group can be determined by

$$P_k^* = (I - H_k)^{-1}\eta_k, \quad \text{for} \quad k = 1, 2, ..., K, \tag{12.8}$$

where $H_k = (h_{ij})_{i,j \in [1,...,N]}$ for $s_i = k, s_j = k$, and $i \neq j$, and η_k is the normalized noise vector for $s_i = k$. The number of the elements of P_k^* is equal to the number of the users who transmit on the same channel.

For a feasible system, P_k^* should be a nonnegative vector, $P_k^*(i) > 0$, $i \in N_k$, with the assumption that the transmission power can be adjusted without limitations. However, in practice, the maximum output power of a transmitter has an upper bound. Taking this limitation into account, the transmission power vector P_k^* can still be determined by Equation (12.8) but with the constraint that

$$0 \leq P_k^*(i) \leq P_{MAX}, \quad i \in N_k, \quad \text{for} \quad k = 1, 2, ..., K, \tag{12.9}$$

where P_{MAX} denotes the maximum transmitter output power depending on the physical device, and/or regulation restrictions. Consequently, the constrained transmission power $\bar{P}_k^*(i) = \min\{P_k^*(i), P_{MAX}\}$, $i \in N_k$, is selected.

12.4 Overhead Analysis for the Cross-Layer Framework

On analyzing the cross-layer Algorithm 12.1 presented in Section 12.2.2, it can be seen that there is no extra control overhead for the route discovery and construction of the AODV routing protocol, since the modification of the routing metric does not impose direct changes on these protocol-specific mechanisms. However, there is an overhead for the cross-layer architecture originating from the following two points:

(1) The route updates triggered by the reconfigurations of transmission channels and power levels at each iteration result in a control overhead for route maintenance. Let O_R be the number of additional packets triggered by the iterative cross-layer implementation at the routing level.

(2) The adaptive channel and power allocation relies on a signaling packet exchange to broadcast the interference distribution information across the network. Let O_{SP} be the number of signaling packets needed by the channel and power assignment algorithms.

The value of the routing overhead O_R cannot be easily assessed analytically, as it depends on the routing protocol implementation. Therefore, we performed a simulation-based estimation of O_R. The worst-case signaling packet overhead O_{SP} can be estimated analytically by a simple analysis of the distributed channel and power allocation scheme, which is presented in the following.

For the channel selection and power control mechanism, the evaluation of the utility function of Equation (12.6) includes two aspects during each iteration:

(1) a measure of the interference created by others for the desired user;

(2) a measure of the interference created by the user for its neighbors' transmissions.

The first part can be estimated at the receiving node, while the second part can only be estimated by the neighboring nodes and has to be communicated to the user via a packet exchange.

It is required that both the transmitter and its one-hop neighbors listen to a common control channel, and that each maintain a channel status table (CST) for all the frequencies, similar to a NAV table in 802.11. The CST of a given node i stores the list of neighbor nodes requesting the same channel, along with their transmission powers. It also includes the estimated link gain between the node i and its intended receiver nodes, and the estimated link gains from the neighbor transmitters to node i, which are used for estimation of the interference level. To update this table, a three-way handshaking procedure is required.[5] The handshaking process and the information carried by the packets between the nodes are illustrated in Figure 12.2 in which each handshaking session carries out a three-packet exchange. In the first **START** packet, every node of the network advertises its chosen power for its current channel plus the interference level sensed at its location. In the second packet, **START_CH**, a neighbor node sends its chosen channel to the transmitting node, which is acknowledged by the **ACK_START_CH** packet. All the nodes that heard the **START_CH** and **ACK_START_CH** packets update their CST accordingly.

For a network of N nodes with a constant traffic pattern, let N_{neb}^i denote the number of neighbors of a particular node i, SP_{ini}^i the number of signaling packets in the initial iteration, and SP_{ite}^i the number of signaling packets for node i to make a channel selection in the following iterations.

The transmitter/receiver should hear at least one packet from each neighboring receiver/transmitter. In other words, all the nodes should send out a **START** packet (in a three-way handshaking) to their neighbors during a certain period in the initial information collection stage. This **START** packet broadcasting should be completed before the transmitting of any **START_CH** packet in the network. After this initial **START** broadcasting period, each node carries out at least one three-way handshaking process (i.e., completes the **START_CH** and **ACK_START_CH** following the initial **START**). Following this procedure, all the nodes can get the required information about their neighbors, and are thus able to complete their CST. Consequently,

$$SP_{ini}^i = 1 + 2N_{neb}^i. \tag{12.10}$$

All the following iterations see a node i carrying out a three-way handshaking procedure with its neighbors to update the information in its CST. Therefore,

$$SP_{ite}^i = 3N_{neb}^i. \tag{12.11}$$

We consider N_{ite} iterations of the cross-layer framework (including the initial iteration with the initial **START** packet exchange). The total number of signaling packets O_{SP} is then calculated as

$$O_{SP} = \sum_{i=1}^{N} \left(SP_{ini}^i + SP_{ite}^i(N_{ite} - 1) \right) = \sum_{i=1}^{N} \left(1 + N_{neb}^i(3N_{ite} - 1) \right). \tag{12.12}$$

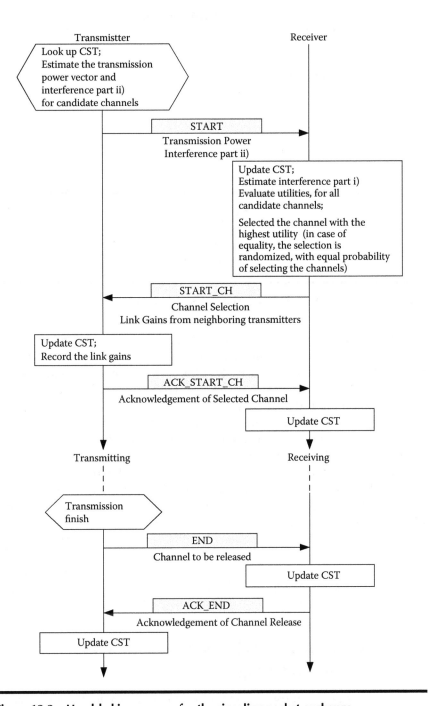

Figure 12.2   Handshaking process for the signaling packet exchange.

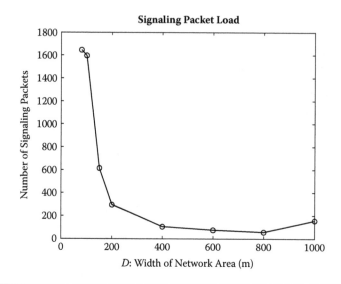

Figure 12.3 Number of signaling packets vs. network area width.

The number of neighboring nodes, N^i_{neb}, depends on the network density and on the receiving threshold. In the worst case (all the nodes are within the sensing range of a node, nodes are neighbors of each other), $N^i_{neb} = N$. In this case, $O_{SP} = (3N_{ite} - 1)N^2 + N$, and we thus have a complexity on the order of $O(N^2)$.

In Figure 12.3, the analytical estimation of the signaling overhead O_{SP} for several network sizes is presented. In this figure, the network is composed of 25 nodes spread across a field of dimension $D \times D$. The plot indicates the number of signaling packets exchanged in the network as the radio nodes collect the required information from their neighboring nodes. In this figure, the node density varies from an average of about 17 neighbors ($D = 100$) to an average of less than 1 neighbor node ($D = 1000$).

The total control overhead of this iterative cross-layer algorithm consists of the signaling packet load O_{SP} and the routing maintenance overhead O_R. Section 12.5 presents an analysis of these values via simulations for various network densities. We will discuss the impact of these overhead figures on the performance of the cross-layer scheme.

12.5 Performance Evaluation and Tradeoffs in Cross-Layering

Numerical results are presented in this section to illustrate the performance of the proposed implementation of our cross-layer framework. These results are compared to a non-cross-layer scenario where no iterative cross-layer optimization is performed. The non-cross-layer algorithm stops once an operating profile (including the channel reuse map, the transmission power assignment, and the corresponding route selection) is determined. This non-cross-layer solution contrasts with the cross-layer algorithm which further searches for a solution (power, channel, and routes) that minimizes the energy consumption of the network by iteratively performing routing and channel/power allocation.

For simulation purposes, we consider a cognitive radio network with N fixed nodes, where $N = 25$, and the width of network area, D, spans from 80 to 1000 meters, which yields various network densities. The message packet length M is 64 bytes, and for simplicity it is assumed that all the bits are information bits, which means $m = M$. The transmission rate R is 11 Mbps. The number of available channels for the radios to share is $K = 4$. The SIR requirement γ^* is 7, and the noise power σ^2 is set to be 10^{-13}. For the numerical results, a path loss coefficient of $\alpha = 2$ is selected. The maximum transmission power at a node P_{MAX} is 10^{-3} W and the initial power of the nodes is set to be P_{MAX} for the power control. To obtain average performance measures, 50 simulation runs are carried out.

An example network topology with a network width $D = 1000$ is illustrated in Figure 12.4. Figure 12.4(a) gives a snapshot of the initial route assignment over the network. The updated route assignment over the network after cross-layering iterations is demonstrated in Figure 12.4(b). The iterative reduction of E_{data} is shown in Figure 12.5. E_{data} is defined by Equation (12.5) as the total energy requirement for every node in the network to successfully deliver one data bit to its destination. It can be seen that the cross-layer iterative algorithm converges to a local minimum in 4 iterations, with about 27% reduction in the energy consumption.

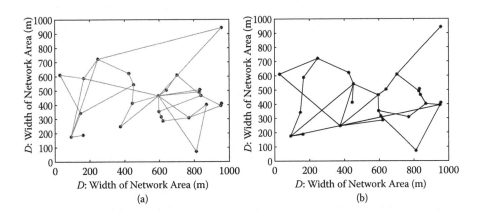

Figure 12.4 Routing example. (a) Initial route assignment over the network. (b) Route assignment over the network after cross-layering iterations.

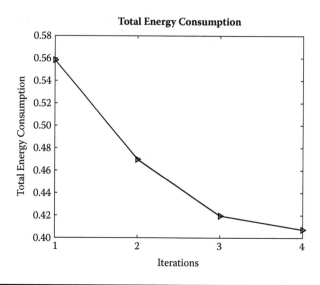

Figure 12.5 Iterative reduction of the total energy consumption for a particular network with $D = 1000$.

In Figure 12.6, we show the average number of iterations carried out by the cross-layer algorithm, with the network width ranging from 80 to 1000 meters. The error bar of each point represents the standard derivation of the number of iterations. It can be seen that, in most of the cases, the algorithm converges in about 2 ± 1 iterations, which corresponds to a fast convergence speed.

To compare the performance of the cross-layer algorithm with that of the non-cross-layer algorithm, several performance measures are considered. First, E_{data}, the total energy requirement for every node in the network to successfully deliver one data bit to its destination, is plotted in Figure 12.7 for several network sizes. The energy consumption in terms of E_{data} is reduced by employing a cross-layer algorithm on the average, but it is for denser networks that cross-layering is the most beneficial. When the nodes are distributed in an area of less than $400 \ m^2$ more energy is saved with the cross-layer scheme, which better mitigates interference for higher density networks. For both cross-layer and non-cross-layer schemes, there is an increase in the energy consumption for dense networks which is impacted by the signaling protocol for power and channel allocation. For lower densities (i.e., larger D values), the gains of cross-layering in terms of energy do not compensate for the cost in energy, as interference is not as limiting as for dense networks.

The performance gain achieved by cross-layering can also be seen in Figure 12.8 with respect to the end-to-end data packet delivery ratio defined

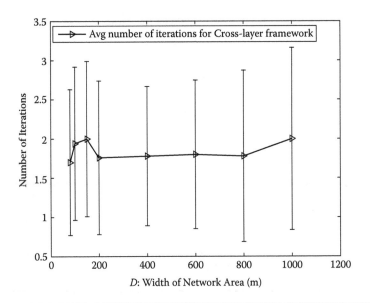

Figure 12.6 **Average and standard deviation for the number of iterations as a function of the network width D.**

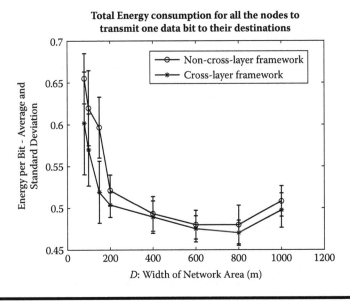

Figure 12.7 **Average and standard deviation for the total data energy consumption as a function of the network width D.**

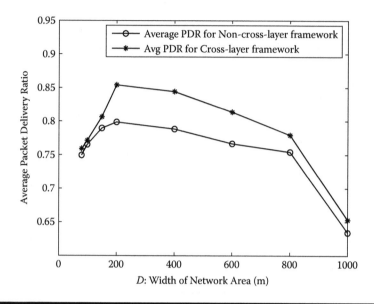

Figure 12.8 **Average data packet delivery ratio as a function of the network width D.**

as the ratio of the number of data packets that are successfully received at their destinations to the number of data packets that are sent out by their sources. The cross-layer framework is best for the scenarios with moderate network density, for example, networks with a width range from 200 to 400 m. In this case, the most efficient routes are obtained with the cross-layer algorithm.

Another performance metric considered in the simulation is end-to-end data packet delay, defined as the average delay for a data packet to be delivered from its source to its destination across the network. Figure 12.9 illustrates that the cross-layer framework results in less delay at higher network density with $D < 400$ m. For the scenarios with lower network density, the cross-layer framework suffers from an increased delay, due to the longer iterative route request process over wider-spread nodes.

As analyzed in Section 12.4, the control overhead introduced by the iterative cross-layer algorithm consists of the signaling packet load and the routing maintenance overhead. In Figure 12.10, the total number of control packets exchanged in the network is demonstrated for both the cross-layer and non-cross-layer frameworks. It is clear that the performance gain of the cross-layer framework is obtained at the cost of extra control overhead in the network. But as highlighted in Figure 12.5, the gain in energy for better coordinating the layers in a cognitive network balances this cost and even provides significant gains for dense networks.

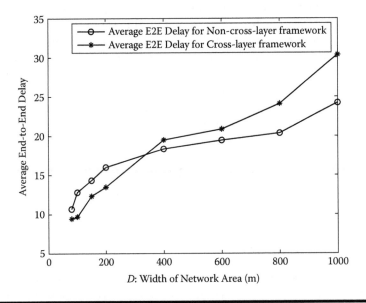

Figure 12.9 **Average end-to-end data delay as a function of the network width D.**

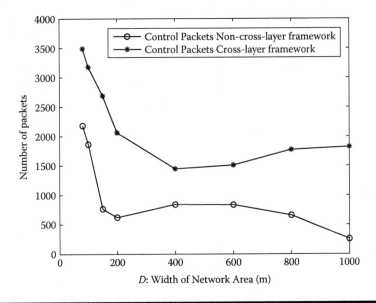

Figure 12.10 **Average number of overall control packets ($O_R + O_{SP}$) as a function of the network width D.**

As illustrated in the simulation results, a cross-layer framework trades off system control overhead for gains on the energy consumption for data transmission. However, for networks with stable topologies and continuous data traffic, this overhead pays off given the long-term benefits of the energy savings, since most of the control overhead is spent at the initial stage when the nodes build up their operating profile (i.e., transmission frequency, power, and data forwarding path).

It also can be seen that the cross-layer framework is beneficial for networks with higher densities if the data services supported by these networks are delay sensitive and throughput tolerant. For networks with lower densities, cross-layer design may be beneficial for delay-tolerant data services since it yields considerable gains in data throughput in terms of improved packet delivery ratio, and also results in significant energy savings.

12.6 Conclusions

In this chapter, we have explored the tradeoffs involved in cross-layering for cognitive radio networks. To this extent we have proposed a hierarchical cross-layer distributed framework for integrating power control, channel allocation and routing. The objective of this framework is to reduce energy consumption, while providing QoS (BER, end-to-end delay, and throughput) for all nodes across the network. Based on this framework, and using a combination of analytical and simulation results, we were able to quantify the gains, as well as the increased overhead, associated with cross-layering for our particular solution. We have shown that cross-layering is particularly beneficial (despite the increased overhead) for dense networks, but can also be beneficial for delay-tolerant traffic in lower-density networks. Overall, the benefits outweigh the increased overhead when networks are quasi-static or static, as the increased overhead is characteristic of the initial setup phase of the network.

References

[1] J. Mitola and G. Maguire, Cognitive radio: making software radios more personal, *IEEE Personal Communications* **6**(4), 13–18 (August, 1999).

[2] FCC. Report and Order, ET Docket No. 03-108 (March, 2005). http://hraunfoss.fcc.gov/edocs_public/attachmatch/FCC-05-57A1.pdf.

[3] S. S. F. Meshkati, H. Poor, and N. Mandayam, An energy efficient approach to power control and receiver design in wireless data networks, *IEEE Trans. Commun.* **53**(11), 1885–1894 (November, 2005).

[4] M. Manshaei, M. Félegyházi, J. Freudiger, J.-P. Hubaux, and P. Marbach, Spectrum Sharing Games of Network Operators and Cognitive Radios, In

eds. F. Fitzek and M. Katz, *Cognitive Wireless Networks: Concepts, Methodologies and Visions Inspiring the Age of Enlightenment of Wireless Communications*, 555–578, Springer, (2007).

[5] N. Nie and C. Comaniciu, Adaptive channel allocation spectrum etiquette for cognitive radio networks, *Mobile Networks and Applications* **11**(6), 779–797 (December, 2006).

[6] H. Mahmood and C. Comaniciu. A cross-layer game theoretic solution for interference mitigation in wireless ad hoc networks, In *Proc. IEEE Military Communications Conference (MILCOM 06)*, 1–7, Washington D.C., USA (October, 2006).

[7] C. Comaniciu, Cooperation for Cognitive Networks: A Game Theoretic Perspective, In eds. F. Fitzek and M. Katz, *Cognitive Wireless Networks: Concepts, Methodologies and Visions Inspiring the Age of Enlightenment of Wireless Communications*, 533–554, Springer, (2007).

[8] V. Srivastava and M. Motani, Cross-layer design: a survey and the road ahead, *IEEE Communications Magazine* **43**(12), 112–119 (December, 2005).

[9] V. Kawadia and P. Kumar, A cautionary perspective on cross-layer design, *IEEE Wireless Communications* **12**(1), 3–11 (February, 2005).

[10] C. Comaniciu and H. Poor, QoS provisioning for wireless ad hoc data networks (invited paper). In *Proc. of 42nd IEEE Conference on Decision and Control* (2003).

[11] D. J. Goodman and N. B. Mandayam, Power control for wireless data, *IEEE Personal Communication* **7**(2), 48–54, (2000).

[12] C. E. Perkins. Ad hoc on-demand distance vector (AODV) routing (1998). RFC 3561, IETF NetworkWorking Group.

[13] N. Nie, C. Comaniciu, and P. Agrawal, A Game Theoretic Approach to Interference Management in Cognitive Networks, In eds. P. Agrawal, D. M. Andrews, P. J. Fleming, G. Yin, and L. Zhang, *The IMA Volumes of Mathematics and its Applications in Wireless Communications*, Springer, (2006).

Chapter 13

Architectures for Cognition in Radios and Networks

Timothy Newman, Muthukumaran Pitchaimani,
Benjamin J. Ewy, and Joseph B. Evans

Contents

13.1 Introduction

Future generations of networks [1,2] have a need for more distributed
and comprehensive management and control functionality for both devices
themselves and the network overall, as the number of devices and complexity of technology continues to expand. These problems are particularly
acute in wireless networks, due to the challenges of the radio environment,
mobility, and varied applications. A long-term vision of a solution for future
network scenarios was presented in [3] to illustrate the issues and objectives
for research. Figure 13.1 represents a proposed architecture [4] based on
cognition that uses distributed sensing and control functionality, a collaborative control mechanism, and learning to support the very large networks
of the future. This approach is applicable to networking in general; wireless
networking in particular can benefit because of the wide range of situations
and dynamic operational environments.

The collaborative control mechanism is expected to incorporate a variety of cognitive [5] approaches and machine learning [6,7] techniques to
provide the needed functionality. This has inspired research [8,9] and calls
[10] for continued research in these areas.

Figure 13.2, derived from [11], shows the range of characteristics and the
wide variety of algorithms that are being brought to bear on these problems.
The time scales of wireless networks dictate that a range of techniques be
used. Short-term wireless environment variations require a tight control
loop with rapid sensing and action, e.g., stimulus/response as is typical in
classical control implementations. More gradual changes, perhaps induced
by local mobility, might be addressed through referral to policy engines
and databases, for example, for a "tactical" response. Large-scale changes,
such as a change of task objective or geographic region, might need to be
addressed through a more considered process of analysis, introspection,
and planning, which might be termed a "strategic" response.

In addition to the development of the cognitive algorithms for performing command and control (both per node as well as collaborative), the

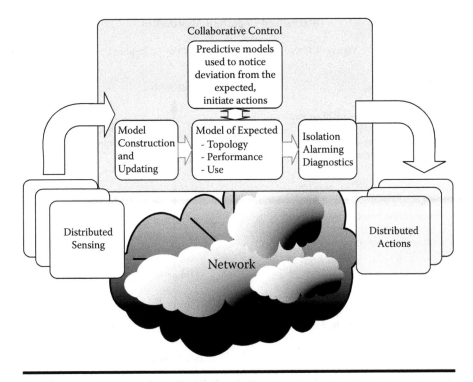

Figure 13.1   Distributed, collaborative architecture for network control and management.

task of relaying measurements to the control logic and its coordination must be carefully considered. A distributed infrastructure must be developed and deployed which scales to millions of nodes, allows end users to automatically share network performance information with other hosts, and operates in diverse environments including wireless. Collaborative control algorithms can then process the information from across the collection of networks in aggregate. The development and deployment of structured peer-to-peer network overlays [12–15] has provided a unique technology for the distribution of network measurements in a scalable fashion.

By developing and deploying these technologies we enable the creation of knowledge-based applications (K-Apps) [3] that can begin to answer *why* a network connection has failed or is performing poorly. For example, actions taken automatically in response to a problem could include

■ Selecting a replica when a web or server site is unavailable
■ Selecting from different local area networks (Ethernet, WiFi, cellular, etc.) when one becomes unreliable or unavailable completely

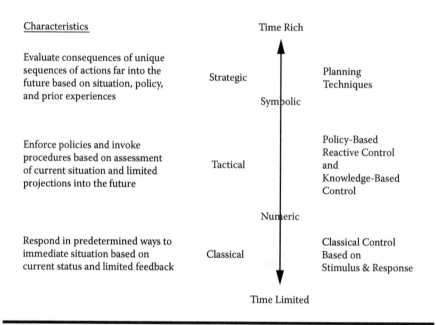

Generalized Cognition Approaches

Variety of Techniques and Technologies May Be Required

Figure 13.2 **Generalized cognition techniques for use in wireless networks.**

- Relaying through an alternate node in a mobile environment
- Adjusting radio parameters for better performance for a particular application communication requirement
- Predicting what should be happening, and focus additional resources on the situation when a deviation from this prediction occurs

We hypothesize that a cognitive and collaborative system performing radio and network configuration and management using distributed reasoning and machine learning techniques will scale to very large network sizes, and address the complexity limitations of a traditional hierarchical network management approach.

The cognitive architectural approach discussed in this chapter extends beyond cognitive radio at layer 1 to encompass the entire networking and communication system. This is the vision of cognitive networking. Because wireless networks provide such a range of challenges, the entire span of the radio, network, and control and management system should be mined for opportunities for improvement. For example, an architecture described in [38] illustrates how the management functions can interact with various

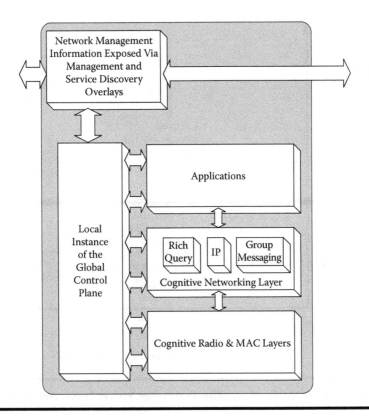

Figure 13.3   Network management and control protocols [38].

layers of the protocol stack. This is depicted in Figure 13.3, where the management and control functions are implemented in the global control plane, which provides interfaces within and beyond the local node. This mechanism allows information to be shared amongst the layers of one node, and with other nodes that have subscribed to this information via various network information sharing options. Each of the layers and the management and control plane might exploit cognition at various time scales (e.g., stimulus/reaction, tactical, and strategic) to improve the overall network performance according to the selected metrics.

The variety of approaches to the application of cognitive techniques at different layers and time scales is illustrated in Figure 13.4. Classical techniques have traditionally been applied at the PHY layers, although these approaches extend up through transport protocols such as TCP. Cognitive techniques such as those we have classified as tactical are also being applied at various layers. Cognitive radio adaptation at moderate time scales and application-focused topology selection (both discussed in this chapter) are

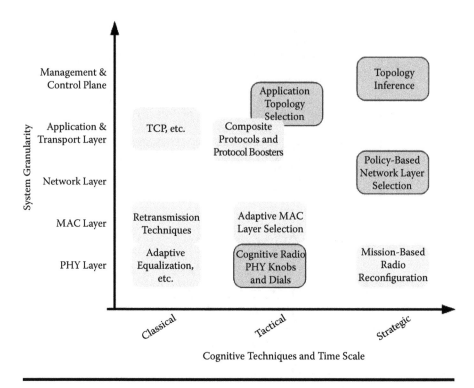

Figure 13.4   Relationship of cognitive approaches, system layers, and time scales.

examples. Other tactical techniques being pursued include adaptive MAC layers and techniques for construction of application-tuned protocol suites. At the strategic time scale, policy-based network selection, management-plane-based topology inference (also both discussed in this chapter), and mission-based radio reconfiguration show promise for emergency response and military use as well as more conventional needs.

13.1.1   Organization

In order to offer insight into these dimensions and their relationships, this chapter will provide examples of the use of cognition at multiple layers and time scales. These are highlighted in Figure 13.4. The remainder of the chapter is organized as follows. In Section 13.2, a tactical layer 1 (cognitive radio) model is presented. Section 13.3 describes a strategic layer 3 (network layer) model. Section 13.4 elaborates on tactical and strategic models for network control and management at tiers, and Section 13.5 provides a brief summary and conclusion.

13.2   Cognitive Radio—Tactical Layer 1

Cognitive radio technology has been receiving significant attention as a way to alleviate the FCC-identified problem of the scarcity of available frequency spectrum [16–19]. However, cognitive radio technology is well equipped to handle more than this single, yet very popular and needed, objective. In the case of frequency adaptation, the physical layer parameters, center frequency and signal bandwidth, can be adapted in real-time to provide for optimal transmission so as not to interfere with other signals within the spectral area. This section introduces several physical and MAC layer parameters than can be adapted using artificial intelligence techniques implemented within a cognitive radio to provide for a fully optimal transmission between a single pair of wireless devices.

Effective wireless communication involves interaction between many transmission parameters spanning several different layers. The primary ability of cognitive radios is to be able to adapt the available transmission parameters in real time in order to provide for optimal communications [20]. The following sections introduce several common transmission and environmental parameters that are common to wireless communication devices. We also introduce a list of performance objectives that are common goals when attempting wireless communication. The cognitive radio transmission parameters are adapted based on the current environment parameters in order to achieve the stated performance objectives. This process of adapting parameters in order to achieve specific goals can be done using one of the many artificial intelligence techniques. We give examples of two different artificial-intelligence-based cognitive engine implementations and describe how they use mathematical relationships between the transmission parameters, environmental parameters, and performance objectives to dynamically adapt to the surrounding wireless environment to achieve optimal communication.

13.2.1   Cognitive Radio Parameters

In developing a cognitive radio system, several inputs should be defined. The accuracy of the decisions made by the cognitive radio engine is based upon the quality and quantity of inputs to the system. More inputs to the system make the radio more informed, thus allowing the decision-making process to generate decisions that are more accurate.

We describe three different categories of inputs used by a cognitive radio system. The transmission parameter set, or "control knobs," represent the set of inputs that can be controlled by the system. The environmental parameters, or "system dials," represent the information about the current wireless environment. In order for the cognitive engine to determine the appropriate transmission parameters to use for communication, the current

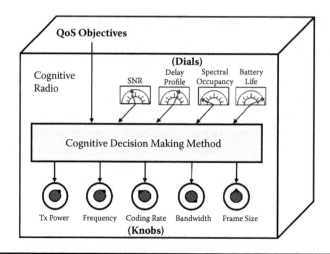

Figure 13.5   Cognitive radio system with knobs and dials.

wireless environment must be modeled internally. This model is created using environmentally-sensed data received by the system using external sensors. In addition to transmission and environmental parameters, performance objectives are used to guide the wireless communication to a specific performance goal. For example, one basic objective is to minimize the bit error rate (BER). This can be done by manipulating the transmission parameters in such a way as to provide for the lowest possible BER given the current environment. The following sections define five performance objectives, ten physical and MAC layer transmission parameters, and six environmental parameters. A visual representation of a cognitive radio system that shows a few common example "knobs" and "dials" and how they are input to the cognitive engine is shown in Figure 13.5.

13.2.2   Transmission Parameters

The transmission parameters described in this section represent parameters that have the largest impact upon the wireless communication operation of a system. We have chosen parameters that would be adjusted on the order of every several minutes in order to keep an optimal communications environment in the face of a more dynamically changing wireless environment. We intentionally do not focus on parameters that change on the order of hours, such as transmission formats (e.g., OFDM or CDMA), encryption (e.g., WEP or PGP), or error control techniques (e.g., Turbo coding or convolutional coding). These parameters are handled by higher layer control systems within the cognitive networking model. The complete list of parameters we have defined is shown in Table 13.1.

Table 13.1   Common Radio Transmission Parameters

Parameter	Description
Transmit Power	Raw transmission power
Modulation Type	Type of modulation
Modulation Index	Number of symbols for a given modulation scheme
Carrier Frequency	Center frequency of the carrier
Bandwidth	Bandwidth of transmission signal
Channel Coding Rate	Specific rate of coding
Frame Size	Size of transmission frame
Time Division Duplexing	Percentage of transmit time
Symbol Rate	Number of symbols per second

13.2.3   Environmental Parameters

Environmental parameters inform the system of the surrounding environment. This environment can include *internal* information about the radio operating state and *external* information representing the wireless channel environment. Both types of information can be used to aid the cognitive controller in making decisions. The complete list of environmental parameters we have defined is shown in Table 13.2.

13.2.4   Performance Objectives

In a wireless communications environment, there are several desirable objectives that the radio system may want to achieve. We define five common objectives that are used to guide the system to an optimal operating state. The five objectives are shown in Table 13.3.

The cognitive engine implementation uses mathematical representations of these performance objectives in order to determine the optimal transmission parameters for communication. These relationships define how each

Table 13.2   Common Radio Environmental Parameters

Parameter	Description
Signal Power	Signal power as seen by the receiver
Noise Power	Noise power density for a given channel
Delay Spread	Variance of the path delays and their amplitudes for a channel
Battery Life	Estimated energy left in batteries
Power Consumption	Power consumption of current configur ation
Spectrum Information	Spectrum occupancy information

Table 13.3 Common Radio Performance Objectives

Objective	Description
Minimize Bit-Error-Rate	Improve the overall BER of the transmission environment
Maximize Data Throughput	Increase the overall data throughput transmitted by the radio
Minimize Power Consumption	Decrease the amount of power consumed by the system
Minimize Interference	Reduce the radios interference contributions
Maximize Spectral Efficiency	Maximize the efficient use of the frequency spectrum

transmission parameter affects the performance objective given a set of environment parameters. We define these relationships as fitness functions. Each performance objective has a fitness function, which takes as inputs both transmission parameters and environment parameters, and outputs a score representing how well a single set of transmission and environment parameters achieves the corresponding objective. The cognitive engine uses these functions to derive the optimal transmission parameter set. The fitness functions are covered in more detail in the following section.

Realistically, all performance objectives will be used simultaneously by the cognitive engine. However, certain objectives will be more important than others in certain scenarios. For example, in an emergency disaster scenario, law enforcement objectives for wireless communication devices may tend to require an emphasis on minimizing interference or minimizing power consumption in a handheld mobile case. However, if the scenario suddenly changes to the need to transfer video of the scene to a local news agency or law enforcement precinct, the objectives may change to emphasis on maximizing data throughput or minimizing BER. This requires the need for preference information attached to each objective. To enable this preference information, we apply weights to each objective, which allows the cognitive engine to determine the appropriate order of importance for simultaneously optimizing multiple objectives.

13.2.5 Fitness Functions

The transmission and environmental parameters are combined with the performance objectives and their weights to create fitness functions. Equation (13.1) gives an example of a multiple-objective fitness function that can be used by an arbitrary cognitive decision-making engine to aid in the parameter selection:

$$f = w_1 * \left(\frac{P_i}{N * P_{\max}} \right) + w_2 * \left(\frac{\log_{10}(0.5)}{\log_{10}(\overline{P_{be}})} \right) + w_3 \left(\frac{M_i}{N * M_{\max}} \right) \quad (13.1)$$

The sample fitness function in Equation (13.1) shows how three performance objectives can be combined to form a single multiple-objective fitness function. Each performance objective has an associated weight, w_i, that denotes the importance of each objective. The inputs to this function include values for both the transmission and environment parameters defined earlier. The output of the function is a score representing how well the transmission parameters achieve the performance objectives in the given environment. Typically this output ranges from 0, the worst case, to 1, the optimal case. Fitness functions can be used in many different ways. We present two different techniques that use the fitness function in drastically different manners in order to implement a cognitive radio engine.

13.2.6 Cognitive Radio Engine Techniques

Several methods are available to implement the cognition engine for a cognitive radio. A wide variety of AI technologies, including neural networks, genetic algorithms, case-based learning, reinforcement learning, fuzzy systems, expert systems, and pattern recognition, exist that can be used as the decision-making engine to determine the transmission parameters in a wireless system. Traditional optimization problems typically consist of derived algorithms that determine the exact optimal parameters for a given problem. Cognitive radio systems operate in a dynamically changing and nonlinear environment requiring nonlinear solutions to determine the proper transmission parameters. In addition to the nonlinearity of the system, the constantly changing environment requires a dynamic solution that can be readily called upon to provide the optimal solutions. Two techniques are presented that are possible candidates for cognitive radio engines that can control physical and MAC layer transmission parameters.

Rule-based systems (RBS) are built upon a set of rules that define how the system operates [21,22]. All possible environments are represented in the rule base along with the optimal decision for each unique environment. This technique requires offline analysis of the problem in order to generate the proper rule set. The advantage of this method is the speed at which a solution is found. However, the high speed solution comes at the price of the large database required to represent the rules.

Fitness functions are used offline to create the rule base used by the RBS. Assumptions are first made about all possible transmission and environmental parameter ranges that may be encountered by the wireless system. These assumptions come down from higher level layers within the cognitive network system; thus the offline processing is dependent on the higher level decisions, limiting the flexibility of this technique. Using these ranges, all possible parameter combinations are input into the fitness function and the optimal solutions are determined and transformed into a usable rule base. This approach uses offline processing to minimize the amount of time spent processing during the radio uptime. The disadvantage of this technique is the loss in flexibility due to the static rule base required. This can be overcome by generating an extremely large rule base; however, memory storage issues can then become a problem.

As an alternative to the RBS, a genetic algorithm (GA)-based solution has also been examined [23]. The advantages of a GA-based solution include increased resilience to convergence at local extrema, unlike traditional optimization techniques. Also, GAs work with a population of possible solutions, not just a single point. This parallel processing allows the GA to explore different portions of the search space simultaneously. Fitness functions are used to score possible solutions and determine which solutions are appropriate for certain environments. In our case, the fitness functions are the mathematical representations of the fitness functions defined in Table 13.3.

The basic idea of genetic algorithms is as follows: the genetic pool of a given population of possible solutions, or chromosomes, potentially contains the optimal solution to a given adaptive problem. The optimal solution is not active in the current population because its genetic combination is split between several other possible solutions. Splitting and combining multiple chromosomes in the population several times can lead to the optimal solution. To perform this procedure, we must define two things. The first is the genetic representation of the solutions, and the second is the fitness function to score the possible solutions. For the genetic representation, the standard representation is an array of bits. The fitness function outputs a score for each chromosome in the current population, and the top-ranking pairs of chromosomes are combined to create a new generation of possible chromosomes. Several techniques exist to combine the chromosomes; one typical approach is called one-point crossover. This technique basically takes two of the high-scoring chromosomes and swaps certain bits between the two, the end objective being that by combining two high-scoring chromosomes, a still higher-scoring chromosome may be produced. This process is repeated for a specific number of generations or until the average score of the population reaches a prespecified limit.

Another advantage of genetic algorithms is the high degree of flexibility with regard to the fitness functions and parameters. In addition, there is no offline processing needed, and in general the ranges of the parameters do not affect the GA processing. This means that there is no dependence on the high layer cognitive network layers. However, genetic algorithms typically require a large amount of processing in order to converge to a near-optimal solution.

13.2.7 Cognitive Radio Conclusions

Autonomous adaptation of the lower level communication layers, specifically the physical and MAC layers, can be achieved using common artificial intelligence techniques. Defining common communication transmission and environment parameters that are shared among the radio devices is essential in creating the model used by the AI engines. Performance objectives, such as minimizing bit error rate or maximizing data throughput, can be represented mathematically and can be used to relate transmission and environmental parameters in order to define a "scoring" function that can rate how well a certain transmission parameter set achieves a performance objective given a set of environment parameters. Cognitive engine decision-making techniques use these fitness functions as the implementation techniques in order to derive the optimal decision regarding transmission parameters. We have shown two different techniques that use fitness functions in different ways. Tradeoffs exist within these techniques, focusing mainly on the tradeoff between memory usage and speed of decision-making. The rule-based system offline approach achieves fast decision-making times at the expense of less flexibility and possibly large memory usage. Genetic algorithms represent a very flexible solution and low memory usage, but suffer from large processing requirements.

13.3 Network Layer Cognition—Strategic Layer 3

Layer 3 cognition is broadly similar to layer 2 cognition and other simple models of learning and adaptation; however, the objectives of the layers are significantly different. The objective of layer 3 is based on end-to-end radio system goals, in contrast to layer 2 goals that are localized to a single radio device. In [24] a cognitive network-like concept was defined as a "pervasive system within the network that builds and maintains high level models of what the network is supposed to do." From this early definition we formally define a *cognitive network* to be *a network that is capable enough to sense its environment and act based upon the information gathered to satisfy overall system objectives.*

A system of cognitive radios without these system objectives will be capable of cognitive processing and communication on an individual basis, but might not be a cognitive network. An example set of objectives for a cognitive network may include policy compliance, security, access control, resource optimization, and similar network goals. At the same time, the cognitive network's capability to adapt depends on underlying elements and the diversity in the configurable "knobs" they provide. Cognition in layer 3 can be termed as providing knobs or modifiable elements that are configured automatically based upon the performance of the network as a whole. This broadly encompasses the issues of routing reconfiguration, addressing-naming control, and many other common network objectives.

13.3.1 Challenges

The unique opportunities offered by a flexible layer 2 provide new challenges for higher layers. Specifically, the issues of routing, naming and addressing need to be examined carefully.

- *Routing*: Routing is an important problem that is not yet understood very well in cognitive networks. The special properties of the open spectrum access paradigm demand new routing algorithms to be developed with multihop communication requirements. There is a need for novel routing algorithms in these environments. Section 13.3.2 outlines a cognitive networking model that makes use of these routing algorithms. An important design decision is the coordination between routing and spectrum management. We explore a design that strives to take advantage of algorithms that are very well understood by modifying their application in the cognitive context, which is intermittent in terms of both time and space [26]. Routing in open spectrum access then involves two approaches, the first being a cross-layered approach that coordinates route construction and spectrum management. The second routing approach is a decoupled approach where route selection is performed independent of spectrum management. The authors of [27] investigate the degree of interdependence between the two routing approaches we identified. While the first approach appears to suit cognitive routing by taking advantage of the unified approach, the latter exploits the available knowledge from the wireless routing world. In this scheme, the route selection is performed independently of the spectrum management by using popular algorithms such as shortest path. The spectrum management is performed on a hop-by-hop basis. Each source node uses methods that individually perform routing decisions using the global view of the network.
- *Naming and addressing*: As discussed in the following section, layer 3 for cognitive networks is envisioned to have a set of protocols, each with a unique name and addressing space. It becomes imperative to preserve a node identity across a myriad of such layers to achieve beneficial cognition. It is expected that these layers

exchange useful knowledge between them in order to have a shorter setup delay, and to reduce the overhead associated with changing layers. The host identity protocol, discussed in Section 13.3.4, provides a consistent naming scheme by introducing a new identity name space across the layers. The upper layer could make use of this space while being transparent to seamless cross-layer relocation.

13.3.2 Cognition Model

Figure 13.6 provides a visual representation of a cognition model at layer 3. The primary objective in developing the model is to strike a balance between two contradictory concepts of generalization and specialization. Communication systems with generalized architectures have been successful and time-tested in being good frameworks for extensible development. For example, the OSI stack has proven to be successful in supporting a wide range of applications at different layers (e.g., TCP or UDP, IP or ATM, voice, video, data, etc.). Models that are specific for particular problems restrict application and implementation. These models try to describe a solution by solving the problem through extending cognition to layer 3, but staying within a generalized framework allowing extension. David Marr described a general cognition model in his work on computer vision [28,29]. It consists of three separate layers: the behavioral, the computational, and the

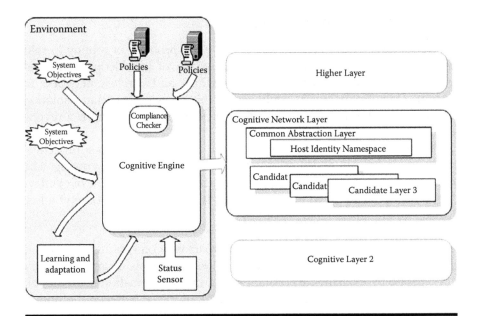

Figure 13.6   Cognition model for layer 3.

neuro-physical. The behavioral layer serves as an environment providing opportunities for cognitive adaptation. The computational layer reasons upon the available inputs from the behavioral layer. The neuro-physical layer activates decisions in the environment.

Cognition at layer 3 is similar to Marr's model, with the environment being the source for system objectives and policies resembling the behavioral layer: the observation and action elements that provide the actual "knobs" for tunable parameters resembling the neuro-physical layer, and the cognitive engine that feeds from the environment layer for cognitive decisions and learning adaptation elements resembling the computational layer, which provides a feedback loop to the engine to retrieve information from previously gathered knowledge. The model follows a conventional layered architecture. The difference lies in the network layer and the new layer above it, but below the transport layer. Essentially the network layer in the model consists of independent, non-overlapping protocols each with its own address space and routing strategy.

The common abstraction layer presents a common name space and other features, such as global network view, that are expected to be beneficial to all the individual protocols. This provides a framework for other features to be incorporated as they continue to evolve. The host identity protocol (HIP) layer separates identity and addressing namespace. The ongoing HIP effort presents necessary naming and state maintenance infrastructure that can be opportunistically used [30,31].

13.3.3 Available Network Layers

Active research in the past decade to solve the problem of routing in self-managed networks has given rise to a wealth of knowledge and good understanding about these kinds of networks and wireless networks in general. Cognitive networks could exploit this knowledge to find a solution that best suits the new challenges they present. The following is a review of routing protocols of interest.

The protocols that we review here fall into four categories: proactive, on-demand, hybrid, and distributed hash table (DHT)-based routing. The third set of protocols (hybrid) are primarily based on the success of overlay-based networks of the wired counterparts wherein routing is based on key lookup among an ordered set of keys.

- *Proactive routing protocols*: Most proactive protocols are based upon the traditional link state routing. These protocols maintain network state by exchanging constant background information even though there is no actual communication taking place between the nodes. The protocols have many desirable properties, especially for applications including real-time communications and quality of service

(QoS) guarantees, such as low latency route access and alternate QoS path support and monitoring. This may be especially desirable in the cognitive context, which has a variability in link properties that makes regular exchange of information of network status desirable and correct.

■ *Reactive routing protocols*: Reactive routing differs from the first approach in that the routes are established when a need for communicating data arises. This requires less space for state maintenance at each node, being inherently scalable to a large set of nodes. Reactive protocols have a route discovery phase where the route query packets are flooded, and the phase is completed when a route is found or all the possible outgoing paths from the source are searched. Routes are discovered using different approaches in these types of protocols. The intermediate nodes use backward learning to populate their forwarding table with a route to the node requesting a route (as in ad hoc on-demand distance vector (AODV) routing [32]). The destination can establish a bidirectional path to the source when it receives the request and traces its origin back to the source. The overhead of new path establishment is reduced when the node stops further flooding of route requests once a route to the destination has been extablished. Routes have a lifetime that is decided by a source taking into account the time a route is needed. A link failure is reported to the source recursively through the intermediate nodes, which in turn triggers another query-response procedure in order to find a new route. Some protocols use source routing where a source indicates in a data packet's header the sequence of intermediate nodes on the routing path. The query packet copies in its header the IDs of the intermediate nodes it has traversed. The destination then retrieves the entire path from the query packet, and uses it to respond to the source, providing the source with the path at the same time. Data packets carry the source route in the packet headers. A node aggressively caches the routes it has learned so far to minimize the cost incurred by the route discovery. When link breakage is detected through passive acknowledgements, route reconstruction can be delayed if the source can use another valid route directly. If no such alternate routes exist, a new search for a route must be re-invoked. The path included in the packet header makes the detection of loops very easy.

■ *Hybrid protocols*: Hybrid routing protocols combine both proactive and on-demand routing strategies and benefits from the advantages of both types. The zone routing protocol (ZRP) [33] is an example of a hybrid protocol that has a predefined zone centered at itself in terms of number of hops. For nodes within the zone, it uses proactive routing protocols to maintain routing information. For those nodes outside of its zone, it adopts on-demand routing methods. The membership of a node within the zone is governed by the availability of a route to that node within a zone. This means that the protocol guarantees a route to a node if it is a member of the zone. When a destination is not known to a node, it must lie outside its own zone.

■ *DHT-based routing protocols*: Virtual ring routing (VRR) uses a ring structure used in DHT overlays, which lie above layer 3, to provide layer 3 routing [34]. Each node has an identity that is unique, random, and ordered in ringlike fashion with wraparound. Every node maintains a route to a set of neighbors in this ring and in either direction. VRR differs from other DHT overlay routing in the way forwarding information is set up. Whereas a DHT-based protocol depends on underlay to perform the forwarding, VRR interacts directly with the link layer. This interaction results in routing table entries to nodes nearby in identity space. Routing is simply carried on by forwarding a packet to the node closest in the identity space.

13.3.4 Host Identity Protocol

Host identity protocol cleanly separates the locator role and the identity role of an IP address by leaving location identity to IP and introducing a new namespace based on host identifiers [31]. This namespace is inserted between the network and transport layers, allowing the transport layer to bind persistent host identities that can migrate among locations, and their corresponding IP addresses, so that higher layers can transparently function across the mobile environment. While HIP is intended for use in both non-mobile and mobile environments, for the purpose of this work we focus on its application in the mobile environment. Host identifiers are the public keys from a public-private key pair, and the public portion can be published using mechanisms such as DNS. In order to allow flexibility in the public-private encryption algorithms used, the host identifier is subjected to a 128-bit hash to create a host identifier tag (HIT) which is exchanged between hosts using a Diffie-Hellman-based mechanism to provide authentication of the endpoints [35]. The HIT is then used by the transport layer to bind sessions to host identities instead of the IP address, decoupling the host's identity from its location. HIP has been extended to provide LO-CATOR records that allow a host to notify its peers about an IP address change, allowing for session preservation as a host moves from one IP address space to another. This method is intended for hosts that are already communicating (i.e., have completed the HIP base exchange). For hosts trying to start a new session, with one or more operating in a changing

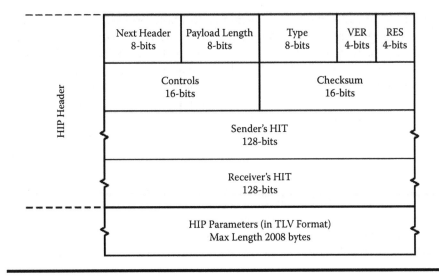

Next Header 8-bits	Payload Length 8-bits	Type 8-bits	VER 4-bits	RES 4-bits
Controls 16-bits		Checksum 16-bits		
Sender's HIT 128-bits				
Receiver's HIT 128-bits				
HIP Parameters (in TLV Format) Max Length 2008 bytes				

Figure 13.7   Host identity protocol packet format.

IP environment, further extensions provide a rendezvous server (RVS) [36]. Using RVS, a host will register its HIT and current IP, and publish its HIT and the RVS's IP address, which is presumed to be fixed, into DNS. When a host wants to initiate a session with the mobile host, it will send its initial packets to the RVS, which will then relay them to the current IP address of the destination host. The mobile host updates the RVS as it moves from one location to another. This allows two hosts who have not communicated to create sessions in a mobile environment. Figure 13.7 depicts the format of an HIP packet [39].

13.3.5   Policy Considerations

Cognitive networks present a wide variety of operating dimensions. For example, frequencies, power, and modulation can all be varied simultaneously, making it difficult to design an optimal algorithm that makes use of available spectrum efficiently and provides an optimal communication environment. A flexible mechanism had to be designed that ensures adherence to regulatory policies at these different operating dimensions. In our model we use a KeyNote-based policy enforcement mechanism that provides system-level policies for various layers of the cognitive protocol stack [37]. The following sections discuss more about KeyNote.

KeyNote is a system with unified notions of security policy, credentials, access control, and authorization. It provides a common domain-independent assertion language that has flexible mechanisms to represent a wide variety of security specifications. KeyNote provides various abstractions to represent entities to which authorization is delegated. These are sources of actions that need to be controlled. The principals are identified by the HITs or names to identify roles. The attribute value set is a common provision to describe actions and policy conditions. A logic engine answers the queries of compliance with an application-specified value. Within our cognitive networking framework, we restrict access to various layer 3 network layers by checking for compliance using KeyNote against the implemented policies. For example, the *join* action that is performed by the node to access a particular layer 3 network is checked against admission control policies. Additionally, nodes that can perform special actions are required to present the authorization information during the request for that operation, such as a relief unit requesting access to a video sensor in a private building. This authorization information must be obtained beforehand and certified by a delegation authority such as a management authority.

13.3.6 An Approach for Dynamic Network Layer Selection

The routing decision made by cognitive radio nodes is based upon information learned from various sections of the environment: for example, communication needs at the application level, security constraints, and lower layer states such as link quality. The decision-making involves both the selection of the appropriate network layer and the routing decisions within the selected layer. The dynamic nature of the delay associated with the decision-making algorithm required to switch to an alternate network layer raises various issues of naming and state maintenance. In HIP there is a natural state maintenance by associating higher layer associations with the host identifier. A host assumes the responsibility of disseminating updates about its new location identifier. This can be used to provide information about the network layer changes and associated name material distribution. The hosts then utilize more appropriate network layers without affecting higher layer states. In particular, the LOCATOR information in the update message can be utilized to inform the peer about the associated changes. We assume here that all the cognitive nodes are able to identify and participate in the desired network layer.

13.4 Cognitive Management and Control—Approaches at Higher Tiers

We continue working our way up the protocol stack by considering a distributed application that allows for the exploration of strategic cognitive processes at the application layer. The use case is based on an application that is tailored to the distribution and searching of rich media content in a military setting. The application distributes multimedia content that is continuously created and updated in response to reports of activity occurring in the field. The application must be able to efficiently search large amounts of data including rich multimedia attachments, as well as perform data transformations to provide bandwidth-efficient representations (e.g., thumbnails) of these media and have the ability to distribute the storage of the full-quality media. To coexist with existing networked applications, the application must efficiently utilize network links with highly variable properties including topology, delay, loss, link bandwidth, available bandwidth, and reachability. Policy-based selection of content through subscriptions, discrete scheduling of media distribution and network utilization, and varying application layer topology provide unique tools for manipulating the network traffic. We use this example to study the strategic management of the topology of the distributed application.

13.4.1 Application Topology Management Using Strategic Cognition Processes

The application layer topology of the distributed system is driven by the nodes available to join the system at any time, the network topology and link capabilities interconnecting those nodes, and the policies of what data is to be made available at each node. Some nodes in the network will be larger infrastructure-type nodes that have a permanent location and are well connected. Other nodes will have reduced capabilities in terms of disk space, available network bandwidth, or even intermittent connectivity, and such nodes can resort to restrictive policy expressions about the amount of data they want distributed to them. The policies used throughout the network, the layer 3 and layer 2 network topology, and the nodes participating in the application generally change on a time scale suitable for strategic cognitive processes. We infer a model of the layer 3 and layer 2 topology of the network between the nodes of the application using a strategic cognitive process based on a distributed measurement process, and then utilize this data and the policy requirements to guide the formation of the application level topology.

The distributed application collaborates on measurement collection to provide more complete coverage and quicker convergence for network topology discovery as compared to using a single measurement point. We present a hybrid approach using multiple viewpoints and both direct and indirect topology observations to discover a more complete topology while using lower aggregate network resources than a single-node system alone.

13.4.2 Data Collection for Inference

Data used for the topology inference comes from two main sources: traceroute-style hop-limited probes, and indirect measurements using single-ended pairwise delay probes. The traceroute technique works by sending hop-limited packets into a network and seeing which nodes return an error message, and in so doing identify themselves. Traceroutes [40] from a set of distributed servers [41] can be stitched together to provide more complete topologies than the tree resulting from a single node's mapping efforts. The hop-count-limited measurements provide layer 3 hop counts, and the unique and potentially correct names for nodes in the topology. This type of traceroute measurement approach for directly measuring topology has been performed by many [42,43], but lacks the ability to discover layer 2 devices, and suffers from problems resolving the aliases that are found due to multiple interfaces on the same physical router responding to probes.

Indirect measurements of network characteristics (network tomography) can be used to solve the alias problem and provide layer 2 details. In order to perform a shared network inference we need to use a metric that is monotonic, such as delay or loss. It can be done with loss, but as the loss rate is generally low, this requires many more samples to perform. Tomography approaches using active probes, including both multicast probes [44] and unicast probes [45,46], have typically required cooperating endpoints, and are done from a single node's point of view, i.e., a tree. Round-trip time tomography using single-ended measurements has been performed [47], but from only a single viewpoint as well. Tomography from two points of view was performed in [48], but required cooperating software on the endpoints, limiting deployability. We use pairwise probes to elicit the degree of shared paths between two nodes, from single and disparate viewpoints. By extending the multiple node probing technique found in [48] to use round-trip time measurements of TCP setup handshakes similar to the single-node approach in [47] but extended to multiple sources for the probing. The round-trip approach allows the measurements to be collected using just the source node, greatly increasing the deployability, and correspondingly the number of hosts, we can collect measurements from.

While performing data collection, a third probe type is used to determine available bandwidth on a link. There are many approaches for attempting to measure this value [49,50], but we utilize an approach similar to abing [51], which uses packet pairs consistent with our delay measurement approach and has a very low overhead associated with its measurement.

13.4.3 Network Topology Inference

We present a hybrid model for inferring the L3 and L2 topology that takes advantage of the information-rich hop-limited probe results to create a

Table 13.4 Measurement data collected

Parameter	Description
ProbeSrcAdr	The IP address of the source of the probe
ProbeDstAdr	The IP address of the destination of the probe
ProbeStart	The start time of the probe
RTT	Round trip time of the probe in ms
ProbeType	Traceroute style, packet pair, available BW
Hops	Intermediate hops and their estimated delay in sequence (for traceroute style)
BW	Available bandwidth on the path
TTL	TTL (time-to-live) found in initial ACK packet

limited layer 3 topology framework, and then uses the pairwise delay co-variance probes and a Markov chain Monte Carlo (MCMC) simulation to address limitations in the topology model. These limitations include collapsing "alias" nodes into a single physical router, identifying layer 2 devices, and filling in the topology around layer 3 devices that are not responsive to hop-limited probes. Despite these limitations, this named node framework and the TTL-based structure from the traceroute data provide an ideal starting point for a more detailed topology inference, and constrains the depths of possible topology trees, making a simulation-based inference more tractable by limiting the search space. MCMC inferences of topology have been performed [52,53], but only active probes and cooperating endpoints have been used.

We begin by assembling a layer 3 framework using the active traceroute data. Developing a model for the topology simulation, we define T as the topology of the network being evaluated, and D as the collection of covariance similarity measures between individual endpoints. We want to maximize

$$P(T|D) \propto P(D|T)P(T) \tag{13.2}$$

We model our topology T as follows: n is a vector describing the number of nodes in a layer of the tree. The interconnection of these nodes, $Z(n)$, is a matrix of indicator vectors from a node to its parent(s), where p is the parent of c for each node defined in n,

$$Z(n) = \begin{pmatrix} p_1 & p_2 & p_3 & p_4 & \cdots & p_n \\ c_1 & c_2 & c_3 & c_4 & \cdots & c_n \end{pmatrix}. \tag{13.3}$$

For the prior, $P(Z(n)|n)$, we assume each node is independent and has a uniform chance of being with each possible parent in the layer above, but start with the hop-limited-probe-derived topology.

The objective function that gives us the best hierarchy given the observed data is then the posterior, $P(n, Z(n)|D)$, of the hierarchy model given the data, varying both size of hierarchy and the interconnections:

$$P(n, Z(n)|D) \propto P(D|Z(n), n)P(Z(n)|n)P(n) \qquad (13.4)$$

The probability of the data D given a particular tree n, $Z(n)$ is a function of a number of terms. In a tree structure, there is only a single path between any two nodes, so the covariance between them is just the sum of the variances of parent nodes until a common one is reached. More complicated topologies require a more detailed method so that we can use the covariance of the delay between two endpoints to suggest whether they have a largely shared topology or not.

Common destinations with shared similarity can be used to decide on the likelihood of a merge of two nodes into a single node to resolve aliases, i.e., shrinking n_i at a particular layer i. Covariance shifts between endpoints that are not reflected by a branch in the layer 3 topology suggest the likelihood of a layer 2 branch, and can be cause for the insertion of a node into the topology, or even increasing the number of layers overall.

We perform a MCMC simulation using Gibbs sampling to maximize the posterior $P(T|D)$. At each step in the sampling we pick one of the nodes and compute the posterior for that node alone given the available covariance measurements. The algorithm holds all other nodes' positions fixed, calculates a single node's position posterior, draws a new sample, and continues. After a burn-in period, the Markov chain's distribution is the desired stationary one, and we converge to a solution.

The algorithm is inherently iterative, and when there are changes or additions to the observations, we can start from the previous accepted best topology and quickly reach an updated one. This property is especially useful in networks with variable topology due to terrestrial wireless or celestial-based links with mobile endpoints and changing uplinks.

Figure 13.8 shows an example of a topology with application nodes N1–N7 with routers and switches, as it goes through the process of starting with hop-limited probes, using inference to refine the topology using the variation in delays on packet pairs, and finally weighting the links via discovered available bandwidth.

13.4.4 Distributed Cognitive Topology Inference

A future task is to extend the cognitive topology inference algorithm to include distributing the inference among the nodes, i.e., not just distributing the collection of data for the inference, but adding collaboration on the inference itself. This will provide a quicker convergence for network topology discovery as compared to a single station performing the inference.

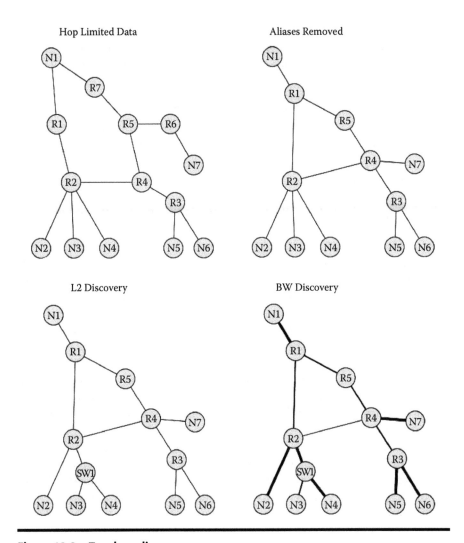

Figure 13.8   Topology discovery.

The decomposition of the problem, and distributed computation, can be performed in such a way as to minimize the data transfer for both the distribution of the data for the inference and the reconstitution of the complete solution, while allowing an overall speedup including all elements of the topological inference. We are able to avoid an explicit parallelization of the inference by taking advantage of the fact that the MCMC inference is a stochastic simulation. The basic idea is to run different simulations simultaneously, but with different Markov chains controlling their search. Higden et al. [54] proposed a method that has each distributed processor running its own Markov chain, exploring different regions of the search space. The approach is to have a range of granularities in the step sizes, and have a coarse-grained simulation quickly find promising regions and swap parameters with fine-grained simulations to allow for detailed searches.

13.4.5 Application Topology Optimization

The goal of the network topology estimation is to guide the formation of the application layer topology of the distributed application. Topologically aware overlays typically add additional complexity to the join and leave steps, as well as additional data structures to make sure the neighbors in terms of the peer-to-peer address space are close in proximity within the network space. Topologically aware systems like Pastry [55] and Tapestry [56] are able to implement data distribution more efficiently than random topologies by using a proximity-based metric (typically round-trip time) to pick the closest neighbor. In general, they look to have topologies with the property that nodes in successively smaller subtrees are increasingly near each other in the network. As a result, the data distribution is more efficient, both with respect to delay and link stress. Another approach to gaining topology information is to use an "oracle" [57] to gain knowledge from an ISP about AS hop counts to help optimize the topology. This approach requires cooperation from an ISP and doesn't handle intra-AS topology optimization, where most of our application is deployed. Our target application's neighbor selection is not dependent on a single network characterization, and extends multicharacteristic distance functions [58] with the addition of policy.

To determine the application topology, we first define C as a vector of all advertised capabilities of nodes in the system as defined by their individual policies. Each node i in the application desires to have a dataset D_i made available to it. The nodes create a list of valid neighbors V by finding the subset of nodes with advertised capabilities C_j for all nodes that have a superset of D_i. We do not explore satisfying D_i by summing multiple j's capabilities in this work, but it is a straightforward extension. Given the list V, node i selects as the best neighbor the node V_j with the best metric found in the physical topology from i to j. This metric is based on an objective function that includes the delay, hop count, and available bandwidth on the path between i and j.

Figure 13.9 shows how the application topology changes as its metric for best neighbor is expanded. First is the closest neighbor by hop count, with the hop count shown on the link. The next graph depicts the case when the nearest neighbor is modified to include bandwidth available on the path between nodes. In particular we see the biasing of the narrow-bandwidth link so that an extra hop is preferred to get to the data for connections between nodes N2 and N7, as well as the direct connection between N4 and N2. Finally, we show how policy can affect the layout; we

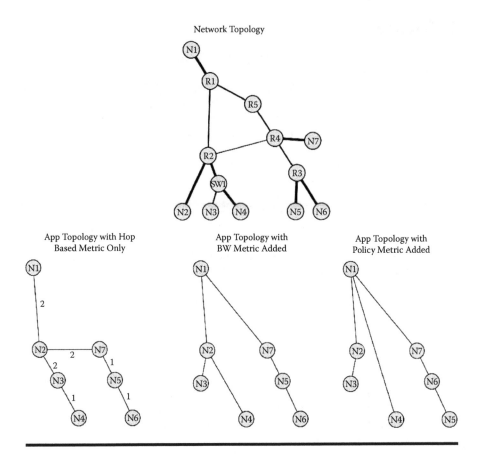

Figure 13.9   Application topology.

assign N1 and N7 to have a complete set of data, while N2 and N6 each have an identical copy of a subset region *A*. N3 and N5 also subscribe to the region *A*, while N4 subscribes to an alternate region *B*, and thus must peer with a further away neighbor in order to fulfill its policy requirements.

13.5   Conclusion

This chapter has discussed a general approach to the use of cognition in wireless networks. Through examples at the cognitive radio layer, at the network layer, and in the control and management plane, we have illustrated how this approach can be instantiated in parts of a comprehensive wireless network architecture. These techniques continue to undergo intensive research that will lead to evaluation of the individual techniques and the overarching approach using system performance metrics.

References

[1] PlanetLab project, http://www.planet-lab.org/

[2] GENI project, http://www.geni.net/

[3] D. D. Clark, C. Partridge, J. C. Ramming, and J. T. Wroclawski, 2003, A Knowledge Plane for the Internet in *Proceedings of the 2003 Conference on Applications, Technologies, Architectures, and Protocols for Computer Communications* (Karlsruhe, Germany, August 25–29, 2003). SIGCOMM '03. New York, NY, ACM Press, 3–10, 2003.

[4] J. Evans, Intelligence in the Network, Invited talk at the 60th Distinguished Professor Lecture Series, University of Kansas.

[5] P. Langley and J. E. Laird, "Cognitive Architectures: Research Issues and Challenges." Draft of October 31, 2002.

[6] D. J. C. MacKay *Information Theory, Inference, and Learning Algorithms*, Cambridge University Press, 2003.

[7] A. Gelman, J. B. Carlin, H. S. Stern, and D. B. Rubin, *Bayesian Data Analysis*, 2nd ed. London: Chapman & Hall, 2004.

[8] M. Zhang, C. Zhang, V. Pai, L. Peterson, and R. Wang, PlanetSeer: Internet Path Failure Monitoring and Characterization in Wide-Area Services, in *Proc. of OSDI*, 2004.

[9] M. Wawrzoniak, L. Peterson, and T. Roscoe, 2004. Sophia: an Information Plane for Networked Systems, in *SIGCOMM Comput. Commun. Rev.* 34, 1, 15–20, (Jan. 2004).

[10] 2006 Workshop on Real-Time Knowledge Processing for Wireless Network Communications, Stanford, CA., March 29 2006.

[11] D. Wilkins, Policy-Defined, Cognitive Radios, 2006 Workshop on Real-Time Knowledge Processing for Wireless Network Communications, Stanford, CA., March 29 2006.

[12] H. Balakrishnan, M. F. Kaashoek, D. Karger, R. Morris, and I. Stoica, Looking Up Data in P2P Systems, in *Commun. ACM* 46, 2, 43–48, Feb. 2003.

[13] E. K. Lua, J. Crowcroft, M. Pias, R. Sharma, and S. Lim. A Survey and Comparison of Peer-to-Peer Network Schemes, in IEEE Communications Tutorials and Surveys, March 2004.

[14] A. Rowstron and P. Druschel, Pastry: Scalable, Decentralized Object Location, and Routing for Large-Scale Peer-to-Peer Systems, in *Lecture Notes in Computer Science*, volume 2218, 329, Jan. 2001.

[15] I. Stoica, R. Morris, D. Karger, M. F. Kaashoek, and H. Balakrishnan Chord: A Scalable Peer-to-Peer Lookup Service for Internet Applications, in *ACM Computer Communication Review*, vol. 31, part 4, 149–160, 2001.

[16] Joint Tactical Radio Systems, Software Communications Architecture Specification, November 2002.

[17] C. J. Rieser, Biologically Inspired Cognitive Radio Engine Model Utilizing Distributed Genetic Algorithms for Secure and Robust Wireless Communications and Networking, Ph.D. dissertation, Virginia Polytechnic Institute and State University, April 2004.

[18] P. R. Etkin and D. Tse, Spectrum Sharing for Unlicensed Bands, in *IEEE International Symposium on New Frontiers in Dynamic Spectrum Access*, 2005.

[19] Spectrum Policy Task Force, Report of the Spectrum Policy Work-Group, November 2002. [Online]. Available: http://www.fcc.gov/sptf/files/SEWGFinalReport 1.pdf.

[20] T. R. Newman, B. A. Barker, A. M. Wyglinski, A. Agah, J. B. Evans, and G. J. Minden, Cognitive Engine Implementation for Wireless Multicarrier Transceivers, in *Wiley Journal on Wireless Communications and Mobile Computing*, November 2007.

[21] J. P. Ignizio, *Introduction to Expert Systems: The Development and Implementation of Rule-Based Expert Systems*, McGraw-Hill, 1991.

[22] B. G. Buchanan and E. H. Shortliffe, *Rule-Based Expert Systems: The MYCIN Experiments of the Stanford Heuristic Programming Project*, Addison-Wesley, 1985.

[23] J. H. Holland, *Adaptation in Natural and Artificial Systems*, MIT Press, 1992.

[24] D. D. Clark, C. Partridge, J. C. Ramming, and J. T. Wroclawski, A Knowledge Plane for the Internet, in *Proceedings of SIGCOMM '03*, (New York, NY, USA), 3–10, ACM Press, 2003.

[25] R. W. Thomas, D. H. Friend, L. A. Dasilva, and A. B. Mackenzie, Cognitive Networks: Adaptation and Learning to Achieve End-to-End Performance Objectives, in *Communications Magazine, IEEE,* vol. 44, 51–57, Dec. 2006.

[26] R. W. Thomas, L. A. DaSilva, and A. B. Mackenzie, Cognitive Networks, in *Proceedings of IEEE DySPAN 2005*, 352–360, November 2005.

[27] Q. Wang, and H. Zheng, Route and Spectrum Selection in Dynamic Spectrum Networks, in *IEEE Consumer Communications and Networking Conference* (CNCC), January 2006.

[28] D. Marr, *Vision: a Computational Investigation into the Human Representation and Processing of Visual Information*, W. H. Freeman, 1982.

[29] P. N. Johnson-Laird, *The Computer and the Mind*, Harvard University Press, 1988.

[30] M. Pitchaimani, B. J. Ewy, and J. B. Evans Evaluating Techniques for Network Layer Independence in Cognitive Networks, in *ICC '07. IEEE International Conference on Communications.*, Glasgow, Scotland, 6527–6531, 2007.

[31] R. Moscowitz, Host Identity Payload and Protocol, Internet Draft: draft-moscowitz-hip-05.txt (work in progress), November 2001, Available online at http://homebase.htt-consult.com/HIP.html.

[32] C. E. Perkins and E. M. Royer, Ad Hoc On-Demand Distance Vector Routing, in *2nd IEEE Workshop on Mobile Computing Systems and Applications*, 90–100, 1999.

[33] Z. Haas and M. R. Pearlman. The Zone Routing Protocol (ZRP) for Ad Hoc Networks, in http://www.ee.cornell.edu/haas/Publications/draft-ietf-manet-zonezrp-02.txt, June 1999.

[34] M. Caesar, M. Castro, E. Nightingale, G. O'Shea, and A. Rowstron, Virtual Ring Routing: Network Routing Inspired by DHTs, in *SIGCOMM Comput. Commun. Rev.* 36, 4, 351–362 Aug. 2006.

[35] W. Diffie and M. E. Hellman, New Directions in Cryptography, in *IEEE Trans. Info. Theory IT-22*, 644–654, November 1976.

[36] T. Koponen, A. Gurtov, P. Nikander, Application Mobility with HIP, in *Proc. of ICT'05*, May 2005.

[37] M. Blaze, J. Feigenbaum, and A. D. Keromytis, KeyNote: Trust Management for Public-Key Infrastructures (position paper), in *Lecture Notes in Computer Science*, 1550:59–63, 1999.

[38] D. Raychaudhuri, N. Mandayam, J. B. Evans, B. J. Ewy, S. Seshan, and P. Steenkiste, CogNet - An Architecture for Experimental Cognitive Radio Networks within the Future Internet, *First ACM/IEEE International Workshop on Mobility in the Evolving Internet Architecture (MobiArch 2006)*, San Francisco, December 2006.

[39] F. Al-Shraideh, Host Identity Protocol, *International Conference on Networking, International Conference on Systems, and International Conference on Mobile Communications and Learning Technologies (ICNICONSMCL'06)*, 203, 2006.

[40] Worldwide Traceroute Servers http://www.traceroute.org/

[41] K. Claffy, T. E. Monk, and D. McRobb, Internet Tomography, in *Nature*, January 1999.

[42] N. Spring, R. Mahajan, D. Wetherall, and T. Anderson, Measuring ISP Topologies with Rocketfuel, in *IEEE/ACM Trans. Netw.* 12, 1, 2–16 Feb. 2004.

[43] B. Huffaker, D. Plummer, D. Moore, and K. Claffy, "Topology Discovery by Active Probing," *Applications and the Internet (SAINT) Workshops*, 90–96, 2002.

[44] N. Duffield, J. Horowitz, F. Lo Presti, and D. Towsley, Multicast Topology Inference from Measured End-to-End Loss, in *IEEE Trans. Inform. Theory* 48, 26–45.

[45] A. Bestavros, J. Byers, and K. Harfoush, Inference and Labeling of Metric-induced Network Topologies, in *Proc. IEEE INFOCOM 2002*, 2, 628–637, IEEE Press, New York 2002.

[46] M. Coates, R. Castro, R. Nowak, M. Gadhiok, R. King, and Y. Tsang, Maximum Likelihood Network Topology Identification from Edge-Based Unicast Measurements, in *Proc. ACM SIGMETRICS 2002*, 11–20, ACM Press, New York, 2002.

[47] Y. Tsang, M. Yildiz, P. Barford, and R. Nowak, Network Radar: Tomography from Round Trip Time Measurements, in *Proceedings of the 4th ACM SIGCOMM Conference on Internet Measurement* (Taormina, Sicily, Italy, October 25–27, 2004). IMC '04. ACM Press, New York, NY, 175–180.

[48] M. Rabbat, R. Nowak, and M. Coates, Multiple Source, Multiple Destination Network Tomography, in *INFOCOM 2004. Twenty-third Annual Joint Conference of the IEEE Computer and Communications Societies*, 3, 1628–1639, 7–11 March 2004.

[49] R. Carter and M. Crovella, Measuring Bottleneck Link Speed in Packet-Switched Networks, in Technical Report 96-006, Boston University, 1996.

[50] V. Ribeiro, R. Riedi, R. Baraniuk, J. Navratil, and L. Cottrell, Pathchirp: Efficient Available Bandwidth Estimation for Network Paths, in *Passive and Active Measurement Workshop*, 2003.

[51] J. Navratil and L. Cottrell, ABwE: A Practical Approach to Available Bandwidth in *Passive and Active Measurement Workshop*, 2003.

[52] R. Castro, M. Coates, and R. Nowak, Likelihood-based Hierarchical Clustering, in *IEEE Trans. Signal Process.* 52, 2308–2321, 2004.

[53] M.-F. Shih and A.O. Hero, Network Topology Discovery Using Finite Mixture Models, in *IEEE International Conference on Acoustics, Speech, and Signal Processing, 2004. Proceedings.* 2, ii-433-ii-436, 17–21 May 2004.

[54] D. Higdon, H. Lee, and C. Holloman, Markov Chain Monte Carlo-based Approaches for Inference in Computationally Intensive Inverse Problems, in *Bayesian Statistics* 7, Oxford University Press, 2003.

[55] A. Rowstron and P. Druschel, Pastry: Scalable, Decentralized Object Location, and Routing for Large-Scale Peer-to-Peer Systems, in *Lecture Notes in Computer Science*, volume 2218, 329, Jan. 2001.

[56] B. Y. Zhao, L. Huang, J. Stribling, S. C. Rhea, A. D. Joseph, and J. D. Kubiatowicz, Tapestry: A Resilient Global-scale Overlay for Service Deployment, in *IEEE Journal on Selected Areas in Communications*, 22(1), 41–53, January 2004.

[57] V. Aggarwal, A. Feldmann, and C. Scheideler, Can ISPs and P2P Users Cooperate for Improved Performance, in *ACM Computer Communication Review*, 37(3), 31–40, 2007.

[58] M. Malli and C. Barakat, Application-Aware Model for Peer Selection, in *Inria Technical Report 5587*, May 2005.

V

ADVANCED TOPICS IN COGNITIVE RADIO NETWORKS

Chapter 14

Ultra-Wideband Cognitive Radio for Dynamic Spectrum Accessing Networks

Honggang Zhang, Xiaofei Zhou, and Tao Chen

Contents

This chapter describes the concept and approach of ultra-wideband cognitive radio (CR-UWB), a wireless technique based on UWB transmission able to self-adapt to the characteristics of the surrounding wireless communications environment. A novel strategy to exploit the advantages and features of integrating cognitive radio (CR) with the UWB technology is investigated with an aim of exploring UWB radio as an enabling technology for implementing cognitive radio. In particular, dynamic spectrum accessing and sharing based upon the spectrum agility and adaptation flexibility is proposed. At first, a common architecture supported by CR-UWB corresponding to heterogeneous networking scenarios is illustrated. Then, a number of concrete technical methods for generation of the transmitter-centric spectrum-agile waveforms, which take advantage of the adaptive combination of a set of orthogonal UWB pulse waveforms complying with the FCC spectral mask at the transmitter side, are discussed. Furthermore, various transmit power control and optimization schemes for improving the system capacity as well as the bit-error-rate (BER) performance are derived. Finally, the receiver-centric spectrum-agile waveform adaptation, which is characterized by generating the spectrum-agile UWB waveforms complying with the receiving interference limit rather than the FCC spectral mask is analyzed.

14.1 Introduction

Spectrum availability in current wireless communication systems is fixed by the regulatory and licensing bodies. To date, static frequency band allocation is widely adopted around the world. However, it has been gradually realized that this static spectrum assignment is one of the main reasons behind the commonly shared concerns of the lack of spectrum resources. This is especially true at frequencies below 3 GHz, with respect to which intense competition for spectral use has occurred over the last few years. At frequencies beyond 3 GHz, the situation is quite different: there is actually very little bandwidth usage in the time and space dimensions. According to the measurements in Ref. 1, in many practical cases, the actual spectral utilization is on the order of a few percent and often a fraction of a percent in large spectrum regions. This is in evident contradiction with the spectrum shortage concerns. It is believed that one of the promising solutions for spectrum abundance is the introduction of cognitive radio and networks.[2,3]

Cognitive radio (CR) is seen as one of the most encouraging approaches to solving the spectrum shortage problem by allowing smart and dynamic spectrum management in future wireless communication systems. CR is characterized as a system able to adapt its transmission or reception functions on the basis of cognitive interaction with the wireless environment in which it operates. This kind of interaction may involve dynamic spectrum sensing or, in general, autonomous communication and negotiation with other spectrum users (e.g., primary users) based on self-organizing procedures of learning and reasoning, which represent the "intelligence" of CR itself. Cognitive radio, by adapting its radio behavior to the local wireless environment, aims at improving spectrum efficiency, capacity, and fairness. As a seismic shift in the wireless communications and networking world, CR could potentially break the ever-serious "bottleneck" of limited spectrum availability and open up new frontiers of opportunities for wireless system designers and application developers. Moreover, with the intelligent capabilities of both radio link and network layers, CR is capable of transmitting information in an optimized way across the available multiple signal dimensions, allowing a huge increase in the prospects for channel capacity, coexistence, compatibility, and interoperability.

Reflecting the ever-growing expectations and needs for smart spectrum management, on December 30, 2003, the FCC (Federal Communications Commission) of the USA released a Notice of Proposed Rulemaking (NPRM) covering various application scenarios for cognitive radio technologies.[4] According to Ref. 4, one of the possible application scenarios of cognitive radio is stated as the following:

> Cognitive radio technologies can be used to enable non-voluntary third party access to spectrum, for instance as an unlicensed device operating at times or in locations where licensed spectrum is not in use.

Therefore, we start from this kind of application scenario and consider it as the most suitable approach for the CR concept to become a reality, an approach that is especially relevant to ultra-wideband (UWB) wireless technology and systems.[5]

UWB technology is currently regarded as a key player for broadband wireless communications in multimedia-rich environments. The UWB signals by definition occupy a bandwidth in excess of 500 MHz, or are such that their fractional bandwidth is greater than 20%. In the context of this chapter, we will concentrate on UWB impulse radio (UWB-IR), based upon the transmission of ultra-short pulses, on the order of a nanosecond, yielding ultra-wide bandwidth. The UWB technology possesses several potential advantages, such as (i) the possibility of achieving high throughputs

(>1 Gbit/s), (ii) immunity to multipath fading, (iii) resilience to interference, (iv) capability of precise ranging and localization (at the centimeter level in the case of UWB impulse radio), (v) enhanced capability to penetrate through obstacles, and (vi) support for the development of small-sized and processing-power-efficient devices. While UWB has generated a great deal of interest, it has also caused a number of controversies among industry, regulation, and standardization bodies. Since UWB signal waveforms are spread over very large bandwidth, they unavoidably overlap with existing and planned (licensed) narrow-band radio systems. In this respect, coexistence and compatibility have become critical issues demanding innovative solutions. In response to this, the FCC released the UWB radio emission mask for the realization of coexistence with traditional and protected wireless services.[6] Similarly, in Europe, the European Conference of Postal and Telecommunications Administrations (CEPT) has provided recommendations for harmonizing radio spectrum usage for UWB.[7] However, it has been widely recognized that UWB signal waveform design is quite a challenging subject, given the need for complying with various spectral masks while still achieving interference avoidance as well as efficient transmission. Taking these facts and requirements into account, our main objective and contribution for coexistence, interference avoidance, and compliance with any regulatory spectral mask is the provision of CR capabilities to UWB devices.

Correspondingly, there exist a number of reasons why it makes sense to promote the use of the UWB technology in the context of cognitive radio and CR networks. The main technical reasons for endowing UWB devices with CR capabilities can be summarized as follows:

- UWB is by definition an *underlay* technology, so it will face severe interference from and cause interference to nearby narrow-band systems; therefore it will surely benefit from utilizing CR techniques implementing collaborative coexistence policies. We believe this is the most realistic and pragmatic scenario for introducing the CR concept.
- UWB devices have inherent capabilities to observe large bandwidths, which is the prerequisite and technical basis for dynamic spectrum sensing.
- There is an intrinsic scalability in the UWB technology, which makes it an ideal candidate for realizing a versatile PHY layer, adaptable to various wireless environment conditions.

Starting from this background, we investigate a novel strategy of utilizing the advantages and features of the integration of cognitive radio with UWB technologies. This strategy exploits UWB radio as an enabling technology for implementing cognitive radio by virtue of the unique attributes of the UWB device endowed with CR-enhanced capabilities. Moreover, through intelligent spectrum planning and appropriate UWB waveform design, such a UWB radio with both spectrum agility and adaptation flexibility is anticipated to improve the prospects for both spectrum compatibility and interoperability among the ever-proliferating wireless communication devices and systems.

14.2 Common Architecture for Dynamic Spectrum Accessing in Heterogeneous Networking Environment

As mentioned before, the FCC has already released the Notice of Proposed Rulemaking (NPRM) covering the various potential applications for cognitive radio technologies. According to this NPRM, four main application scenarios are introduced as possible targets for cognitive radio systems:

- First, a licensee can employ cognitive radio technologies internally within its own network to increase the efficiency of spectrum use.
- Second, cognitive radio technologies can facilitate secondary markets in spectrum use, implemented by voluntary agreements between licensees and third parties. For instance, a licensee and a third party could sign an agreement allowing secondary spectrum uses made possible only by deployment of cognitive radio technologies. Ultimately, cognitive radio devices could be developed that "negotiate" with a licensee's system and use spectrum only if agreement is reached between a device and the system.
- Third, cognitive radio technologies can facilitate automated frequency coordination among licensees of co-primary services. Such coordination could be done voluntarily by the licensees under more general coordination schemes imposed by the FCC rules, or the FCC could require the use of an automated coordination mechanism.
- Fourth, cognitive radio technologies can be used to enable nonvoluntary third party access to spectrum, for instance as an unlicensed device operating at times or in locations where licensed spectrum is not in use.

Similarly, across the whole European area there has been a drive for more innovative use of radio spectrum as a result of the European Commission's strategy for introducing more flexible radio frequency usage in reaction to evolving market demands. This was reflected in the relevant

Communications Paper of the European Commission.[8] In this Communications Paper, it was proposed by the European Commission that a

> flexible, non-restrictive approach to the use of radio resources for electronic communications services, which allows the spectrum user to choose services and technology, should from now on be the rule, as opposed to the restrictive approach which is often still used today....

Furthermore, the European Commission indicated that

> Avoiding interference remains a key element of spectrum management, but the way it can be achieved has evolved due to technological progress. This progress means that the traditional spectrum management approach should be replaced by a more flexible one, which not only facilitates technical efficiency, but also economic efficiency in spectrum use.

Cognitive radio techniques underpin the drive to move towards dynamic, service-and-technology-neutral spectrum allocations, improve all levels of spectrum efficiency, and maximize economies of scale without the inefficiencies associated with spectrum management. In particular, the UWB-based cognitive radio approach may have an impact on spectrum management and use by addressing the critical coexistence issue in frequency, time, and space domains. Given the diversity and complexity of existing radio spectrum allocations around the world, it is likely that cognitive radio and the associated dynamic spectrum access possibilities will initially be verified in a small subset of the whole range of spectrum bands, such as those identified as the ultra-wideband frequency bands allocated within 3.1–10.6 GHz.

By taking into account the technical aspects and basic functionality of UWB radio in a heterogeneous networking environment, a general architecture of cognitive radio networks targeting dynamic spectrum access can be constructed by including two distinct application scenarios as illustrated in Figure 14.1, namely,

1. Autonomous cognitive radio networks
2. Infrastructure-based cognitive radio networks

The first scenario corresponds to networks created by end-user devices with cognitive functionality and capability, each operating autonomously. Each cognitive radio device should be sufficiently sophisticated to assess how its own communications affects the static and dynamic spectral use (in frequency, time, and space) in the communications of licensed systems. Based on this information, each cognitive device makes autonomous decisions on the best available spectrum to access in order to satisfy the end

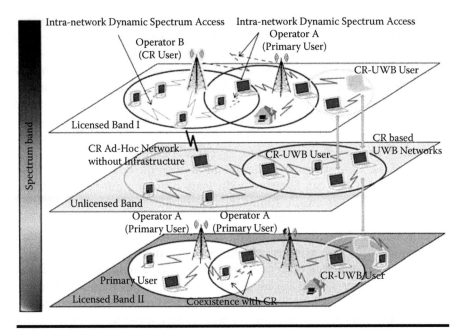

Figure 14.1 Common architecture for dynamic spectrum accessing and sharing in heterogeneous networking environment.

user's requirement for a given service or range of services. In this process, the cognitive radio device may engage in negotiation with the surrounding systems, including both licensed systems and other neighbouring cognitive radios, and may form ad hoc networks for a range of tasks including cooperative spectrum sensing, interference avoidance, spectrum pooling, node clustering, and cognitive packet routing.

In the second scenario, sophisticated cognitive functionality in the end-user devices will exist, but it is provided via the base station (or the access point) and the infrastructure made available by the base station (or the access point). Subscriber devices usually are software-defined radio (SDR) and are spectrum-agile. They may become more versatile and more fully cognitive, but it is likely that their intelligence will be activated or downloaded mainly as necessary to enable the selected features. The actual reasoning, spectral estimation, negotiation, and network management will be carried out by the base station (or the access point). The base station (or the access point) will convert the logic into controls and software modules to control the end-user units.

No matter what kind of application scenario is involved (autonomous or infrastructure-based), the CR-UWB user first needs to guarantee smooth coexistence with other CR or non-CR devices. Apart from coexistence improvements, the CR-UWB devices may further provide additional benefits to infrastructure-based networks (e.g., cellular networks) in the heterogeneous networking environment shown in Figure 14.1. The CR-UWB devices could sense a huge collection of frequency bands and forward a "radio exposition picture" of a certain area to the cellular base station (potentially via a cellular terminal station) depending upon the information on the relative position/ranging of the UWB devices. This is expected to serve as an incentive to open even a dedicated communication channel (e.g., cognitive pilot channel) between the CR-UWB and cellular devices, which could potentially provide another unique "diversity" based upon a common estimation of the local radio environment. Moreover, the concept of cognitive pilot channel (CPC) can facilitate coordination among the heterogeneous networks, as well as ensure future interoperability for smooth end-to-end reconfigurability. The common information provided by the CR-UWB devices with regard to frequency bands, services, location situation, etc., can be broadcast to other wireless networks or terminals via CPC. With such information, the mobile terminals can initiate a communication session in an optimized way, taking advantage of situation and location information. In particular, since in a dynamic spectrum accessing environment, due to dynamic frequency relocation mechanisms, the mobile terminals may not know the available spectrum in a timely way, it becomes more critical to adopt the UWB-enabled CPC to broadcast such information to the terminals who want to set up a communication session.

14.3 Transmitter-centric Spectrum-Agile Waveform Adaptation: Matching with UWB Spectral Mask

Essentially, cognitive radio functionality requires the capability of accurately ascertaining the surrounding wireless environment by sensing the spectrum over wide bandwidths, adapting the air-interface to the environmental conditions in a timely way while coping with various interferences, and achieving the overall goal of optimizing the whole system operation. Accordingly, one of the basic ideas behind cognitive radio is that performance can be improved by reducing interference if wireless systems are aware of other radio signals in their operating environment.

As illustrated in Figure 14.2, since UWB signals are permitted to spread over a broad spectral region at low power levels near the noise floor, they will unavoidably overlap with the existing and dedicated narrow radio systems, such as the global positioning system (GPS), cellular mobile communications, and wireless LAN systems (e.g., IEEE 802.11a/b/g). There have been some concerns among the incumbent users of the radio spectrum that UWB would artificially boost the interference floor and degrade the

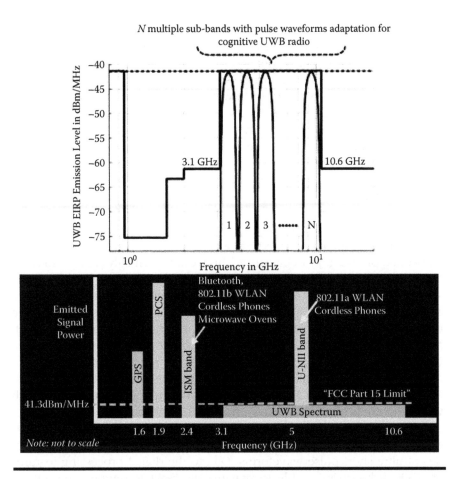

Figure 14.2 **Spectral characteristics of UWB wireless system and the flexible band plan of CR-UWB matching with the FCC spectral mask and the neighboring wireless environment (multi-band division with signal waveform adaptation).**

performances of the primary users. Therefore, how to achieve interference avoidance and coexistence is the common challenge for cognitive radio as well as UWB radio. That is also the reason why the FCC approved the UWB radio emission limit, which not only defined the UWB signals as noted previously, but also defined the spectral mask that specifies the amount of power that can be radiated by any UWB device.

From the perspective of technology evolution, cognitive radio actually roots its adaptability and reconfigurability features in an advanced software-defined radio, and further adds to itself the capability to sense its wireless environment and take action in the most intelligent manner. Such a cognitive radio should help to fully utilize the available radio resources by dynamically adapting to the available spectrum in real time, and should be able to offer the bandwidth on demand, accordingly choosing the most appropriate air-interface based upon perception of wireless channels, locations, spectrum rules, spectral occupancy, and other parameters. Therefore, the problem of how to generate a set of adaptive transmitting signals corresponding to the local spectrum status in a dynamic manner, and then intelligently choose the most appropriate air-interface, which is a prerequisite for realizing cognitive radio, calls for fundamentally novel solutions.

Correspondingly, a CR-UWB capable of adaptively reacting to the spectral environment by seamlessly implementing any suitable waveform through the modification of its generation algorithm is one of the primary goals in this chapter. Here we consider a variety of impulsive UWB waveforms, and their associated adaptation schemes, entirely in a software-embedded process and architecture. This would essentially create a software-based impulse radio, capable of changing its waveform based upon any spectral need, thus reducing the UWB physical layer interface to a "software abstraction." It would also be capable of implementing different impulsive physical layers corresponding to specific operating environments in which the CR-UWB could find itself. We focus on the key issue, namely, adaptive UWB pulse waveform generation for spectrum sharing with interference avoidance, which may also potentially provide us with the required physical basis if we want to further implement dynamic spectrum access with varying degrees of cognitive awareness and intelligence.

As we know, the UWB radio is characterized by its unique modulation scheme of directly conveying digital binary information over a series of ultra-short pulse waveforms. Due to the digital processing features of UWB radio, it is feasible to dynamically react to various spectral requirements by seamlessly implementing a number of specific pulse waveforms through the design of their generation algorithms. Therefore, pulse waveform adaptation and cooperative utilization of multiple UWB frequency bands are the key solutions, adjustable as they are to the actual interference environment and the spectral mask requirements. As described in Figure 14.2, the flexible and adaptive pulse waveform design is essential here in order to provide the expected radio interface, even if the local wireless channel (e.g., spectral mask) is changed, meanwhile still maintaining coexistence and interoperability among the different systems.

Taking these requirements into account, we aim at designing a number of adaptive pulse waveforms, corresponding to the expected UWB spectrum features and aiming at achieving the required spectral agility for effective coexistence with interference avoidance. This kind of adaptive pulse waveform is preferably expressed by means of a limited linear combination of a series of core UWB pulse shapes, namely, taking advantage of orthogonal kernel functions and their related auxiliary functions and factors. These core pulse shapes (i.e., kernel functions) are expected to possess the following properties:

- The kernel function's spectrum should be contained in the desired frequency band allocated by the FCC spectral mask; namely, it should be bandwidth-limited.
- The kernel function's waveform should be limited to short duration with efficient energy concentration, realizing a data rate as high as possible and interpulse interference (IPI) as low as possible; namely, it should be time-limited.
- The kernel functions are preferred to be orthogonal to each other, allowing linear combinations of them to serve as the basis for further designing much more complex pulse waveforms.
- The kernel function set is preferred to be flexibly combinable or extendable, being thereby capable of fitting any further wireless channel change, as well as spectral mask modification by other regional regulatory bodies around the world.

Following this kind of philosophy and technical requirements, it is necessary for the CR-UWB to make the pulse waveforms as adaptive as possible to satisfy arbitrary spectral requirements for spectrum sharing and interference avoidance. Fortunately, we have found the expected UWB core pulse shapes, rooted in orthogonal pulse waveforms based on prolate spheroidal wave functions (PSWF).[9] As illustrated in Figure 14.3, the PSWF-based UWB waveforms present several advantages: (i) the pulse waveforms with different orders are mutually orthogonal to each other within the same band, (ii) the pulse width and bandwidth can be simultaneously set, and (iii) the pulse width and bandwidth can be kept constant for the different orders of the waveforms. Moreover, these core pulse shapes allow the defining of nonoverlapping spectral subbands and the building of additional orthogonal waveforms within such subbands, leaving great freedom in the spectrum agility and adaptation flexibility.

For instance, in the extreme case of radio emission prohibition like that imposed in the frequency band of radio astronomy, any kind of UWB radio emission should be avoided. Considering this kind of special spectral requirement and still complying with the FCC emission mask (3.1–10.6 GHz), we have designed a unique adaptive pulse waveform as shown in Figure 14.4, in which a limited linear combination of PSWF-based orthogonal

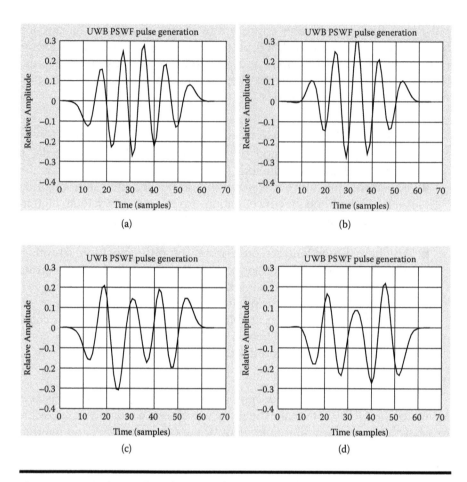

Figure 14.3 Orthogonal pulse waveforms generation based on PSWF (3.1–10.6 GHz), (a) order = 1, (b) order = 2, (c) order = 3, and (d) order = 4.

pulse wavelets and their related auxiliary functions is utilized.[10] As shown in Figure 14.4, the pulse waveform is located in the lower band of 3.1–6.4 GHz, in which several spectral notches (with notch depth more than 20 dB) are generated, in order to avoid any possible severe interference to the existing narrow-band wireless services, such as radio astronomy, the fixed satellite services, and the 3G wireless communications services.

Furthermore, future work can be pursued with the aim of allowing each CR-UWB system to build its own "spectral signature," describing which waveforms in a given set are being used and how information is associated to them. Such a "spectral signature" will be considered as a kind of "chromosome" and allowed to evolve in time to adapt to the surrounding environmental changes.

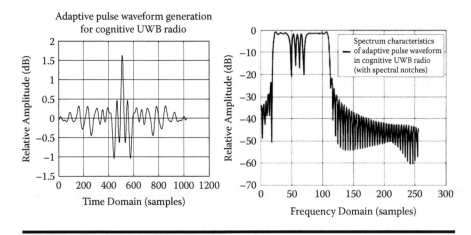

Figure 14.4 **Adaptive pulse waveform generation and its spectrum-agile characteristics.**

14.4 Transmit Power Control and Optimization in Ultra-wideband Cognitive Radio

As noted previously, the UWB communication system represents an ideal "landing platform" for feasible technical approaches suitable for the realization of cognitive radio. In the earlier sections, we have analyzed the spectrum-agile UWB waveform generation methods, based on the linear combination of orthogonal core pulse wavelets to achieve a "proof of concept" for the cognitive UWB radio. Such a CR-UWB system would have cognitive radio capabilities in terms of spectral shaping and dynamically adapting to its available spectrum sources, i.e., generating several unique spectral notches to mitigate interference in the victim receiver's band of interest, as well as satisfying any region-dependent UWB spectral emission masks around the world.

Recently, an M-ary pulse shape modulation (PSM) scheme using a set of orthogonal UWB waveforms (e.g., PSWF-based UWB waveforms) has also been investigated for the expected spectrum agility and adaptation flexibility.[11] In the PSM scheme, a sequence of binary information bits can be directly modulated and carried by the specific shapes of the multiple orthogonal pulses, and thus the combined multiple UWB waveforms can transmit additional bits due to their orthogonality and overlapping in the time domain. It has been verified that this PSM scheme is also capable of achieving higher data throughput even in severe multipath fading environments.

In regard to the CR-UWB scenario involving multiple users with each user adopting multiple orthogonal pulse waveforms and pulse shape modulation, we hereafter focus on developing several transmit power control and optimization schemes that either maximize the data throughput or minimize the average bit error rate, under the stringent restrictive conditions on the UWB radio transmission—the emission limits imposed by the regulatory agencies (e.g., FCC spectral mask).

14.4.1 System Model

As described in Section 14.3, four typical examples of the PSWF-based UWB pulse waveforms are presented in Figure 14.3. For more detailed information on the generation process of those waveforms, readers are suggested to refer to Ref. [11]. Here, in a similar approach to that found in Ref. [11], the typical transmit structure for the M-ary pulse shape modulation is depicted in Figure 14.5(a), where one binary information bit is modulated and carried by one of the orthogonal pulses, the PSM scheme thus needing only K pulses for a symbol of K bits, while the conventional PSM scheme generally needs $2^K - 1$ pulses.

In general, the transmitted UWB signal with the M-ary ($M = 2^K$) PSM for user n is given as[11]

$$s^{(n)}(t) = \sqrt{p} \sum_{i=-\infty}^{\infty} \sum_{l=0}^{K-1} b^{(n)}_{\lfloor i/K \rfloor + l} \psi_l^{(n)} \left(t - \lfloor i/K \rfloor T_f - l T_l^{(n)} - c^{(n)}_{\lfloor i/K \rfloor} T_c \right),$$

$$(14.1)$$

where p is the transmit signal power; $b^{(n)}_{\lfloor i/K \rfloor + l}$ is the binary information data sequence for user n, consisting of a train of i.i.d. data bits that take the values of ± 1 with equal probability; and $\psi_l^{(n)}(t)$, of width T_p, are the PSWF-based pulse waveforms applied to the M-ary pulse shape modulation. All these pulse shapes are normalized to have totally combined unit power. In addition, T_l is the average period between two consecutive pulses ($T_l \geq T_p$), namely the pulse repetition interval (PRI); T_f is the frame repetition time of the pulse sequences for all users; and $\lfloor \cdot \rfloor$ denotes the integer part. To avoid catastrophic collisions due to multiple accesses, the ith orthogonal pulse group undergoes an additional time shift of $c^{(n)}_{\lfloor i/K \rfloor} T_c$, in which $c^{(n)}_{\lfloor i/K \rfloor}$ is the user-specific time-hopping (TH) code and T_c is the chip width of each TH code.

At the receiver, as shown in Figure 14.5(b), the received pulse sequence signals for the nth user are generally distorted by the transmission multipath channel and can be given as

$$r^{(n)}(t) = s^{(n)}(t) + n(t), \qquad (14.2)$$

where $n(t)$ is the zero mean additive white Gaussian noise (AWGN). Assuming that the ith orthogonal pulse sequence is transmitted, the jth decision

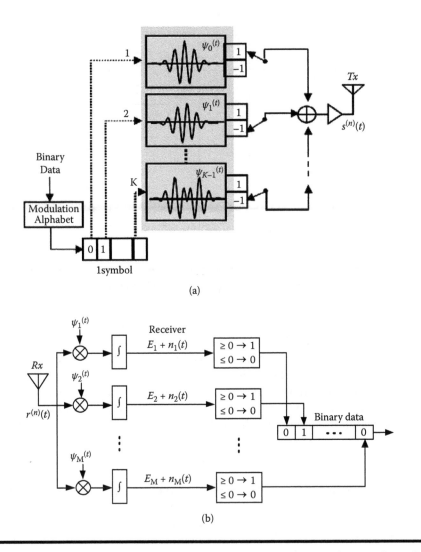

Figure 14.5 *M*-ary PSWF-based PSM communication system: (a) transmitter, (b) receiver. (Note: the superscript *n* is omitted for expression simplicity.)

statistics after a coherent pulse shape detection and demodulation at the receiver can be expressed as

$$D_j^{(n)} = \int_{-\infty}^{\infty} r^{(n)}(t)\psi_j^{(n)}(t)dt = \begin{cases} E_i^{(n)} + n_i(t), & i = j, \\ n_j(t), & i \neq j, \end{cases} \tag{14.3}$$

where $n_j(t) = \int_{-\infty}^{\infty} n(t)\psi_j^{(n)}(t)dt$ and $E_i^{(n)}$ is the symbol energy for the received PSWF pulse.

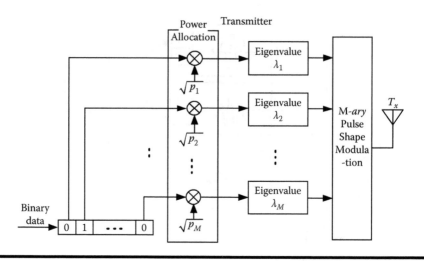

Figure 14.6 Transmit structure with optimal transmit power allocation scheme.

From now on, for simplicity, we focus on a one-user communication system that adopts the M-ary PSWF-based pulse shape modulation scheme with optimal transmit power allocation among the M pulse waveforms; the transmit structure is as shown in Figure 14.6.

Taking into account the multipath fading factors, in a general way, we describe the output of an arbitrary pulse waveform (e.g., mth pulse waveform) through the multipath channel as

$$y_m = \sqrt{p_m}h_m b_m + n_m, \qquad m = 1, 2, \ldots, M, \qquad (14.4)$$

where h_m denotes the multipath fading coefficient for the mth pulse waveform, n_m represents the additive noise, and these noises are assumed to be i.i.d. Gaussian random variables with zero mean and variance of σ^2. The data bit b_m is carried by one orthogonal pulse ψ_m in the pulse shape modulation scheme. Here, we just present b_m with a simpler subindex than that in Equation (14.1), in order to underline the $m = 1, 2, \ldots, M$ pulses. In Equation (14.4), p_m represents the transmit power allocated to the mth data bit, with the total power constraint given as

$$\sum_{m=1}^{M} p_m = M\overline{P}, \qquad (14.5)$$

where $\overline{P}$ denotes the average transmit power per pulse, when the total transmit power is equally distributed into the M pulse waveforms. At the receiver part, ψ_m represents the corresponding template pulse for coherent detection.

It is assumed that at the receiver, a selective Rake combiner is employed to effectively collect the multipath signals. After the Rake reception, the SNR γ_m for the mth pulse is calculated as

$$\gamma_m = \alpha_m p_m, \tag{14.6}$$

where $\alpha_m \triangleq |b_m|^2/\sigma^2$ is the ratio of the combined multipath gain of the mth pulse to the noise power, representing the overall channel state for the mth pulse. We further express α_m by deriving the eigenvalue of the mth pulse after its transmission through the multipath channel. The channel state information α_m is assumed to be known in advance by the transmitter, which is required to determine the optimal transmit powers $\{p_m\}$.

14.4.2 Transmit Power Allocation Schemes

Starting from the aforementioned system model, we first develop the transmit power allocation scheme that is optimal in terms of channel data throughput in Section 14.4.2.1. Then the transmit power allocation scheme that is optimal in terms of BER performance is investigated in Section 14.4.2.2, and finally a case study of the BER function optimization in the context of 8-ary pulse shape modulation is provided.

14.4.2.1 Optimal Power Allocation for Channel Data Throughput Improvement

Rooted in information theory, the water-filling technique has been recognized as an optimal method for transmit power allocation in multiple bands or multiuser communication systems.[12] As we agree,

> In mathematical terms, the essence of transmit power control for a noncooperative multiuser radio environment is stated as: Given a limited number of spectrum holes, select the transmit-power levels of n unserviced users so as to jointly maximize their data-transmission rates, subject to the constraint that the interference-temperature limit is not violated.[2]

Accordingly, in our simplified system model, we would like to state our problem as follows: *In an M-ary pulse shape modulation transmission scheme, how to control the transmit power levels of the M pulse waveforms of a specific user so as to jointly maximize its whole data transmission rates, subject to the constraint that the UWB emission limit is not violated.*

To solve this kind of optimization problem, we apply the water-filling algorithm to these PSWF-based pulse waveforms in order to optimally allocate the transmit power, as illustrated in Figure 14.6. Generally, the achievable throughput per hz of the mth waveform is bounded by[12]

$$C_m = \log_2(1 + \gamma_m) = \log_2(1 + \alpha_m p_m), \tag{14.7}$$

where the denotations of γ_m, α_m, p_m are the same as we described in equation (14.6). Hence, the throughput of the channel per hz is

$$C = \sum_{m=1}^{M} \log_2(1 + \alpha_m p_m). \tag{14.8}$$

Then, the optimization problem can be described as

$$\text{maximize } \sum_{m=1}^{M} \log_2(1 + \alpha_m p_m) \tag{14.9a}$$

$$\text{subject to } \sum_{m=1}^{M} p_m = M \overline{P} \tag{14.9b}$$

$$p_m \geq 0. \tag{14.9c}$$

The solution to such an objective function can be obtained by citing the ideas of optimization processing as in Ref. 13. We first transform (14.9) to a standard convex optimization problem,

$$\text{minimize } - \sum_{m=1}^{M} \log_2(1 + \alpha_m p_m) \tag{14.10a}$$

$$\text{subject to } \sum_{m=1}^{M} p_m = M \overline{P} \tag{14.10b}$$

$$-p_m \leq 0. \tag{14.10c}$$

By introducing Lagrange multipliers λ_l for the inequality constraint $-p_m \leq 0$, and a multiplier ν for the equality constraint $\sum_{m=1}^{M} p_m = M\overline{P}$, we obtain the *Karysh-Kuhn-Tucker* (KKT) conditions

$$-p_m \leq 0 \tag{14.11a}$$

$$\sum_{m=1}^{M} p_m = M\overline{P} \tag{14.11b}$$

$$\lambda_l \geq 0 \tag{14.11c}$$

$$\lambda_l \cdot p_m = 0 \tag{14.11d}$$

$$-\frac{\alpha_m}{1+\alpha_m p_m} - \lambda_l + \nu = 0. \tag{14.11e}$$

From (14.11c) and (14.11e) we can derive

$$\lambda_l = \nu - \frac{\alpha_m}{1 + \alpha_m p_m} \tag{14.12}$$

and

$$\nu \geq \frac{1}{\beta_m + p_m}, \tag{14.13}$$

where $\beta_m = 1/\alpha_m$.

Continuing, from (14.11d) and (14.12) we can achieve

$$\lambda_l \cdot p_m = \left(\nu - \frac{1}{\beta_m + p_m} \right) \cdot p_m = 0. \tag{14.14}$$

If $\nu < 1/\beta_m$, the condition (14.13) can only hold when $p_m > 0$, which by the condition (14.14) implies that

$$\nu = \frac{1}{\beta_m + p_m}. \tag{14.15}$$

Solving this equation, we can obtain

$$p_m = \frac{1}{\nu} - \beta_m. \tag{14.16}$$

If $\nu \geq 1/\beta_m$, we assume $p_m \neq 0$, which implies that $p_m > 0$, and then $\nu \geq 1/\beta_m > 1/(\beta_m + p_m)$. However, from (14.14) one can conclude that $p_m = 0$, and this is a contradiction to our assumption. So, $p_m = 0$ if $\nu \geq 1/\beta_m$.

Thus we have the solution as

$$p_m = \begin{cases} \dfrac{1}{\nu} - \beta_m, & \nu < 1/\beta_m, \\ 0, & \nu \geq 1/\beta_m. \end{cases} \tag{14.17}$$

It can be expressed more concisely as

$$p_m = \left(\frac{1}{\nu} - \beta_m \right)^+, \tag{14.18}$$

where $(\cdot)^+$ denotes $\max(0, \cdot)$, and ν is determined in order to satisfy the power constraint $\sum_{m=1}^{M} p_m = M\overline{P}$.

Substituting this expression for p_m into the condition $\sum_{m=1}^{M} p_m = M\overline{P}$, we obtain

$$\sum_{m=1}^{M} \max \left\{ 0, \frac{1}{\nu} - \beta_m \right\} = M\overline{P}. \qquad (14.19)$$

The left-hand side is a piecewise linear increasing function of $1/\nu$, with breakpoints at $1/\alpha_m$, so the equation has a unique solution which is readily determined.

This algorithm can be implemented as follows.

(1) Set a small positive termination threshold p_{thr} and step size ξ, initialize ν and p_{total} to be zero.
(2) While $|p_{total} - M\overline{P}| > p_{thr}$, do
for $m = 1 : M$
{
    if $\nu \geq 1/\beta_m$, $p_m = 0$;
    else
        find p_m that satisfies $\frac{1}{\beta_m + p_m} = \nu$ by the following steps:
        (a) Set a small positive termination threshold ν_{thr} and step size μ;
        (b) Initialize $p_m = 0$;
        (c) While $\left| \nu - \frac{1}{\beta_m + p_m} \right| > \nu_{thr}$, do
        find derivative δ of $(\nu - \frac{1}{\beta_m + p_m})^2$ with respect to p_m:

$$\delta = \frac{1}{(\beta_m + p_m)^2} \left(\nu - \frac{1}{\beta_m + p_m} \right).$$

        Update $p_m = p_m - \mu\delta$.

}
Compute $p_{total} = \sum_{m=1}^{M} p_m$;
Update ν as $\frac{1}{\nu} = \frac{1}{\nu} + \xi(1 - p_{total})$.
(3) Output all p_m.

14.4.2.2 Optimal Power Allocation for BER Performance Improvement

Continuing, we now discuss the optimal power allocation scheme for BER performance improvement in Section 14.4.2.2.1. Then, the case study of improving the BER performance in the context of M-ary pulse shape modulation is performed in Section 14.4.2.2.2.

14.4.2.2.1 Optimal Power Allocation Scheme

To derive the optimal power allocation scheme, we first express the overall BER as a function of the transmit power for M waveforms, $\{p_m|m = 1, 2, \ldots, M\}$, and then find $\{p_m\}$ that minimizes the overall BER. The BER for the mth wavefrom is generally a function of the SNR γ_m, and thus, the BER $p_b(e|\alpha_m)$ for a given channel state α_m may be expressed as

$$p_b(e|\alpha_m) = f(\gamma_m) = f(\alpha_m p_m), \qquad m = 1, 2, \ldots, M, \qquad (14.20)$$

where $f(\cdot)$ is a function determined by a specific modulation scheme. In the PSM scheme, data streams are carried and transmitted by mutual orthogonal waveforms with equal rate constraint, and since the multiple waveforms are statistically independent, the joint BER for given channel states of $\{\alpha_m|m = 1, 2, \ldots, M\}$ can be calculated as an arithmetic mean of $p_b(e|\alpha_m)$,

$$p_b(e|\alpha_1, \alpha_2, \ldots, \alpha_M) = \frac{1}{M} \sum_{m=1}^{M} p_b(e|\alpha_m) = \frac{1}{M} \sum_{m=1}^{M} f(\alpha_m p_m). \qquad (14.21)$$

Note that the average BER becomes minimal when the BER in (14.21) is minimized for each waveform. To find the optimal $\{p_m\}$ that minimizes (14.21), we use the Lagrange multiplier method with the same total power constraint as in (14.5). The Lagrangian function may be expressed as

$$J(p_1, p_2, \ldots, p_M) = \frac{1}{M} \sum_{m=1}^{M} f(\alpha_m p_m) + \lambda \left(\sum_{m=1}^{M} p_m - M\overline{P} \right), \qquad (14.22)$$

where λ denotes the Lagrange multiplier. By differentiating (14.22) with respect to p_m and setting it to zero, we obtain a set of equations as

$$\frac{1}{M} \frac{d}{dp_m} f(\alpha_m p_m) + \lambda = 0, m = 1, 2, \ldots, M. \qquad (14.23)$$

By simultaneously solving the equations in (14.5) and (14.23), we can calculate the optimal set of the transmit power $\{p_m\}$.

As mentioned above, the BER function in (14.20) is a function determined by a specific modulation scheme. For a binary differential phase-shift keying (DPSK), for example, the BER function may be expressed as an exponential function,[14] and a closed-form solution of (14.5) and (14.23) may be easily found. For M-ary orthogonal signals, however, the BER function is expressed as an integration function,[14] and it is difficult to find a closed-form solution. In this case, an adaptive method, such as the steepest-descent algorithm,[15] may be employed to find a solution in an iterative manner as follows.

(1) Initialization: Set an iteration number $i = 1$, a step size $\mu(1) = \mu_1$, and an arbitrary initial positive power set $\{p_m\}$ satisfying (14.5).

(2) Power set update: For $m = 1, 2, \ldots, M$, update the transmit power $p_m(i)$ as

$$p_m(i+1) = p_m(i) - \mu(i)\frac{\partial}{\partial p_m(i)}J\left(p_1(i), p_2(i), \ldots, p_M(i)\right)$$

$$= p_m(i) - \mu(i)\left(\frac{1}{M}\frac{d}{dp_m(i)}f\left(\alpha_m p_m(i)\right) + \lambda(i)\right),$$

$$\text{(14.24)}$$

where $\lambda(i)$ is determined from the power constraint in (14.5) and is updated as

$$\lambda(i) = -\frac{1}{M^2}\sum_{m=1}^{M}\frac{d}{dp_m(i)}f(\alpha_m p_m(i)). \qquad \text{(14.25)}$$

(3) Step-size adjustment: If all components of the updated power set $\{p_m(i+1)\}$ in (2) are positive, then go to (4) with $\mu(i+1) = \mu_1$. Otherwise, compute $\tilde{\mu}_m(i) \triangleq p_m(i)/((1/M)(df(\alpha_m p_m(i)))/(dp_m(i))+\lambda(i))$ for m, which is associated with $p_m(i+1) \leq 0$, set the step size $\mu(i)$ to $\rho \cdot \min_{m: p_m(i+1) \leq 0} \tilde{\mu}_m(i)$, where ρ is a positive scaling factor smaller than one, and return to (2).

(4) Repetition or termination: If more iterations are required for convergence, increase by one and go to (2). Otherwise, terminate the adaptive procedure.

The adaptive algorithm described above converges to the global optimum solution for the convex BER function. Note that the BER function of M-ary orthogonal signals may be approximated by a Q-function upper bound, which is a convex function.[14]

14.4.2.2.2 Case Study

In order to discuss the BER performance of the M-ary PSM by executing the above optimization algorithm, we first develop the Lagrangian function for the M-ary orthogonal signals, and then the steepest-descent algorithm as mentioned above is performed to calculate the optimal set of the transmit power $\{p_m\}$, since we cannot get a closed-form solution.

According to the theoretical calculations for the M-ary orthogonal modulation from Ref. 14, the BER function is expressed as

$$p_b(e) = \frac{1}{2} \cdot \frac{M}{M-1} p_M(e), \tag{14.26}$$

where $p_M(e)$ is the symbol error probability (SEP) and

$$p_M(e) = \frac{1}{\sqrt{2\pi}} \int_{-\infty}^{\infty} \left[1 - \left(\frac{1}{\sqrt{2\pi}} \int_{-\infty}^{y} e^{-\frac{x^2}{2}} dx \right)^{M-1} \right]$$

$$\cdot \exp\left[-\frac{1}{2} \left(y - \sqrt{2 \sum_{m=1}^{M} \alpha_m p_m \Big/ N_0} \right)^2 \right] dy. \tag{14.27}$$

In our case study, we set $M = 8$. Accordingly,

$$J(p_1, p_2, \ldots, p_8) = \frac{1}{8} \sum_{m=1}^{8} \left[\frac{4}{7\sqrt{2\pi}} \cdot \int_{-\infty}^{\infty} \left[1 - \left(\frac{1}{\sqrt{2\pi}} \int_{-\infty}^{y} e^{-\frac{x^2}{2}} dx \right)^7 \right] \right.$$

$$\times \exp\left[-\frac{1}{2} \left(y - \sqrt{2 \sum_{m=1}^{8} \alpha_m p_m \Big/ N_0} \right)^2 \right] dy \Bigg]$$

$$+ \lambda \left(\sum_{m=1}^{8} p_m - 8\bar{p} \right). \tag{14.28}$$

By differentiating (14.28) with respect to p_m and setting it to zero, we can obtain

$$\frac{1}{14\sqrt{2\pi}} \cdot \frac{d}{dp_m} \left\{ \int_{-\infty}^{\infty} \left[1 - \left(\frac{1}{\sqrt{2\pi}} \int_{-\infty}^{y} e^{-\frac{x^2}{2}} dx \right)^7 \right] \right.$$

$$\left. \times \exp\left[-\frac{1}{2} \left(y - \sqrt{2 \sum_{m=1}^{8} \alpha_m p_m \Big/ N_0} \right)^2 \right] dy \right\} + \lambda = 0. \tag{14.29}$$

To simplify the mathematical deduction process in solving the above equation, we transform $\left(\frac{1}{\sqrt{2\pi}} \int_{-\infty}^{y} e^{-\frac{x^2}{2}} dx \right)$ into the Q-function

$$\left(\frac{1}{\sqrt{2\pi}} \int_{-y}^{\infty} e^{-\frac{t^2}{2}} dt \right) = Q(-y).$$

As we have $Q(y) < e^{-y^2/2}$, thus (14.29) can be further simplified by the lower bound $Q(-y)$, as

$$
\frac{1}{14\sqrt{2\pi}} \frac{d}{dp_m} \left\{ \int_{-\infty}^{\infty} \left[1 + \left(e^{\frac{-y^2}{2}} \right)^7 \right] \right.
$$

$$
\left. \exp\left[-\frac{1}{2} \left(y - \sqrt{2 \sum_{m=1}^{8} \alpha_m p_m \bigg/ N_0} \right)^2 \right] dy \right\} + \lambda = 0.
$$

(14.30)

Continuing, by differentiating the later part of the above equation, we can further express it as

$$
\frac{1}{14\sqrt{2\pi}} \int_{-\infty}^{\infty} \left[1 + \left(e^{\frac{-y^2}{2}} \right)^7 \right] \exp\left[-\frac{1}{2} \left(y - \sqrt{2 \sum_{m=1}^{8} \alpha_m p_m \bigg/ N_0} \right)^2 \right] \cdot
$$

$$
\left(\sqrt{2 \sum_{m=1}^{8} \alpha_m p_m \bigg/ N_0} - y \right) \cdot \left(-1 \bigg/ \sqrt{2 N_0 \cdot \sum_{m=1}^{8} \alpha_m p_m} \right) \cdot \alpha_m dy + \lambda = 0.
$$

(14.31)

14.4.3   Simulation Results and Discussion

In this section, the performance of the two proposed optimization schemes is evaluated in a direct-sequence UWB (DS-UWB) system, where the information bit is modulated and transmitted over the parallel sequences of orthogonal pulses (e.g., PSWF-based UWB pulses). In order to express the two schemes clearly in the following figures, we hereby denote the scheme for data throughput improvement as scheme 1, and the scheme for BER performance improvement as scheme 2. In the simulation, we generate eight PSWF-based orthogonal pulse waveforms and compare the data throughput and the BER performance of the 8-ary PSM case in an indoor UWB multipath fading environment. Here, the standard IEEE 802.15.3a S-V UWB channel model is considered.[16] There are four different multipath fading channels defined in this model, namely, CM1, CM2, CM3, and CM4. The simulation procedures are conducted in CM1 and CM3 as the representative cases. At the transmission side, we also adopt an eight-by-eight orthogonal ternary complementary code set as the direct-sequence codes, which are assigned to the eight different PSWF-based orthogonal pulses. Signal energy is allocated among these 8 pulses according to the proposed scheme. At the receiver side, we employ the selective Rake receiver to collect the multipath signals, and the Rake fingers are set to 5 and 3 in scheme 1 and scheme 2, respectively.

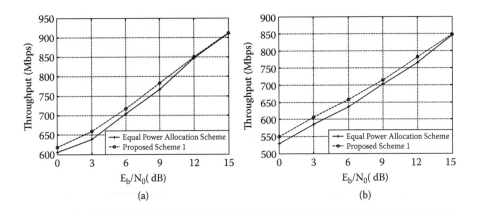

Figure 14.7   Throughput of the proposed scheme 1 and the equal power allocation scheme, (a) in CM1, (b) in CM3.

Figure 14.7 (a) and (b) show the attainable throughput of scheme 1 and the conventional equal power allocation scheme. Hereinafter, the chip pulse duration is set to 1 μs, which means the maximal transmission data rate of this DS-UWB system can be as high as 1 Gbps. From Figure 14.7 (a) and (b), it can be observed that with the proposed power allocation scheme (scheme 1), the channel data throughput is significantly improved by about 20 Mbps when SNR is below 10 dB, which represents 2% of the total capacity. When SNR is above 10 dB, the proposed scheme still performs similarly to the equal power allocation scheme. The reason behind the decreased difference is that the capacity is a logarithmic function of SNR, and therefore the data rate is usually insensitive to the exact power allocation, except when the SNR is in the low range.[17]

As we know, the water-filling algorithm generally allots more power to the channels with lower noise. In our DS-UWB simulation model, in regard to each PSWF-based pulse waveform, the eigenvalue is the ratio of the combined multipath gain of the corresponding pulse to the noise power, which represents the overall channel state for that specific pulse. The lower eigenvalue means that pulse has a relatively poorer combined gain. Consequently, the water-filling algorithm allocates more energy to the pulse waveform with a bigger eigenvalue, aiming at maximizing the data rate.

Figure 14.8 (a) illustrates the BER performance achieved by scheme 2 and the equal power allocation scheme in CM1. As expected, the proposed scheme outperforms the other one by 1.5–2.5 dB in the high SNR (>10 dB) range, and by 1–2 dB in the low SNR range. Similarly, as shown in Figure 14.8 (b), for which the simulation is performed in CM3, the SNR gain of the proposed scheme is about 1.0–2.5 dB in the high SNR range and 0.5–1.0 dB in the low SNR range.

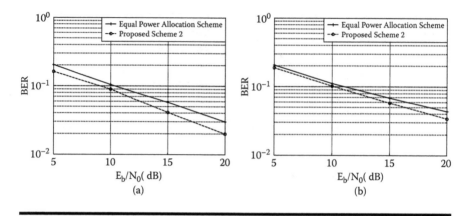

Figure 14.8   BER performance of the proposed scheme 2 and the equal power allocation scheme, (a) in CM1, (b) in CM3.

It is worthwhile to note that the characteristics of this optimal transmit power allocation scheme (scheme 2) derived in Section 14.4.2.2 differ from those of scheme 1 in Section 14.4.2.1, which improves the capacity rather than the BER performance. According to the simulation results, from the power set $\{p_m\}$, we can observe that in the high SNR range, the optimal power allocation scheme 2 tends to allocate more transmit power to the more attenuated waveforms, which is contrary to the behavior of scheme 1.

In this chapter, following our previous work in Refs. [18] and [19], we are mainly focusing on a peer-to-peer transmission scenario. However, the proposed schemes can be extended to multiuser cognitive radio environments, where the CR-UWB nodes may operate in a decentralized manner, and each node may cooperate with every other node to control its transmit power level, and thus maintain the cumulative RF energy under the victim receiver's interference limit.

14.5   Receiver-Centric Spectrum-Agile Waveforms Adaptation: Matching with Interference Temperature Limit

Traditionally, interference avoidance and coexistence issues are regulated in the transmitter-centric mode, based upon frequency band allocation and coordination, transmit power limitation, out-of-band emission control, and so on. Recently, the FCC Spectrum Policy Task Force[20] has introduced a paradigm shift in interference regulation, namely a shift from the transmitter-centric mode toward the one of "real-time interactions between the transmitter and receiver in an adaptive manner." As opposed to the widely adopted transmitter-centric approach, this model attempts to assess interference at the receiver, which has given rise to a new metric named the *interference temperature* for measuring and managing the interference sources in a radio environment. Although in May 2007 the FCC temporarily terminated the interference temperature approach,[21] responding to several radio spectrum regulatory issues, the basic philosophy of the interference temperature is still extremely valuable and meaningful for research on cognitive radio networks. Therefore, more and more academic efforts on how to effectively implement the technical solution based on the interference temperature can be expected. According to the principle of the interference temperature, a maximum acceptable interference level would be set in a specific frequency band and at a particular geographic location, while any new wireless device could be put into service if it could be expected not to "overflow" the *interference temperature limit*. Here, the interference temperature model provides an advanced measurement for the cumulative radio frequency energy from all kinds of transmissions and sets a maximum "cap" on their aggregate level in the frequency band of interest. Therefore, given a specific frequency band in which the interference temperature limit is not exceeded, that spectrum resource could be made available to any unserviced user. Moreover, it is interesting to note that the interference temperature model is somewhat analogous to the UWB spectral mask, in which the UWB transmit signal power is limited low enough so that harmful interference would not take place. If we follow this kind of paradigm shift from the transmitter-centric way to the receiver-centric way recommended by the FCC, then it makes sense for us to further relocate the UWB emission limit (i.e., UWB spectral mask) from the transmitter side to the receiver side. If this concept is approved by the regulatory bodies and really works well in actual wireless environments, a promising scenario for high-data-rate UWB communication applications can be anticipated, with radio links extended from "short-range" to "long-range," since the "short-range" transmission power limitation would no longer be a key restriction factor.

Corresponding to this paradigm shift, we may further consider a receiver-centric spectrum-agile waveform adaptation approach. Since UWB radio is naturally aware of its transmit power level and radiated signal waveforms, with the assistance of its position-aware capability it could potentially know the precision location as well as the through-the-air flying distance to the target receiver. By virtue of this kind of information, a spectrum-agile CR-UWB can effectively estimate the cumulative probability that its transmission would cause significant interference to the neighboring receiver. Therefore, we may further exploit the new dimension of realizing the spectrum-agile waveform adaptation by matching with the interference temperature limit at the receiver rather than with the UWB

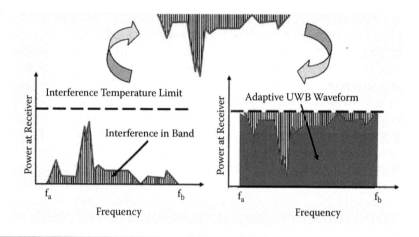

Figure 14.9 Receiver-centric spectrum-agile CR-UWB waveform adaptation matching with the interference temperature limit.

spectral mask at the transmitter. The basic concept of the proposed receiver-centric spectrum-agile waveform adaptation is illustrated as in Figure 14.9. In regard to Figure 14.9, if the CR-UWB knows in advance the interference temperature conditions at the target receiver based on the channel state information and iterative feedbacks between them, the transmitter can generate a spectrum-adaptive UWB waveform that is matched with the interference temperature limit correspondingly. The receiver-centric spectrum-agile waveform could also be designed by employing the well-known water-filling scheme, which consists of various spectral components with varying spectral power density (PSD) indicated by the "dark-shadowed" area shown in Figure 14.9. It should be noted that although there is no practical method for a spectrum-agile radio to measure the individual effect of its transmission on all possible receivers or assess the interference-tolerating capability of those receivers, the proposed receiver-centric UWB waveform adaptation approach may stimulate further research efforts and exploration.

14.6 Conclusions

In summary, the chapter has presented novel solutions that aim at effectively integrating the advantages and features in the field of UWB wireless transmission with the emerging paradigm of cognitive radio. We have explored UWB radio as an enabling technology for implementing cognitive radio by virtue of the unique attributes of the UWB device with CR-enhanced capabilities. In particular, following the philosophy of both spectrum agility and adaptation flexibility, we have presented concrete technical approaches for the design of transmitter-centric spectrum-agile waveform adaptation in the UWB cognitive radio environment, which were based on the linear combination of a set of orthogonal UWB pulse waveforms complying with the FCC spectral mask at the transmitter side. Furthermore, we have tested various transmit power control and optimization methods for improving the system capacity as well as the bit-error-rate performance of the CR-UWB. By adopting the water-filling algorithm, we optimized the transmit power allocation, achieving the target of total channel throughput improvement. Moreover, in the scheme for BER performance improvement, we developed the Lagrangian function from the BER functions of the M-ary orthogonal signals, and simplified the mathematical expression in the typical case of 8-ary PSM. Numerical results show that both of our proposed schemes improved the data throughput and the system's BER performance respectively, compared with the conventional equal power allocation scheme.

As a brainstorming approach, another kind of spectrum-agile waveform adaptation has been proposed, namely, generating receiver-centric spectrum-agile UWB waveforms complying with the interference temperature limit rather than the FCC spectral mask.

Finally, it can be concluded that through dynamic spectrum accessing and appropriate UWB waveform design, such a CR-UWB with both spectrum agility and adaptation flexibility is surely capable of improving the prospects for both spectrum compatibility and interoperability among the ever-proliferating wireless communication systems.

References

[1] D. Cabric, S. M. Mishra, D. Willkomm, R. W. Broderson, and A. Wolisz, "A cognitive radio approach for usage of virtual unlicensed spectrum," *Proc. of 14th IST Mobile Wireless Communications Summit 2005*, Dresden, Germany, June 2005.

[2] S. Haykin, "Cognitive radio: Brain-empowered wireless communications," *IEEE J. Select. Areas Commun. (JSAC)*, vol. 23, no. 2, Feb. 2005.

[3] J. Mitola and G. Maguire, "Cognitive radio: Making software radios more personal," *IEEE Pers. Commun.*, vol. 6, no. 6, pp. 13–18, Aug. 1999.

[4] Federal Communication Commission, "Facilitating opportunities for flexible, efficient, and reliable spectrum use employing Cognitive Radio technologies," NPRM & Order, ET Docket No. 03-108, FCC 03-322, Dec. 30, 2003.

[5] M. Win and R. Scholtz, "Ultra-wide bandwidth time-hopping spread-spectrum impulse radio for wireless multiple-access communications," *IEEE Trans. Communications*, vol. 48, pp. 679–689, April 2000.

[6] Federal Communication Commission, "Revision of Part 15 of the commission's rules regarding ultra-wideband transmission systems," First Report and Order, ET Docket 98-153, FCC 02-48, Feb. 2002.

[7] European Conference of Postal and Telecommunications Administrations, "Radio spectrum decision," European Parliament and Council, 676/2002/EC, March 7, 2002.

[8] European Commission, available: http://eur-lex.europa.eu/LexUriServ/site/en/com/2007/com2007_0050en01.pdf.

[9] D. Slepian, "Some comments on Fourier analysis, uncertainty and modeling," *SIAM Review*, vol. 25, issue 3, pp. 379–393, July 1983.

[10] H. Zhang, X. Zhou, K. Y. Yazdandoost, and I. Chlamtac, "Multiple signal waveforms adaptation in cognitive ultra-wideband radio evolution," *IEEE J. Select. Areas Commun (JSAC)*, vol. 24, no. 4, pp. 878–884, April 2006.

[11] K. Usuda, H. Zhang, and M. Nakagawa, "*M*-ary pulse shape modulation for PSWF-based UWB systems in multipath fading environment," *Proc. IEEE Globecom 2004*, pp. 3498–3504, Dallas, Dec. 2004.

[12] T. M. Cover and J. A. Thomas, *Elements of Information Theory*. New York: Wiley, 1991.

[13] S. Boyd and L. Vandenberghe, *Convex Optimization*. Cambridge, UK, Cambridge University Press, 2004.

[14] J. G. Proakis, *Digital Communications*, Fourth Edition, Chapter 5, McGraw Hill, 2001.

[15] B. Widrow and S. D. Stearns, *Adaptive Signal Processing*, Englewood Cliffs, NJ: Prentice-Hall, 1985.

[16] A. F. Molisch, "Ultrawideband propagation channels-Theory, measurement and modeling," *IEEE Trans. Veh. Technol*, vol. 54, issue 5, pp. 1528–1545, Sep. 2005.

[17] W. Yu and J. M. Cioffi, "On constant power water-filling," in *Proc. IEEE Int. Conf. Communications 2001 (ICC'01)*, vol. 44, pp. 1665–1669, June 2001.

[18] X. Zhou, H. Zhang, and I. Chlamtac, "Transmit power allocation among PSWF-based pulse wavelets in cognitive UWB radio," *Proc. First International Conference on Cognitive Radio Oriented Wireless Networks and Communications (CrownCom2006)*, Mykonos Island, Greece, June 2006.

[19] X. Zhou, H. Zhang, and I. Chlamtac, "Transmit power allocation among orthogonal pulse wavelets for BER performance improvement in cognitive UWB radio," *Proc. the First IEEE Workshop on Cognitive Radio Networks (CCNC 2007 Workshop)*, January 11–13, 2007, Las Vegas, USA.

[20] FCC Spectrum Policy Task Force (SPTF), available: http://www.fcc.gov/sptf/.

[21] Federal Communication Commission, "Re: Establishment of an Interference Temperature Metric to Quantify and Manage Interference and to Expand Available Unlicensed Operation in Certain Fixed, Mobile, and Satellite Frequency Bands (ET Docket No. 03-237)," Order FCC-07-78, May 2007.

Chapter 15

Analytical Models for Multihop Cognitive Radio Networks

Yi Shi and Y. Thomas Hou

Contents

15.1 Introduction

A cognitive radio (CR) is a frequency-agile data communication device with a rich control and monitoring (spectrum sensing) interface [11]. It capitalizes on advances in signal processing and radio technology, as well as recent changes in spectrum policy. A CR node constantly senses the spectrum to detect any change in white space; its frequency-agile radio module is capable of reconfiguring RF and switching to newly-selected frequency bands. A CR can be programmed to tune to a wide spectrum range and operate on any frequency band in the range.

The need for CR functionality is indeed compelling, as it affects many important wireless communications applications, including public safety, military, and wireless applications for the general public. Since its conceptual inception, there has been growing interest in the research community to develop cognitive radio platforms. As research in CR continues to intensify, it is expected that future truly CR networks will ultimately have the processing power and flexibility needed to support spectrum sensing, flexible waveform configuration, spectrum negotiation in dynamic environments, and many other functions. Also, it is not hard to foresee that CR will eventually play a pivotal role in multihop wireless networking, such as mesh and ad hoc networks, which are currently mostly based on 802.11 technology. Under CR-enabled wireless networks, each node individually detects the white bands at its particular location, and the spectrum that can be used for communication can be different from node to node. It is not unusual to find that although each node may have some white bands to access, there still may not exist a common frequency band shared by all the nodes in the network.

To better understand the unique challenges associated with a CR-enabled wireless network, we compare it with multichannel multiradio (MC-MR) wireless networks (e.g., [1,7,14]), which have been an area of active research in recent years. First, an MC-MR platform employs traditional *hardware-based* radio technology (i.e., channel coding, modulation, etc. are all implemented in hardware). Thus, each radio can only operate on a single channel at a time and there is no switching of channels on a per-packet basis. The number of concurrent channels that can be used at a wireless node is limited by the number of hardware-based radios. In contrast, the radio technology in CR is software-based; a CR is capable of switching frequency bands on a per-packet basis and over a wide range of spectrum. As a result, the number of concurrent frequency bands that can be shared by a single CR is typically much larger than what can be supported by MC-MR. Second, due to the nature of hardware-based radio technology in MC-MR, a common assumption in MC-MR is that there is a set of *common* channels available for every node in the network; each channel typically has the same bandwidth. Such an assumption may not hold for CR networks, in which each node may have a different set of frequency bands, each of potentially unequal size. A CR node is capable of working on a set of "heterogeneous" channels that are scattered across widely separated slices of the frequency spectrum with different bandwidths. An even more profound advance in CR technology is that there is no requirement that a CR needs to select contiguous frequencies/channels for transmission/reception—the radio can send packets over noncontiguous frequency bands. These important differences between MC-MR and CR warrant that the algorithm design for a CR network be substantially more complex than that under MC-MR. In some sense, an MC-MR-based wireless network can be considered as a special (simple) case of a CR-based wireless network. Thus, algorithms designed for a CR network can be tailored to address MC-MR networks while *the converse is not true*.

Due to the unique characteristics associated with CR networks, problems for CR networks are expected to be much more challenging and interesting. As a first step toward studying CR networks systematically, it is important to develop analytical models. Such models should capture characteristics of CR across multiple layers, such as power control, scheduling, and routing. Performance objectives should correlate with spectrum usage and its occupancy in space, which are unique to CR networks. The goal of this chapter is to present a unified mathematical model for the physical (i.e., power control), link (i.e., scheduling), and network (i.e., routing) layers in a multihop cognitive radio network environment.

The foundation of our analytical model is built upon the so-called interference modeling. There are two popular approaches to model interference in wireless networks, namely, the *physical model* and the *protocol model*. Under the *physical model* (see, e.g., [3,5,6,8,10]), a transmission is considered successful if and only if the signal-to-interference-and-noise ratio (SINR) exceeds a certain threshold, where the interference includes all other concurrent transmissions. Since the calculation of a link's capacity involves not only the transmission power on this link, but also the transmission power on interference links, it is difficult to develop a tractable optimal solution whenever link capacity is involved. Another approach to model interference is called the *protocol model* [10], whereby a transmission is considered successful if and only if the receiving node is in the transmission range of the corresponding transmission node and is out of the interference range of all other transmission nodes. This model is easy to understand and facilitates the building of tractable models. It has been used successfully to address various hard problems in wireless networks (see, e.g., [1,4,13–15,19]); results from these efforts have already offered many important insights. In this chapter, we will follow the protocol model in our analytical modeling.

We organize this chapter as follows. In Section 15.2, we present an analytical model for a multihop CR network at multiple layers. Within this section, we start in Section 15.2.1 with the modeling problem of spectrum band division and allocation, which is unique to CR networks. Then in Section 15.2.2, based on the protocol interference model, we develop an analytical model in the form of a set of constraints for power control and scheduling. Finally, in Section 15.2.3, we present models for flow routing and link capacity. To show the application of the mathematical models in Section 15.2, we apply these models to solving some real problems for multihop CR networks in Section 15.3. We first present a new objective function called the *bandwidth footprint product* (BFP), which uniquely characterizes the spectrum and space occupancy for a node in a CR network. Using this objective function, we study two specific problems in a multihop CR network. In Section 15.3.1, we study the subband division and allocation problem under fixed transmission power (i.e., no power control). The goal is to explore how the different-sized available bandwidth in the network should be optimally divided and allocated among the nodes. In Section 15.3.2, we study the power control problem when all the subbands are identical (i.e., fixed channel bandwidth). The goal there is to explore how to optimally adjust the power level for each node in the network so that the total BFP is minimized. The results from these two case studies not only validate the practical utility of the analytical models, but also help gain some theoretical understanding of a multihop CR network.

15.2 Analytical Models at Multiple Layers

We consider an ad hoc network consisting of a set of $\mathcal{N}$ nodes. Among these nodes, there is a set of $\mathcal{L}$ unicast communication sessions. Denote $s(l)$ and $d(l)$ the source and destination nodes of session $l \in \mathcal{L}$, and $r(l)$ the rate requirement (in b/s) of session l. Table 15.1 lists all notation used in this chapter.

15.2.1 Modeling of Available Spectrum Allocation

This part of mathematical modeling is unique to CR networks and does not exist in MC-MR networks. In a multihop CR network, the available spectrum bands at one node may be different from those at another node in the network. Given a set of available frequency bands at a node, the sizes (or bandwidths) of each band may differ drastically. For example, among the least-utilized spectrum bands found in [17], the bandwidth between [1240, 1300] MHz (allocated to amateur radio) is 60 MHz, while the bandwidth between [1525, 1710] MHz (allocated to mobile satellites, GPS systems, and meteorological applications) is 185 MHz. Such large differences in bandwidths among the available bands suggests the need for further division of the larger bands into smaller subbands for more flexible and efficient frequency allocation. Since equal subband division of the available spectrum band is likely to yield suboptimal performance, an unequal division is desirable.

More formally, we model the union of the available spectrum among all the nodes in the network as a set of M unequally sized bands (see Figure 15.1). $\mathcal{M}$ denotes the set of these bands and $\mathcal{M}_i \subseteq \mathcal{M}$ denotes the set of available bands at node $i \in \mathcal{N}$, which is possibly different from that at another node, say $j \in \mathcal{N}$, i.e., $\mathcal{M}_i \neq \mathcal{M}_j$. For example, at node i,

Table 15.1   Notation

Symbol	Definition
d_{ij}	Distance between nodes i and j
$d(l)$	Destination node of session l
$f_{ij}(l)$	Data rate that is attributed to session l on link (i, j)
g_{ij}	Propagation gain from node i to node j
h_{ij}^m	$\in \{0, 1, 2, \ldots, H\}$, the integer power level for q_{ij}^m
H	The total number of power levels at a transmitter
$\mathcal{I}_j^m$	The set of nodes that can use band m and are within the interference range of node j
$K^{(m)}$	The maximum number for subband division in band m
$\mathcal{L}$	The set of active user sessions in the network
$\mathcal{M}_i$	The set of available bands at node $i \in \mathcal{N}$
$\mathcal{M}$	$= \bigcup_{i \in \mathcal{N}} \mathcal{M}_i$, the set of available bands in the network
M	$= |\mathcal{M}|$, the number of available bands in the network
$\mathcal{M}_{ij}$	$= \mathcal{M}_i \bigcap \mathcal{M}_j$, the set of available bands on link (i, j)
n	Path loss index
$\mathcal{N}$	The set of nodes in the network
$q_{ij}^{(m,k)}$	The transmission power spectral density from node i to node j in subband (m, k)
Q	Maximum transmission power spectral density at a transmitter
Q_T	The minimum threshold of power spectral density to decode a transmission at a receiver
Q_I	The maximum threshold of power spectral density for interference to be negligible at a receiver
$r(l)$	Rate of session $l \in \mathcal{L}$
$R_T(q), R_I(q)$	Transmission range and interference range under q, respectively
R_T^{max}, R_I^{max}	Transmission range and interference range under Q, respectively
$s(l)$	Source node of session l
T_i^m	The set of nodes that can use band m and are within the transmission range of node i
T_i	$= \bigcup_{m \in \mathcal{M}_i} T_i^m$, the set of nodes within the transmission range of node i
$u^{(m,k)}$	The fraction of bandwidth for the k-th subband in band m
$W^{(m)}$	Bandwidth of band $m \in \mathcal{M}$
$x_{ij}^{(m,k)}$	Binary indicator to mark whether or not subband (m, k) is used by link (i, j).
β	An antenna-related constant
η	Ambient Gaussian noise density

$\mathcal{M}_i$ may consist of bands I, III, and V, while at node j, $\mathcal{M}_j$ may consist of bands I, IV, and VI. $W^{(m)}$ denotes the bandwidth of band $m \in \mathcal{M}$. For more flexible and efficient bandwidth allocation and to overcome the disparity in bandwidth size among the spectrum bands, we assume that band m can be

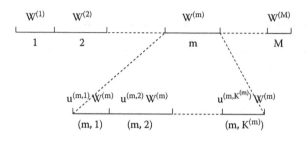

Figure 15.1   A schematic illustrating the concept of bands and subbands in spectrum sharing.

further divided into up to $K^{(m)}$ subbands, each of which may be of *unequal* bandwidth. $u^{(m,k)}$ denotes the fraction of bandwidth for the k-th subband in band m. Then we have

$$\sum_{k=1}^{K^{(m)}} u^{(m,k)} = 1.$$

As an example, Figure 15.1 shows M bands in the network and for a specific band m, it displays a further division into $K^{(m)}$ subbands. Then the M bands in the network are effectively divided into $\sum_{m=1}^{M} K^{(m)}$ subbands, each of which may be of different size. Note that for a specific optimization problem, some $u^{(m,k)}$s can be 0 in the final optimal solution. This suggests that we will have fewer subbands than $K^{(m)}$ in the optimal solution.

15.2.2   Modeling of Power Control and Scheduling

In this section, we show analytical models for power control and scheduling for multihop CR networks. We will examine the notion of transmission and interference ranges in a wireless network, as well as the necessary and sufficient condition for successful transmission.

Transmission and Interference Ranges.   We follow the protocol model for a wireless network [10], where each transmitting node is associated with a transmission range and an interference range. Both transmission and interference ranges directly depend on a node's transmission power and propagation gain. For transmission from node i to node j, a widely used model for power propagation gain g_{ij} is

$$g_{ij} = \beta d_{ij}^{-n}, \tag{15.1}$$

where β is an antenna-related constant, d_{ij} is the physical distance between nodes i and j, and n is the path loss index. Note that we are considering a uniform gain model here and assuming the same gain model on all frequency bands. The case of a nonuniform gain model or a band-dependent gain behavior can be extended without much technical difficulty.

In this context, we assume a data transmission from node i to node j is successful only if the received power spectral density at node j exceeds a threshold, say Q_T. Suppose node i's transmission power spectral density is q and denote the transmission range of this node as $R_T(q)$. Then based on $g_{ij} \cdot q \geq Q_T$ and (15.1), we can calculate the transmission range of this node as follows:

$$R_T(q) = \left(\frac{q}{Q_T} \right)^{1/n}. \tag{15.2}$$

Similarly, we assume that an interference is non-negligible only if it exceeds a power spectral density threshold, say Q_I, at a receiver. Denote the interference range of a node by $R_I(q)$. Then following the same reasoning as in the derivation for the transmission range, we can obtain the interference range of a node as follows:

$$R_I(q) = \left(\frac{q}{Q_I} \right)^{1/n}.$$

Note that since $Q_I < Q_T$, the interference range is greater than the transmission range at a node, i.e., $R_I(q) > R_T(q)$.

Necessary and Sufficient Condition for Successful Transmission. In a CR network, each node has a set of frequency subbands that it may use for transmission and reception. Suppose that subband (m, k) is available at both node i and node j, and $q_{ij}^{(m,k)}$ denotes the transmission power from node i to node j in subband (m, k). Then to schedule a successful transmission from node i to node j, the following necessary and sufficient condition, expressed as two constraints, must be met. The first constraint (C-1) is that receiving node j must be physically within the transmission range of node i, i.e.,

$$(\text{C-1}) \qquad d_{ij} \leq R_T(q_{ij}^{(m,k)}) = \left(\frac{q_{ij}^{(m,k)}}{Q_T} \right)^{1/n}.$$

The second constraint (C-2) is that the receiving node j must not fall into the interference range of any other node p ($p \in \mathcal{N}$, $p \neq i$) that is transmitting on the same subband, i.e.,

$$(\text{C-2}) \qquad d_{jp} \geq R_I(q_{pz}^{(m,k)}) = \left(\frac{q_{pz}^{(m,k)}}{Q_I} \right)^{1/n},$$

where z is the intended receiving node of transmitting node p.

Now we formalize the necessary and sufficient condition for successful transmission into a mathematical model in the general context of multihop CR networks. Suppose that subband (m, k) is available at both node i and node j, i.e., $m \in \mathcal{M}_{ij}$, where $\mathcal{M}_{ij} = \mathcal{M}_i \cap \mathcal{M}_j$. Define

$$
x_{ij}^{(m,k)} = \begin{cases} 1 & \text{if node } i \text{ transmits data to node } j \text{ on subband } (m, k), \\ 0 & \text{otherwise.} \end{cases}
$$

We consider scheduling in the frequency domain. Thus, once a subband (m, k) is used by node i for transmission to node j, $m \in \mathcal{M}_{ij}$, this subband cannot be used again by node i to transmit to a different node. That is,

$$
\text{(C-3)} \qquad \sum_{j \in T_i^m} x_{ij}^{(m,k)} \leq 1,
$$

where T_i^m is the set of nodes that are within the transmission range of node i under full power spectral density Q on band m.

$R_T^{\max}$ denotes the maximum transmission range of a node when it transmits at full power. Then based on (15.2), we have

$$
R_T^{\max} = R_T(Q) = \left(\frac{Q}{Q_T} \right)^{1/n}.
$$

Thus, we have

$$
Q_T = \frac{Q}{(R_T^{\max})^n}.
$$

Then for a node transmitting at a power $q \in [0, \ Q]$, its transmission range is

$$
R_T(q) = \left(\frac{q}{Q_T} \right)^{1/n} = \left[\frac{q(R_T^{\max})^n}{Q} \right]^{1/n} = \left(\frac{q}{Q} \right)^{1/n} R_T^{\max}. \qquad (15.3)
$$

Similarly, $R_I^{\max}$ denotes the maximum interference range of a node when it transmits at full power. Then by the same token we have

$$
R_I^{\max} = R_I(Q) = \left(\frac{Q}{Q_I} \right)^{1/n},
$$

$$
Q_I = \frac{Q}{(R_I^{\max})^n}.
$$

For a node transmitting at a power $q \in [0, Q]$, its interference range is

$$R_I(q) = \left(\frac{q}{Q}\right)^{1/n} R_I^{\max}. \qquad (15.4)$$

Recall that T_i^m denotes the set of nodes that are within the transmission range from node i under full power spectral density Q on band m. More formally, we have $T_i^m = \{j : d_{ij} \leq R_T^{\max}, j \neq i, m \in \mathcal{M}_j\}$. Similarly, $\mathcal{I}_j^m$ denotes the set of nodes that can cause interference to node j on band m under full power spectral density Q, i.e., $\mathcal{I}_j^m = \{p : d_{jp} \leq R_I^{\max}, m \in \mathcal{M}_p\}$. Note that the definitions of T_i^m and $\mathcal{I}_j^m$ are both based on full transmission power spectral density Q. When the power spectral density level q is below Q, the corresponding transmission and interference ranges will be smaller. As a result, it is necessary to keep track of the set of nodes that fall into the transmission range and the set of nodes that can produce interference whenever transmission power changes at a node.

Applying (15.3) and (15.4) to the two constraints (C-1) and (C-2) for successful transmission from node i to node j, we have

$$d_{ij} \leq R_T\left(q_{ij}^{(m,k)}\right) = \left(\frac{q_{ij}^{(m,k)}}{Q}\right)^{1/n} R_T^{\max}, \qquad (15.5)$$

$$d_{jp} \geq R_I\left(q_{pz}^{(m,k)}\right) = \left(\frac{q_{pz}^{(m,k)}}{Q}\right)^{1/n} R_I^{\max} \quad (p \in \mathcal{I}_j^m, \, p \neq i, \, z \in T_p^m). \qquad (15.6)$$

Based on (15.5) and (15.6), we have the following requirements for the transmission link $i \to j$ and interfering link $p \to z$:

$$q_{ij}^{(m,k)} \begin{cases} \in \left[\left(\frac{d_{ij}}{R_T^{\max}}\right)^n Q, \, Q\right] & \text{if } x_{ij}^{(m,k)} = 1, \\ = 0 & \text{if } x_{ij}^{(m,k)} = 0, \end{cases}$$

$$q_{pz}^{(m,k)} \leq \begin{cases} \left(\frac{d_{kj}}{R_I^{\max}}\right)^n Q & \text{if } x_{ij}^{(m,k)} = 1, \\ Q & \text{if } x_{ij}^{(m,k)} = 0, \end{cases} \quad (p \in \mathcal{I}_j^m, \, p \neq i, \, z \in T_p^m).$$

Mathematically, these requirements can be rewritten as

$$(C\text{-}1') \qquad q_{ij}^{(m,k)} \in \left[\left(\frac{d_{ij}}{R_T^{\max}}\right)^n Q x_{ij}^{(m,k)}, \, Q x_{ij}^{(m,k)}\right],$$

$$(C\text{-}2') \qquad q_{pz}^{(m,k)} \leq Q - \left[1 - \left(\frac{d_{pj}}{R_I^{\max}}\right)^n\right] Q x_{ij}^{(m,k)} \quad (p \in \mathcal{I}_j^m, \, p \neq i, \, z \in T_p^m).$$

Recall that we consider scheduling in the frequency domain, and that in (C-3) we state that once a subband (m, k) is used by node i for transmission to node j, this subband cannot be used again by node i to transmit to a different node. In addition, for successful scheduling in the frequency domain, the following two constraints must also hold:

(C-4) For a subband (m, k) that is available at node j, this subband cannot be used for both transmission and receiving. That is, if subband (m, k) is used at node j for transmission (or receiving), then it cannot be used for receiving (or transmission).

(C-5) Similarly to constraint (C-3) on transmission, node j cannot use the same subband (m, k) for receiving from two different nodes.

Note that (C-4) can be viewed as a "self-interference" avoidance constraint where at the same node j, its transmission to another node z on band m interferes with its reception from node i on the same band. It turns out that the above two constraints are mathematically *embedded* in (C-1′) and (C-2′). That is, once (C-1′) and (C-2′) are satisfied, then both constraints (C-4) and (C-5) are also satisfied. This result is formally stated in the following lemma. Its proof can be found in [20].

Lemma 15.1
If transmission powers on every transmission link and interference link satisfy (C-1′) and (C-2′) in the network, then (C-4) and (C-5) are also satisfied.

The significance of Lemma 15.1 is that since (C-4) and (C-5) are embedded in (C-1′) and (C-2′), they can be removed from the list of scheduling constraints. That is, it is sufficient to consider constraints (C-1′), (C-2′), and (C-3) for scheduling and power control.

15.2.3 Flow Routing and Link Capacity Model

Recall that we consider an ad hoc network consisting of a set of $\mathcal{N}$ nodes. Among these nodes, there is a set of $\mathcal{L}$ active user communication (unicast) sessions. $s(l)$ and $d(l)$ denote the source and destination nodes of session $l \in \mathcal{L}$ and $r(l)$ denotes the rate requirement (in b/s) of session l. To route these flows from their respective source nodes to their destination nodes, it is necessary to employ multihop due to limited transmission range of a node. Further, to have better load balancing and flexibility, it is desirable to employ multipath routing (i.e., allow flow splitting) between a source node and its destination node. This is because a single path is overly restrictive and usually does not yield an optimal solution.

Mathematically, this can be modeled as follows. $f_{ij}(l)$ denotes the data rate on link (i, j) that is attributed to session l, where $i \in \mathcal{N}$,

$j \in T_i = \bigcup_{m \in M_i} T_i^m$. If node i is the source node of session l, i.e., $i = s(l)$, then

$$\sum_{j \in T_i} f_{ij}(l) = r(l). \tag{15.7}$$

If node i is an intermediate relay node for session l, i.e., $i \neq s(l)$ and $i \neq d(l)$, then

$$\sum_{\substack{j \in T_i}}^{j \neq s(l)} f_{ij}(l) = \sum_{\substack{k \in T_i}}^{k \neq d(l)} f_{ki}(l). \tag{15.8}$$

If node i is the destination node of session l, i.e., $i = d(l)$, then

$$\sum_{k \in T_i} f_{ki}(l) = r(l). \tag{15.9}$$

It can be easily verified that once (15.7) and (15.8) are satisfied, (15.9) must also be satisfied. As a result, it is sufficient to have (15.7) and (15.8) in the formulation.

In addition to the above flow balance equations at each node $i \in \mathcal{N}$ for session $l \in \mathcal{L}$, the aggregated flow rates on each radio link cannot exceed this link's capacity. Under $q_{ij}^{(m,k)}$, we have

$$\sum_{\substack{l \in \mathcal{L}}}^{s(l) \neq j, d(l) \neq i} f_{ij}(l) \leq \sum_{m \in M_{ij}} \sum_{k=1}^{K^{(m)}} c_{ij}^{(m,k)}$$

$$= \sum_{m \in M_{ij}} \sum_{k=1}^{K^{(m)}} W^{(m)} u^{(m,k)} \log_2 \left(1 + \frac{g_{ij}}{\eta W} q_{ij}^{(m,k)} \right), \tag{15.10}$$

where η is the ambient Gaussian noise density. Note that the denominator inside the log function contains only ηW. This is due to the use of protocol interference modeling, i.e., when node i is transmitting to node j on band m, then the interference range of all other nodes on this band should not contain node j.

15.3 Case Study

In the last section, we presented analytical models for power control, scheduling, and flow routing. In this section, we show how to apply these analytical models to solve some real problems for multihop CR networks.

First and foremost, we should have an objective function. For CR networks, a number of objective functions can be considered for problem formulation. A commonly used objective is to maximize network capacity, which can be expressed as maximizing a scaling factor for all the rate requirements of the communication sessions in the network (see, e.g., [1,14]). For CR networks, we consider an objective called the bandwidth footprint product (BFP), which characterizes the spectrum and space occupancy for the nodes in a CR network. The BFP was first introduced by Liu and Wang in [16]. The so-called footprint refers to the interference area of a node under a given transmission power, i.e., $\pi \cdot (R_I(q))^2$. Since each node in the network will use a number of bands for transmission and each band will have a certain footprint corresponding to its transmission power, an important objective is to minimize the network-wide BFP, which is the sum of BFPs among all the nodes in the network. That is, our objective is to minimize

$$\sum_{i \in \mathcal{N}} \sum_{m \in \mathcal{M}_i} \sum_{j \in T_i^m} \sum_{k=1}^{K^{(m)}} W^{(m)} u^{(m,k)} \cdot \pi \left(R_I \left(q_{ij}^{(m,k)} \right) \right)^2,$$

which is equal to

$$\pi (R_I^{\max})^2 \sum_{i \in \mathcal{N}} \sum_{m \in \mathcal{M}_i} \sum_{j \in T_i^m} \sum_{k=1}^{K^{(m)}} W^{(m)} u^{(m,k)} \left(\frac{q_{ij}^{(m,k)}}{Q} \right)^{2/n}.$$

Since $\pi (R_I^{\max})^2$ is a constant factor, we can remove it from the objective function.

Using this objective function, we study two specific problems for CR networks. In Section 15.3.1, we study the subband division and allocation problem under fixed transmission power (i.e., no power control). The goal is to explore how the different-sized available bandwidth in the network should be optimally divided and allocated among the nodes. In Section 15.3.2, we study the power control problem when all the subbands are identical (i.e., fixed channel bandwidth). The goal there is to explore how to optimally adjust the power level at each node in the network so that the objective function is minimized.

15.3.1 Case A: Subband Division and Allocation Problem

Problem Formulation.   In this case study, we focus on subband division and spectrum sharing for CR networks. We assume the transmission power spectral density is Q and is fixed, i.e.,

$$q_{ij}^{(m,k)} = \begin{cases} Q & \text{if } x_{ij}^{(m,k)} = 1, \\ 0 & \text{if } x_{ij}^{(m,k)} = 0, \end{cases}$$

which is equivalent to

$$q_{ij}^{(m,k)} = Q x_{ij}^{(m,k)}. \tag{15.11}$$

As a result, the objective function of minimizing the network-wide BFP is equivalent to minimizing the network-wide bandwidth usage, i.e., $\sum_{i \in \mathcal{N}} \sum_{m \in \mathcal{M}_i} \sum_{j \in T_i^m} \sum_{k=1}^{K^{(m)}} W u^{(m,k)} (x_{ij}^{(m,k)})^{2/n}$. Since $x_{ij}^{(m,k)}$ is a binary variable (0 or 1), this is equivalent to minimizing

$$\sum_{i \in \mathcal{N}} \sum_{m \in \mathcal{M}_i} \sum_{j \in T_i^m} \sum_{k=1}^{K^{(m)}} W u^{(m,k)} x_{ij}^{(m,k)}.$$

We now examine the power control and scheduling constraints (C-1′) and (C-2′) under fixed transmission power. Due to (15.11), (C-1′) always holds and thus can be removed from problem formulation. Due to fixed transmission power, (C-2′) can be rewritten as

$$Q x_{pz}^{(m,k)} \le Q - \left[1 - \left(\frac{d_{pj}}{R_I^{\max}} \right)^n \right] Q x_{ij}^{(m,k)} \quad (p \in \mathcal{I}_j^m, \ p \ne i, \ z \in T_p^m),$$

which is equal to

$$x_{pz}^{(m,k)} \begin{cases} = 0 & \text{if } x_{ij}^{(m,k)} = 1. \\ \le 1 & \text{if } x_{ij}^{(m,k)} = 0 \end{cases} \quad (p \in \mathcal{I}_j^m, \ p \ne i, \ z \in T_p^m),$$

since $x_{pz}^{(m,k)}$ and $x_{ij}^{(m,k)}$ are binary variables. Thus, $x_{pz}^{(m,k)} + x_{ij}^{(m,k)} \le 1$ for $p \in \mathcal{I}_j^m, \ p \ne i, \ z \in T_p^m$. On the other hand, $\sum_{z \in T_p^m} x_{pz}^{(m,k)} \le 1$ due to (C-3). Putting these together into a more compact form, we have

$$\sum_{z \in T_p^m} x_{pz}^{(m,k)} + x_{ij}^{(m,k)} \le 1 \quad (p \in \mathcal{I}_j^m, p \ne i).$$

Under fixed transmission power, the link capacity constraint (15.10) is

$$\sum_{l \in \mathcal{L}}^{s(l) \ne j, d(l) \ne i} f_{ij}(l) \le \sum_{m \in \mathcal{M}_{ij}} \sum_{k=1}^{K^{(m)}} W^{(m)} u^{(m,k)} \log_2 \left(1 + \frac{g_{ij} Q}{\eta W} \right) x_{ij}^{(m,k)}.$$

Now we have the following optimization problem for subband division and allocation:

$$\text{Min} \quad \sum_{i \in \mathcal{N}} \sum_{m \in \mathcal{M}_i} \sum_{j \in T_i^m} \sum_{k=1}^{K^{(m)}} W^{(m)} x_{ij}^{(m,k)} u^{(m,k)}$$

$$\text{s.t.} \qquad \sum_{k=1}^{K^{(m)}} u^{(m,k)} = 1 \qquad (m \in \mathcal{M}),$$

$$\sum_{j \in T_i^m} x_{ij}^{(m,k)} \leq 1 \qquad \left(i \in \mathcal{N},\, m \in \mathcal{M}_i,\, 1 \leq k \leq K^{(m)}\right),$$

$$x_{ij}^{(m,k)} + \sum_{z \in T_p^m} x_{pz}^{(m,k)} \leq 1 \qquad \left(i \in \mathcal{N},\, m \in \mathcal{M}_i,\, j \in T_i^m,\right.$$

$$\left. 1 \leq k \leq K^{(m)},\, p \in \mathcal{I}_j^m,\, p \neq i\right),$$

$$\sum_{l \in \mathcal{L}}^{s(l) \neq j,\, d(l) \neq i} f_{ij}(l) - \sum_{m \in \mathcal{M}_{ij}} \sum_{k=1}^{K^{(m)}} W^{(m)} \log_2 \left(1 + \frac{g_{ij} Q}{\eta}\right) x_{ij}^{(m,k)} u^{(m,k)}$$

$$\leq 0 \qquad\qquad (i \in \mathcal{N},\, j \in T_i),$$

$$\sum_{j \in T_i} f_{ij}(l) = r(l) \qquad (l \in \mathcal{L},\, i = s(l)),$$

$$\sum_{j \in T_i}^{j \neq s(l)} f_{ij}(l) - \sum_{p \in T_i}^{p \neq d(l)} f_{pi}(l) = 0 \qquad (l \in \mathcal{L},\, i \in \mathcal{N},\, i \neq s(l), d(l)),$$

$$x_{ij}^{(m,k)} = 0 \text{ or } 1,\, u^{(m,k)} \geq 0 \qquad \left(i \in \mathcal{N},\, m \in \mathcal{M}_i,\, j \in T_i^m,\, 1 \leq k \leq K^{(m)}\right),$$

$$f_{ij}(l) \geq 0 \qquad (l \in \mathcal{L},\, i \in \mathcal{N},\, i \neq d(l),\, j \in T_i,\, j \neq s(l)),$$

where $W^{(m)}$, g_{ij}, Q, η, and $r(l)$ are constants, while $x_{ij}^{(m,k)}$, $u^{(m,k)}$, and $f_{ij}(l)$ are optimization variables.

The above optimization problem is a *mixed-integer nonlinear programming* (MINLP) problem, which is NP-hard in general [9]. Although existing software (e.g., BARON [2]) can solve very small-sized network instances (e.g., several nodes), the time complexity becomes prohibitively high for large-sized networks.

Solution Approach. Here we show a promising approach to solving this problem. Under this approach, we first find a lower bound for the objective via a linear relaxation, which is done by relaxing the integer variables and using a linearization technique. Using this lower bound as a performance benchmark, we develop a highly effective algorithm based on a so-called *sequential fixing* (SF) procedure [12]. The main idea of SF is to determine the binary variables in the scheduling decisions iteratively via the relaxed formulation. Specifically, since the integer variables $x_{ij}^{(m,k)}$ can only have binary values of either 0 or 1, we can set $x_{ij}^{(m,k)}$ iteratively based on their closeness to either 0 or 1 in the solution to the relaxed formulation. That is, if during an iteration $x_{ij}^{(m,k)}$ has a value close to 1, then we can make a scheduling decision by fixing this $x_{ij}^{(m,k)}$ to 1. In the next iteration, by updating the relaxed problem with these newly fixed values, we can continue the same process to determine (fix) other $x_{ij}^{(m,k)}$ variables. The iteration continues until we fix all the $x_{ij}^{(m,k)}$s to binary values. The details of this approach can be found in [12].

Simulation Results.   The performance of the SF algorithm can be substantiated by simulation results. Specifically, we will show that the solution obtained via the SF algorithm is very close to the lower bound. Since the optimal objective value lies between the lower bound and the solution obtained by the SF algorithm, the solution given by the SF algorithm must be even closer to the true optimum.

We consider $|\mathcal{N}| = 20$ nodes in a 500×500 area (in meters). Among these nodes, there are $|\mathcal{L}| = 5$ active sessions, each with a random rate within [10, 100] Mb/s. We assume that there are $M = 5$ bands that can be used for the entire network (see Table 15.2). Recall that the set of available bands at each CR node is a *subset* of these five bands depending on the node's location, and the sets of available bands at any two nodes in the network may not be identical. In the simulation, this aspect is modeled by randomly selecting a subset of bands from the pool of five bands for each node. Further, we assume that bands I to V can be divided into 3, 5, 2, 4, and 4 subbands, although other desirable divisions can be used. Note that the size of each subband may be unequal and is part of the optimization problem. We assume that the transmission range at each node is 100 m and that the interference range is 150 m. The path loss index n is assumed to be 4, and $\beta = 62.5$. The threshold Q_T is assumed to be

Table 15.2   Available Bands $\mathcal{M}$ in the Network in the Simulation Study

Band Index	Spectrum Range (MHz)	Bandwidth (MHz)
I	[1240, 1300]	60.0
II	[1525, 1710]	185.0
III	[902, 928]	26.0
IV	[2400, 2483.5]	83.5
V	[5725, 5850]	125.0

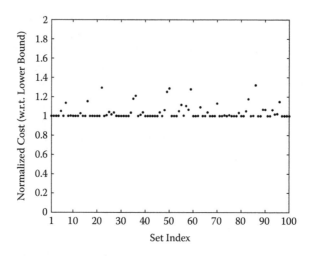

Figure 15.2 **Normalized costs (with respect to lower bound) on 100 sets of randomly generated 20-node networks.**

10η. Thus, we have $Q_I = \left(\frac{100}{150}\right)^n Q_T$ and the transmission power spectral density $Q = (100)^n Q_T / \beta = 1.6 \cdot 10^7 \eta$.

We present simulation results for 100 randomly generated data sets for 20-node networks. Figure 15.2 shows the normalized costs for 100 data sets. The running time for each point is less than 10 seconds on a Pentium 3.4 GHz machine. For each point, we use the SF algorithm to determine the cost, which is the total required bandwidth in the objective function. Then we normalize this cost with respect to the lower bound obtained by linear relaxation. The average normalized cost among the 100 simulations is 1.04 and the standard deviation is 0.07. There are two observations that can be made from this figure. First, since the ratio of the solution obtained by SF (upper bound of optimal solution) to the lower bound solution is close to 1 (in many cases, they coincide with each other), the lower bound must be very tight. Second, since the optimal solution (unknown) is between the solution obtained by the SF algorithm and the lower bound, *the SF solution must be even closer to the optimum.*

15.3.2 Case B: Power Control Problem

Problem Formulation. In this case study, we focus on power control for CR networks. We assume that $\mathcal{M}$ bands are divided equally, each with a bandwidth W. The set of available bands at each node i is $\mathcal{M}_i$.

For power control, we allow the transmission power spectral density to be adjusted between 0 and Q. In practice, the transmission power spectral density can only be tuned into a finite number of discrete levels between 0 and Q. To model this discrete version of power control, we introduce an integer parameter H that represents the total number of power levels to which a transmitter can be adjusted, i.e., $0, \frac{1}{H}Q, \frac{2}{H}Q, \dots, Q$. Denote $b_{ij}^m \in \{0, 1, 2, \dots, H\}$ the integer power level for q_{ij}^m, i.e., $q_{ij}^m = \frac{b_{ij}^m}{H}Q$.

Under such discrete levels of power control, constraints (C-1'), (C-2'), and (15.10) can be rewritten as follows:

$$b_{ij}^m \in \left[\left(\frac{d_{ij}}{R_T^{\max}} \right)^n H x_{ij}^m, \; H x_{ij}^m \right],$$ (15.12)

$$b_{pz}^m \leq H - \left[1 - \left(\frac{d_{pj}}{R_I^{\max}} \right)^n \right] H x_{ij}^m \quad (p \in \mathcal{I}_j^m, \; p \neq i, \; z \in T_p^m),$$ (15.13)

$$\sum_{l \in \mathcal{L}}^{s(l) \neq j, d(l) \neq i} f_{ij}(l) \leq \sum_{m \in \mathcal{M}_{ij}} W \log_2 \left(1 + \frac{g_{ij} Q}{\eta W H} b_{ij}^m \right).$$

For the objective function, again we consider the bandwidth footprint product (BFP), which is

$$\pi (R_I^{\max})^2 \sum_{i \in \mathcal{N}} \sum_{m \in \mathcal{M}_i} \sum_{j \in T_i^m} W \left(\frac{q_{ij}^m}{Q} \right)^{2/n}$$

$$= \pi (R_I^{\max})^2 \sum_{i \in \mathcal{N}} \sum_{m \in \mathcal{M}_i} \sum_{j \in T_i^m} W \left(\frac{b_{ij}^m}{H} \right)^{2/n}.$$

Since $\pi (R_I^{\max})^2$ is a constant factor, we can remove it from the objective function.

Putting all these together, we have the following formulation:

$$\text{Min} \qquad \sum_{i \in \mathcal{N}} \sum_{m \in \mathcal{M}_i} \sum_{j \in T_i^m} W \left(\frac{b_{ij}^m}{H} \right)^{2/n}$$

$$\text{s.t.} \qquad \sum_{j \in T_i^m} x_{ij}^m \leq 1 \qquad (i \in \mathcal{N}, \; m \in \mathcal{M}_i),$$

$$b_{ij}^m - \left(\frac{d_{ij}}{R_T^{\max}} \right)^n H x_{ij}^m \geq 0 \qquad (i \in \mathcal{N}, \; m \in \mathcal{M}_i, \; j \in T_i^m),$$ (15.14)

$$b_{ij}^m - H x_{ij}^m \leq 0 \qquad (i \in \mathcal{N}, \; m \in \mathcal{M}_i, \; j \in T_i^m),$$ (15.15)

$$\sum_{z \in T_p^m} b_{pz}^m + \left(1 - \left(\frac{d_{pj}}{R_I^{\max}}\right)^n\right) Hx_{ij}^m \le H$$

$$(i \in \mathcal{N}, m \in \mathcal{M}_i, j \in T_i^m, p \in \mathcal{I}_j^m, p \ne i), \quad (15.16)$$

$$\sum_{l \in \mathcal{L}}^{s(l) \ne j, d(l) \ne i} f_{ij}(l) - \sum_{m \in \mathcal{M}_{ij}} W \log_2 \left(1 + \frac{g_{ij}Q}{\eta WH} b_{ij}^m\right) \le 0, \quad (i \in \mathcal{N}, j \in T_i),$$

$$\sum_{j \in T_i} f_{ij}(l) = r(l) \qquad (l \in \mathcal{L}, i = s(l)),$$

$$\sum_{j \in T_i}^{j \ne s(l)} f_{ij}(l) - \sum_{p \in T_i}^{p \ne d(l)} f_{pi}(l) = 0 \qquad (l \in \mathcal{L}, i \in \mathcal{N}, i \ne s(l), d(l)),$$

$$x_{ij}^m \in \{0, 1\}, b_{ij}^m \in \{0, 1, 2, \ldots, Q\} \qquad (i \in \mathcal{N}, m \in \mathcal{M}_i, j \in T_i^m),$$

$$f_{ij}(l) \ge 0 \qquad (l \in \mathcal{L}, i \in \mathcal{N}, i \ne d(l), j \in T_i, j \ne s(l)),$$

where $W, g_{ij}, R_T^{\max}, R_I^{\max}, Q, \eta, r(l)$, and H are constants, and x_{ij}^m, b_{ij}^m, and $f_{ij}(l)$ are optimization variables. In this formulation, (15.14) and (15.15) come from (15.12), while (15.16) is based on (15.13) by noting that in (C-3) there is at most one $z \in T_p^m$ such that $x_{pz}^m = 1$. As a result, by (15.15) there is at most one $z \in T_p^m$ such that $b_{pz}^m > 0$. Thus, (15.13) can be rewritten as

$$\sum_{z \in T_p^m} b_{pz}^m \le H - \left(1 - \left(\frac{d_{pj}}{R_I^{\max}}\right)^n\right) Hx_{ij}^m \qquad (p \in \mathcal{I}_j^m, p \ne i),$$

which is equivalent to (15.16).

This optimization problem is in the form of a *mixed-integer nonlinear programming* (MINLP) problem, which is NP-hard in general [9]. In [20], we developed a solution procedure based on branch-and-bound approach [18] and convex hull relaxation. The details of this solution approach are quite mathematically involved, and we refer readers to [20].

Simulation Results. To offer some insights into the power control problem, we present some simulation results from [20]. We consider a 20-node ad hoc network with each node randomly placed in a 500 × 500 area (in meters). An instance of network topology is given in Figure 15.3 with each node's location listed in Table 15.3. We assume there are $|\mathcal{M}| = 10$ frequency bands in the network and each band has a bandwidth of

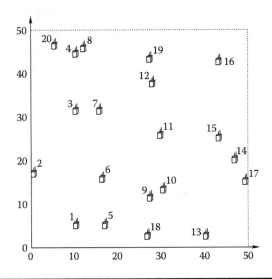

Figure 15.3   A 20-node ad hoc network.

Table 15.3   Each Node's Location and Available Frequency Bands for the 20-node Network

Node Index	Location	Available Bands
1	(10.5, 4.3)	I, II, III, IV, V, VI, VII, VIII, IX, X
2	(1.7, 17.3)	II, III, IV, V, VI, VII, X
3	(10.7, 30.8)	I, III, IV, V, VI, VII, VIII, IX, X
4	(10.2, 45.3)	I, III, IV, V, VI, VII, VIII, IX, X
5	(17.8, 4)	I, II, V, VI, VII, VIII, IX
6	(17.2, 15.2)	I, II, IV, VIII
7	(16.9, 30.8)	I, II, III, IV, V, VI, VII, VIII, IX, X
8	(12.3, 47.3)	I, III, IV, V, VII, VIII, IX
9	(28.2, 11.5)	I, III, V, VII
10	(32.1, 13.8)	I, II, III, IV, VI, VII, VIII, IX, X
11	(30.4, 25.6)	I, II, III, V, VI, VIII, IX, X
12	(29.7, 36)	I, II, III, IV, VI, VI
13	(41.7, 3.1)	I, II, III, V, VI, VIII, IX, X
14	(41.7, 3.1)	I, IV, V, VIII, IX, X
15	(43.3, 25.3)	II, III, IV, V, VI, VII, VIII, IX, X
16	(44.1, 42.7)	I, II, IV, VI, VII, VIII, IX, X
17	(49.6, 15.8)	I, II, III, IV, V, VI, VII, VIII
18	(28.7, 2.5)	I, II, III, VI, VII, VIII, IX, X
19	(28, 43.5)	II, IV, V, VI, VIII
20	(5, 46.9)	II, IV, V, VI, VII

Table 15.4 Source Node, Destination Node, and Rate Requirement of the Five Active Sessions

Source Node	Destination Node	Rate Requirement
7	16	28
8	5	12
15	13	56
2	18	75
9	11	29

$W = 50$ MHz. Each node may only have a subset of these frequency bands. In the simulation, this is modeled by randomly selecting a subset of bands for each node from the pool of 10 bands. Table 15.3 shows the available bands for each node.

We assume that, under maximum transmission power, the transmission range of each node is 100 m and the interference range is twice the transmission range (i.e., 200 m). Both transmission range and interference range will be smaller when transmission power is less than maximum. The path loss index n is assumed to be 4, and $\beta = 62.5$. The threshold Q_T is assumed to be 10η. Thus, we have $Q_I = \left(\frac{100}{200}\right)^n Q_T$ and the maximum transmission power spectral density $Q = (100)^n Q_T/\beta = 1.6 \cdot 10^7 \eta$.

Within the network, we assume there are $|\mathcal{L}| = 5$ user communication sessions, with source node and destination node randomly selected and the rate of each session randomly generated within [10, 100]. Table 15.4 specifies an instance of the source node, destination node, and rate requirement for the 5 sessions in the network.

We apply the solution procedure to the 20-node network described above for different levels of power control granularity (H). Under the branch-and-bound solution procedure we set the desired approximation error bound ε to be 0.05, which guarantees that the obtained solution is within 5% of the optimum [20].

Note that $H = 1$ corresponds to the case that there is no power control, i.e., a node uses its peak power spectral density Q for transmission. When H is sufficiently large, the discrete nature of power control diminishes and power control becomes continuous between [0, Q].

Figure 15.4 shows the results of our solution procedure. First, we note that power control has a significant impact on BFP. Comparing the case when there is no power control ($H = 1$) and the case of $H = 15$, we find that there is nearly a 40% reduction in the total cost (objective value). Second, the cost (objective) is a nonincreasing function of H. However, when H becomes sufficiently large (e.g., 10 in this network setting), further increases in H will not yield much reduction in cost. This suggests that for practical purposes, the number of power levels to achieve a reasonably good result does not need to be a large number.

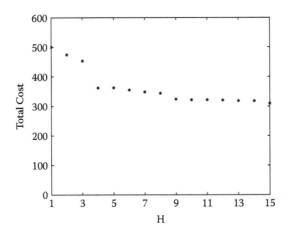

Figure 15.4 Total cost as a function of the number of power levels.

For $H = 10$, the transmission power levels are

$$b^1_{9,11} = 3,$$
$$b^2_{12,16} = 4,$$
$$b^3_{7,12} = 3,$$
$$b^4_{2,1} = 4,$$

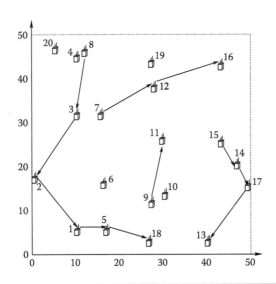

Figure 15.5 Flow routing topology for the five communication sessions in the 20-node network ($H = 10$).

$$b_{2,1}^5 = 4,$$
$$b_{5,18}^6 = 2,$$
$$b_{17,13}^7 = 4,$$
$$b_{8,3}^8 = 5, b_{14,17}^8 = 1,$$
$$b_{1,5}^9 = 1, b_{15,14}^9 = 1,$$
$$b_{3,2}^{10} = 5.$$

For each $b_{ij}^m > 0$, the corresponding scheduling variable $x_{ij}^m = 1$, otherwise $x_{ij}^m = 0$. The flow routing topology for $H = 10$ is shown in Figure 15.5. The corresponding flow rates are:

$$f_{7,12}(1) = 28, f_{12,16}(1) = 28,$$
$$f_{8,3}(2) = 12, f_{3,2}(2) = 12, f_{2,1}(2) = 12, f_{1,5}(2) = 12,$$
$$f_{15,14}(3) = 56, f_{14,17}(3) = 56, f_{17,13}(3) = 56,$$
$$f_{2,1}(4) = 75, f_{1,5}(4) = 75, f_{5,18}(4) = 75,$$
$$f_{9,11}(5) = 29.$$

Note that a link may be used by multiple sessions. For example, link $(2, 1)$ is used by sessions 2 and 4. As a result, the total data rate on link $(2, 1)$ is $f_{2,1}(2) + f_{2,1}(4) = 12 + 75 = 87$.

15.4 Conclusion

Cognitive radio (CR) is a revolution in radio technology that promises unprecedented flexibility in radio communications and is viewed as an enabling technology for future wireless networks. Due to the unique characteristics associated with CR networks, problems for CR networks are expected to be much more challenging and interesting. In this chapter, we presented an analytical model for a multihop CR network at multiple layers. The basic building block in our analytical model is the protocol interference model. Building upon this protocol model, we presented models for power control, scheduling, and routing at physical, link, and network layers, respectively. To demonstrate the practical utility of these analytical models, we studied two practical problems in a multihop CR network as case studies. The first case study addressed the subband division and allocation problem, and the second addressed the power control problem. The results from these two case studies not only validated the practical utility of the analytical models, but also helped gain some theoretical understanding of a multihop CR network.

Acknowledgment

This work has been supported in part by NSF Grant CNS-0721570 and ONR Grant N00014-05-0481.

References

[1] M. Alicherry, R. Bhatia, and L. Li, "Joint channel assignment and routing for throughput optimization in multi-radio wireless mesh networks," in *Proc. ACM Mobicom,* pp. 58–72, Cologne, Germany, Aug. 28–Sep. 2, 2005.

[2] BARON Global Optimization Software, http://www.andrew.cmu.edu/user/ns1b/baron/baron.html.

[3] A. Behzad and I. Rubin, "Impact of power control on the performance of ad hoc wireless networks," in *Proc. IEEE Infocom,* pp. 102–113, Miami, FL, March 13–17, 2005.

[4] R. Bhatia and M. Kodialam, "On power efficient communication over multi-hop wireless networks: joint routing, scheduling and power control," in *Proc. IEEE Infocom,* pp. 1457–1466, Hong Kong, China, March 7–11, 2004.

[5] C.C. Chen and D.S. Lee, "A joint design of distributed QoS scheduling and power control for wireless networks," in *Proc. IEEE Infocom,* Barcelona, Catalunya, Spain, April 23–29, 2006.

[6] R.L. Cruz and A.V. Santhanam, "Optimal routing, link scheduling and power control in multi-hop wireless networks," in *Proc. IEEE Infocom,* pp. 702–711, San Francisco, CA, March 30–April 3, 2003.

[7] R. Draves, J. Padhye, and B. Zill, "Routing in multi-radio, multi-hop wireless mesh networks," in *Proc. ACM Mobicom,* pp. 114–128, Philadelphia, PA, Sep. 26–Oct. 1, 2004.

[8] T. Elbatt and A. Ephremides, "Joint scheduling and power control for wireless ad hoc networks," in *Proc. IEEE Infocom,* pp. 976–984, New York, NY, June 23–27, 2002.

[9] M.R. Garey and D.S. Johnson, *Computers and Intractability: A Guide to the Theory of NP-completeness,* W.H. Freeman and Company, pp. 245–248, New York, NY, 1979.

[10] P. Gupta and P.R. Kumar, "The capacity of wireless networks," *IEEE Transactions on Information Theory,* vol. 46, no. 2, pp. 388–404, March 2000.

[11] S. Haykin, "Cognitive radio: Brain-empowered wireless communications," *IEEE Journal on Selected Areas in Communications,* vol. 23, no. 2, pp. 201–220, Feb. 2005.

[12] Y.T. Hou, Y. Shi, and H.D. Sherali, "Optimal spectrum sharing for multi-hop software defined radio networks," in *Proc. IEEE Infocom,* pp. 1–9, Anchorage, AL, May 6–12, 2007.

[13] K. Jain, J. Padhye, V. Padmanabhan, and L. Qiu, "Impact of interference on multi-hop wireless network performance," in *Proc. ACM Mobicom,* pp. 66–80, San Diego, CA, Sep. 14–19, 2003.

[14] M. Kodialam and T. Nandagopal, "Characterizing the capacity region in multi-radio multi-channel wireless mesh networks," in *Proc. ACM Mobicom,* pp. 73–87, Cologne, Germany, Aug. 28–Sep. 2, 2005.

[15] P. Kyasanur and N.H. Vaidya, "Capacity of multi-channel wireless networks: impact of number of channels and interfaces," in *Proc. ACM Mobicom,* pp. 43–57, Cologne, Germany, Aug. 28–Sep. 2, 2005.

[16] X. Liu and W. Wang, "On the characteristics of spectrum-agile communication networks," in *Proc. IEEE DySpan,* pp. 214–223, Baltimore, MD, Nov. 8–11, 2005.

[17] M. McHenry and D. McCloskey, "New York City Spectrum Occupancy Measurements September 2004," available at http://www.sharedspectrum.com/inc/content/measurements/nsf/NYC_report.pdf.

[18] G.L. Nemhauser and L.A. Wolsey, *Integer and Combinatorial Optimization,* John Wiley & Sons, New York, NY, 1999.

[19] K. Ramachandran, E. Belding-Royer, K. Almeroth, and M. Buddhikot, "Interference-aware channel assignment in multi-radio wireless mesh networks," in *Proc. IEEE Infocom,* Barcelona, Spain, April 23–29, 2006.

[20] Y. Shi and Y.T. Hou, "Optimal power control for multi-hop software defined radio networks," in *Proc. IEEE Infocom,* pp. 1694–1702, Anchorage, AL, May 6–12, 2007.

Chapter 16

Cognitive Radio for Wireless Regional Area Networks

Ying-Chang Liang, Yonghong Zeng, Anh Tuan Hoang, Chee Wei Ang, and Edward Peh

Contents

16.1   Introduction

With the proliferation of wireless communications technology in the last couple of decades, in many countries most of the available spectrum has been allocated. This results in radio spectrum scarcity, which poses a serious problem for the future development of the wireless communications industry. On the other hand, careful studies of the usage pattern reveal that most of the allocated spectrum experiences low utilization. Recent measurements by the FCC show that 70% of the allocated spectrum in the United States is not utilized. Furthermore, the time scale of spectrum occupancy varies from milliseconds to hours.[1] This motivates the concept of *opportunistic spectrum access* that allows secondary networks to borrow unused radio spectrum from primary licensed networks.

The core technology behind opportunistic spectrum access is *cognitive radio*,[3] which consists of the following three components:

(1) *Spectrum sensing*: The wireless devices can sense the radio spectrum environment within their operating range to detect frequency bands that are not occupied by primary users.

(2) *Dynamic spectrum management*: Cognitive radio networks dynamically select the best available bands for communications and monitor the radio environment in order to protect primary users.

(3) *Adaptive communications*: A cognitive radio device can configure its transmission parameters to opportunistically make the best use of the ever-changing available spectrum.

In December 2003, the FCC issued a Notice of Proposed Rule Making that identifies cognitive radio as the candidate for implementing negotiated/opportunistic spectrum sharing.[2] In response to this, the IEEE has formed the 802.22 Working Group to develop a standard for wireless

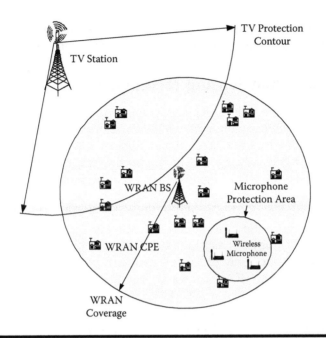

Figure 16.1   An IEEE 802.22 WRAN deployment scenario. Each WRAN system consists of a base station (BS) serving fixed wireless subscribers called customer premise equipment (CPE). Incumbent users are TV receivers and the FCC's Part 74 wireless microphones. Note the difference in the protection ranges for TV and wireless microphone operations.

regional area networks (WRAN) based on cognitive radio technology.[5] WRAN systems operate on the VHF/UHF TV bands, which range from 54 MHz to 862 MHz. While doing so, they must ensure that no harmful interference is caused to the incumbent users, which, for the VHF/UHF bands, include TV receivers and the FCC's Part 74 wireless microphones.[5]

Figure 16.1 illustrate the topology of WRAN where the primary users include TV transmitters and wireless microphones, and the secondary users include both WRAN base stations (BS) and WRAN customer premise equipment (CPEs). These systems will provide wireless broadband access to rural and suburban areas, with an average coverage radius of 33 km and coverage that can extend up to 100 km. The operating principle of WRAN is based on making use of opportunistic unlicensed access to temporary unused TV spectrum. To achieve that, WRAN systems will be able to sense the spectrum, identify unused TV channels, and utilize these channels for supporting broadband connections.

WRAN is expected to be one of the first commercial systems that utilizes cognitive radio technology. Since WRAN reuses TV spectrum, the success of WRAN will provide evidence to the FCC to further open up other licensed bands for frequency reuse. In this chapter, we will illustrate how the three key components of cognitive radio are implemented in WRAN systems.

16.2 Spectrum Sensing Techniques

Unique to cognitive radio operation is the requirement that the radio be able to sense the environment over a huge swath of spectrum and adapt to it, since the radio does not have primary rights to any preassigned frequencies. That is, it is necessary to dynamically detect the existence of signals of primary users or other cognitive users. Furthermore, a cognitive radio should not only be able to sense the existence of the signals, it should also have the capability to identify the type of the signals. There are several factors which make the sensing task difficult. First, the available signal-to-noise ratio (SNR) may be very low. In a WRAN deployment, some Part 74 wireless microphones only transmit signal with power less than 50 mW in a 200 kHz bandwidth. If the sensor is several hundred meters away from the devices, the received SNR may go below −20 dB. Second, wireless fading and multipath propagation can significantly complicate the problem. Fading will cause the signal power to fluctuate dramatically (the magnitude of the effect can be 10 dB or even higher), while unknown multipath interactions will make coherent detection methods unreliable. Third, there exists noise uncertainty, which includes receiver noise uncertainty and environment noise uncertainty. There are several sources of receiver noise uncertainty,[6,19,21] including (a) nonlinearity of components and (b) thermal noise in components (nonuniform, time-varying). The environment noise uncertainty may be caused by transmissions of other users unintentionally (close by) or intentionally (far away). Due to noise uncertainty, in practice it is very difficult (virtually impossible) to obtain accurate noise power measurements. In addition, it is usually required that sensing time be limited and complexity be low. Finally, there may be hidden incumbents whose transmit power may be very low. This makes it very difficult for a far away sensor to detect these low-power incumbent users.

16.2.1 General Model of Sensing

Assume that we are interested in the frequency band with central frequency f_c and bandwidth W. We sample the received signal at a sampling rate higher than the Nyquist rate. The discrete signal can be generally written as

$$y(n) = x(n) + w(n), \tag{16.1}$$

where $x(n)$ represents the signal samples and $w(n)$ represents the noise samples. We define two-hypothesis testing as follows:

$$H_0 : y(n) = w(n) \text{ (signal does not exist)},$$
$$H_1 : y(n) = x(n) + w(n) \text{ (signal exists)}.$$

Let P_d be the probability of detection, that is, on hypothesis H_1, the probability of the algorithm having detected the signal. Let P_{fa} be the probability of false alarm, that is, on hypothesis H_0, the probability of the algorithm having detected the signal. Obviously, for a good detection algorithm, P_d should be high and P_{fa} should be low. The requirements for P_d and P_{fa} depend on the applications. As an example, according to the IEEE 802.22 requirements, WRAN systems must be able to detect DTV signal strength of -116 dBm at $P_d \geq 90\%$ and $P_{fa} \leq 10\%$.

16.2.2 Sensing Algorithms

There are various sensing algorithms. We summarize some of them in the following.

16.2.2.1 Energy Detection

Energy detection was proposed in Ref. [18]. It does not require any information of the source signal, and is robust to unknown dispersed channels and fading. If the noise power is known, it is optimal for detecting independent and identically distributed (i.i.d.) signals.[20]

Let $T_y(N)$ be the average power of the received signal, that is,

$$T_y(N) = \frac{1}{N} \sum_{n=0}^{N-1} |y(n)|^2, \tag{16.2}$$

where N is the number of available samples. Let λ be the threshold. The decision is: if $T_y(N) > \lambda$, signal exists; otherwise, signal does not exist.

For given P_{fa}, the threshold can be found as (for large N)

$$\lambda = \left(\sqrt{\frac{2}{N}} Q^{-1}(P_{fa}) + 1 \right) \sigma_w^2, \tag{16.3}$$

where

$$Q(t) = \frac{1}{\sqrt{2\pi}} \int_t^{+\infty} e^{-u^2/2} du \tag{16.4}$$

and σ_w^2 is the noise power.

Energy detection solely uses the received signal power, which is the sum of the source signal power and the noise power. The receiver does not know the source signal power (it does not know if there is signal or not) and only has some knowledge of noise power. To guarantee reliable detection, the threshold must be set according to the noise power and the

Table 16.1   Noise Uncertainty

method	EG-2 dB	EG-1.5 dB	EG-1 dB	EG-0.5 dB	EG-0dB
P_{fa}	0.489	0.487	0.482	0.461	0.098

number of samples. Accurate knowledge of the noise power is the key to the method. Unfortunately, in practice, noise uncertainty is always present. We can model this by assuming that the noise power is uncertain and can take any value in an interval $[\sigma_1, \sigma_2]$, where σ_1 and σ_2 are related to the receiver devices and the environment. Obviously, if the source signal power is less than $\sigma_2 - \sigma_1$, it is impossible to detect the signal while keeping the P_{fa} low.

Let the estimated noise power be $\hat{\sigma}_w^2 = \alpha\sigma_w^2$, where σ_w^2 is the real noise power. We define the noise uncertainty factor (in dB) as

$$B = \max\{10\log_{10}\alpha\}. \tag{16.5}$$

It is assumed that α (in dB) is evenly distributed in the interval $[-B, B]$.[6,22] In practice, the noise uncertainty factor of a receiving device is normally 1 to 2 dB.[6,22] The environment/interference noise uncertainty can be much higher.

When there is noise uncertainty, if we still use the threshold formula (16.2) based on the estimated noise power $\hat{\sigma}_w^2$, the real P_{fa} will far surpass the targeted value. For example, if we set the target $P_{fa} = 0.1$, the simulated values at different noise uncertainties are shown in Table 16.1 at $N = 50,000$. In Table 16.1 and in the following, "EG-x dB" means the energy detection with x dB noise uncertainty. The noise uncertainty also greatly affects the P_d. Figure 16.2 shows the simulation results when the source signal samples are i.i.d. with Gaussian distribution. The P_d is dramatically reduced when there is noise uncertainty. At low SNR and with noise uncertainty, both the P_d and P_{fa} approach 0.5, which means that the detection is like a random choice between "signal present" and "signal not present." The noise uncertainty problem cannot be solved by increasing the number of signal samples.[21,22] Hence the energy detection algorithm is very unreliable in practical situations with noise uncertainty.

We can change the threshold to guarantee that the P_{fa} is the targeted probability, but then the P_d will be further reduced. Simulations for this are given in Ref. [22].

16.2.2.2   Eigenvalue-Based Detection

Let us consider L consecutive samples and define the following vectors:

$$\mathbf{x}(n) = [x(n)\,x(n-1)\cdots x(n-L+1)]^T, \tag{16.6}$$

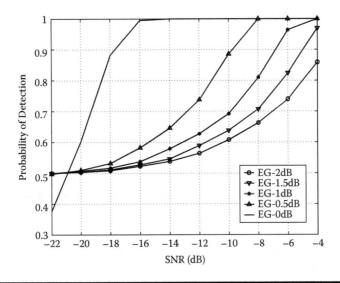

Figure 16.2   Probability of detection for energy detector.

$$\mathbf{y}(n) = [y(n)\ y(n-1)\ \cdots\ s(n-L+1)]^{T}, \tag{16.7}$$

$$\boldsymbol{w}(n) = [w(n)\ w(n-1)\ \cdots\ w(n-L+1)]^{T}, \tag{16.8}$$

where L is called the smoothing factor. Considering the statistical covariance matrices defined as

$$\mathbf{R}_x = \mathrm{E}[\mathbf{x}(n)\mathbf{x}^T(n)], \tag{16.9}$$

$$\mathbf{R}_y = \mathrm{E}[\mathbf{y}(n)\mathbf{y}^T(n)], \tag{16.10}$$

we can verify that

$$\mathbf{R}_y = \mathbf{R}_x + \sigma_w^2 \mathbf{I}_L. \tag{16.11}$$

Let λ_{max} and λ_{min} be the maximum and minimum eigenvalue of $\mathbf{R}_y$, and ρ_{max} and ρ_{min} be the maximum and minimum eigenvalue of $\mathbf{R}_x$. Then $\lambda_{max} = \rho_{max} + \sigma_w^2$ and $\lambda_{min} = \rho_{min} + \sigma_w^2$. Obviously, $\rho_{max} = \rho_{min}$ if and only if $\mathbf{R}_x = \delta \mathbf{I}_L$, where δ is a positive number. In practice, when signal is present, it is very unlikely that $\mathbf{R}_x = \delta \mathbf{I}_L$ (due to dispersive channel and/or oversampling and/or correlation among the signal samples). Hence, if there is no signal, $\lambda_{max}/\lambda_{min} = 1$; otherwise, $\lambda_{max}/\lambda_{min} > 1$. The ratio $\lambda_{max}/\lambda_{min}$ can be used to detect the presence of signal.

In practice, the statistical covariance matrix can only be calculated by using a limited number of signal samples. Define the sample autocorrelations

414 ■ Cognitive Radio Networks

of the received signal as

$$\lambda(l) = \frac{1}{N} \sum_{m=0}^{N-1} y(m)y(m-l), \quad l = 0, 1, \ldots, L-1. \quad (16.12)$$

The statistical covariance matrix $\mathbf{R}_y$ can be approximated by the sample covariance matrix defined as

$$\hat{\mathbf{R}}_y(N) = \begin{bmatrix} \lambda(0) & \lambda(1) & \cdots & \lambda(L-1) \\ \lambda(1) & \lambda(0) & \cdots & \lambda(L-2) \\ \vdots & \vdots & \vdots & \vdots \\ \lambda(L-1) & \lambda(L-2) & \cdots & \lambda(0) \end{bmatrix}. \quad (16.13)$$

Note that the sample covariance matrix is symmetric and Toeplitz. Based on the sample covariance matrix, we propose the following signal detection method.

Algorithm 16.1 *Maximum-minimum eigenvalue (MME) detection.*

Step 1: *Compute the sample autocorrelations in (16.12) and form the sample covariance matrix defined in (16.13).*

Step 2: *Obtain the maximum and minimum eigenvalue of the matrix $\hat{\mathbf{R}}_y(N)$, that is, $\hat{\lambda}_{max}(N)$ and $\hat{\lambda}_{min}(N)$.*

Step 3: *Decision: if $\hat{\lambda}_{max}(N) > \gamma\hat{\lambda}_{min}(N)$, signal exists ("yes" decision); otherwise, signal does not exist ("no" decision), where $\gamma > 1$ is a threshold given below.*

Based on random matrix theory,[23–26] it is shown in Ref. 31 that for given P_{fa}, the threshold should be chosen as

$$\gamma = \frac{(\sqrt{N}+\sqrt{L})^2}{(\sqrt{N}-\sqrt{L})^2}\left(1 + \frac{(\sqrt{N}+\sqrt{L})^{-2/3}}{(NL)^{1/6}}F_1^{-1}(1-P_{fa})\right), \quad (16.14)$$

where F_1 is the cumulative distribution function (CDF) (sometimes simply called distribution function) of the Tracy-Widom distribution of order 1. The Tracy-Widom distributions were found by Tracy and Widom (1996) as the limiting law of the largest eigenvalue of certain random matrices.[27,28] It is generally difficult to evaluate them. Fortunately, tables have been made for the functions[24] and Matlab codes exist to compute them.[29] Table 16.2 gives some values of the function.

Note that, unlike the case of energy detection, here the threshold is not related to noise power. The threshold can be precomputed based only on N, L, and P_{fa}, irrespective of signal and noise.

Table 16.2   CDF of the Tracy-Widom Distribution of Order 1

t	-3.90	-3.18	-2.78	-1.91	-1.27	-0.59	0.45	0.98	2.02
$F_1(t)$	0.01	0.05	0.10	0.30	0.50	0.70	0.90	0.95	0.99

We can also derive other methods using the eigenvalues. For example, we have the energy with minimum eigenvalue (EME) detection as follows.

Algorithm 16.2 *Energy with minimum eigenvalue (EME) detection.*

Step 1: *The same as that in Algorithm 16.1.*

Step 2: *Compute the average power of the received signal $T_y(N)$ (defined in (16.2)), and the minimum eigenvalue $\hat{\lambda}_{min}(N)$ of the matrix $\hat{R}_y(N)$.*

Step 3: *Decision: if $T_y(N)/\hat{\lambda}_{min}(N) > \gamma$, signal exists ("yes" decision); otherwise, signal does not exist ("no" decision), where $\gamma > 1$ is a threshold.*

The difference between the energy detection algorithm and EME is that energy detection compares the signal energy to the noise power, while EME compares the signal energy to the minimum eigenvalue of the sample covariance matrix (online computed).

Remark: Similarly to energy detection, both MME and EME only use the received signal samples for detections (no information on the transmitted signal and channel is needed). Such methods can be called blind methods. The major advantage of the proposed eigenvalue methods is that energy detection needs the noise power for decision, while the proposed methods do not need it (the decision is based on two online computed statistics).

Here simulations are given for wireless microphone signal detection. FM modulated analog wireless microphones are widely used in the United States and elsewhere. They operate on TV bands and typically occupy about 200 kHz (or less) of bandwidth.[30] The detection of the signal is one of the major challenge in 802.22 WRAN. In this simulation, a wireless microphone soft speaker signal[30] is used. The sampling rate at the receiver is 6 MHz (the same as the TV bandwidth in the United States). The smoothing factor is chosen as $L = 10$ and sensing time is 10 ms (corresponding to 60,000 samples). Simulation results are shown in Figure 16.3. From the figure, we see that the MME is much better than the ideal energy detection. EME is usually worse than the ideal energy detection. However, if noise uncertainty is considered, EME can be much better.

More details and simulation results can be found in Refs. [31–33].

16.2.2.3 Covariance-Based Detection

In addition to using the eigenvalues of the covariance matrix described above, we can also use other properties of the matrix for detection. From

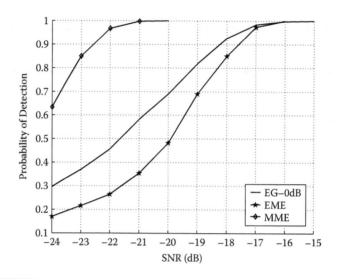

Figure 16.3   Probability of detection (wireless microphone signal, $P_{fa} = 0.1$).

(16.11), if the signal $x(n)$ is not present, $\mathbf{R}_y = \sigma_w^2 \mathbf{I}_L$. Hence the off-diagonal elements of $\mathbf{R}_y$ are all zeros. If there is signal and the signal samples are correlated, $\mathbf{R}_y$ is not a diagonal matrix. Hence, some of the off-diagonal elements of $\mathbf{R}_y$ should be nonzeros. Denote r_{nm} as the element of matrix $\mathbf{R}_y$ at the nth row and mth column, and let

$$T_1 = \frac{1}{L} \sum_{n=1}^{L} \sum_{m=1}^{L} |r_{nm}|, \qquad (16.15)$$

$$T_2 = \frac{1}{L} \sum_{n=1}^{L} |r_{nn}|. \qquad (16.16)$$

Then, if there is no signal, $T_1/T_2 = 1$. If the signal is present, $T_1/T_2 > 1$. Hence, the ratio T_1/T_2 can be used to detect the presence of the signal.

In practice, the statistical covariance matrix can only be calculated using a limited number of signal samples. Based on the sample covariance matrix, we propose the following signal detection method.

Algorithm 16.3 *Covariance absolute value (CAV) detection.*

  Step 1: Compute the autocorrelations of the received signal $\lambda(l)$, $l = 0$, 1, ..., $L - 1$, and form the sample covariance matrix.

Step 2: *Compute*

$$\hat{T}_1(N) = \frac{1}{L} \sum_{n=1}^{L} \sum_{m=1}^{L} |\hat{r}_{nm}(N)|, \qquad (16.17)$$

$$\hat{T}_2(N) = \frac{1}{L} \sum_{n=1}^{L} |\hat{r}_{nn}(N)|, \qquad (16.18)$$

where $\hat{r}_{nm}(N)$ are the elements of the sample covariance matrix $\hat{\mathbf{R}}_y(N)$.

Step 3: *Determine the presence of the signal based on $\hat{T}_1(N)$, $\hat{T}_2(N)$ and the threshold γ, i.e., if $\hat{T}_1(N)/\hat{T}_2(N) > \gamma$, signal exists; otherwise, signal does not exist.*

It is proved in Ref. 34 that for a given P_{fa}, the threshold should be chosen as

$$\gamma = \frac{1 + (L-1)\sqrt{\frac{2}{N\pi}}}{1 - Q^{-1}(P_{fa})\sqrt{\frac{2}{N}}}. \qquad (16.19)$$

Simulations show that the method has similar performance to the MME. An advantage of the CAV is the reduction of computational complexity, because CAV does not need to compute the eigenvalues of the covariance matrix. More details of the method can be found in Refs. [34–36].

16.2.2.4 Feature-Based Detection

Practical communication signals are not purely random signals. They have some nonrandom components such as double-sidedness due to sinewave carrier and keying rate due to symbol period. Such signals are usually cyclostationary, that is, their statistical parameters vary in time with single or multiple periodicities. The cyclostationarity of the signal can be extracted by the spectral correlation density (SCD) function.[8] For a cyclostationary signal, its SCD function is not zero at some nonzero cycle frequencies. On the other hand, noise does not have any cyclostationarity at all; that is, the SCD function of a noise signal is always zero at nonzero cycle frequencies.[8] Hence, we can distinguish signal from noise by analyzing the SCD function. Furthermore, it is possible to distinguish the signal type because different signals may have different nonzero cycle frequencies. For the case of 802.22 WRAN sensing, analog TV signal has cycle frequencies at multiples of the TV-signal horizontal line-scan rate, which is 15.75 kHz in the United States and 15.625 kHz in Europe.[8]

We list the cyclic frequencies of some signals of interest in the following.[8–10]

- *Analog TV signal*: It has cycle frequencies at multiples of the TV-signal horizontal line-scan rate (15.75 kHz in the United States, 15.625 kHz in Europe).
- *AM signal*: $x(t) = a(t)\cos(2\pi f_0 t + \phi_0)$. It has cycle frequencies at $\pm 2 f_0$.
- *PM and FM signal*: $x(t) = \cos(2\pi f_0 t + \phi(t))$. It usually has cycle frequencies at $\pm 2 f_0$. The characteristics of the SCD function at cycle frequency $\pm 2 f_0$ depend on $\phi(t)$.
- *Digital modulated signals*:
  - *Amplitude-shift keying*: $x(t) = [\sum_{n=-\infty}^{\infty} a_n p(t - nT_0 - t_0)] \cos(2\pi f_0 t + \phi_0)$. It has cycle frequencies at k/T_0, $k \neq 0$, and $\pm 2 f_0 + k/T_0, k = 0, \pm 1, \pm 2, \ldots$.
  - *Phase-shift keying*: $x(t) = \cos[2\pi f_0 t + \sum_{n=-\infty}^{\infty} a_n p(t - nT_0 - t_0)]$. For BPSK, it has cycle frequencies at k/T_0, $k \neq 0$, and $\pm 2 f_0 + k/T_0, k = 0, \pm 1, \pm 2, \ldots$. For QPSK, it has cycle frequencies at $k/T_0, k \neq 0$.
- *Multipath signal*: After the transmitted signal $x(t)$ passes a multipath fading channel $h(t)$, the received signal is $y(t) = x(t) * h(t)$. Its SCD function is $S_y^\alpha(f) = H(f + \alpha/2)H^*(f - \alpha/2)S_x^\alpha(f)$, where $H(f)$ is the Fourier transform of the channel $h(t)$ and $S_x^\alpha(f)$ is the SCD function of $x(t)$. Usually the multipath channel does not change the cycle frequencies of the transmitted signal.

Although cyclostationarity detection has advantages, it also has some disadvantages: (1) it needs oversampling, thus requiring a large number of samples, and has high computational complexity; (2) it is still unknown how to set the threshold to meet required performance; (3) it is mathematically intractable to obtain analytic expressions for the P_d and P_{fa} (therefore difficult to predict the detection reliability); and (4) if the noise is not stationary, such as impulse noise, the SCD function of the noise may not be zero at some cycle frequencies. Hence, the noise also contributes to the SCD at some non-zero cyclic frequencies and invalidates the theory.

There is another popular method called matched filtering (MF). It uses the known signal signature to match the received signal. The MF-based method requires perfect knowledge of the channel responses from the primary user to the receiver and accurate synchronization (otherwise its performance will be reduced dramatically).[16,17] In practice, it is hard to obtain the channel responses and get synchronized if the primary users do not cooperate with the secondary users.

16.2.3 Cooperative Sensing

In a cognitive network, cooperation among multiple secondary users within a certain region can achieve higher sensing reliability.[7] There are various different cooperation schemes. For example, each user can send its test statistics to a specified user, which then processes the data collected. This scheme is called data fusion. Alternatively, multiple secondary nodes sense independently and then send their decisions to a specific user. This is called decision fusion.

16.2.3.1 Optimal Decision Fusion Rule

Let us define $P(\mathcal{H}_1)$ as the activity probability for which the primary user is active in the frequency band of interest, and $P(\mathcal{H}_0) = 1 - P(\mathcal{H}_1)$ as the inactivity probability. Let I_i be the binary decision from the ith time slot, where $I_i \in (0, 1)$ for $i = 1, \ldots, M$. The optimal decision fusion rule is the Chair-Varshney fusion rule,[14] which is a threshold test of the following statistic:

$$\Lambda_0 = \sum_{i=1}^{M} \left[I_i \log \frac{P_{d,i}}{P_{f,i}} + (1 - I_i) \log \frac{1 - P_{d,i}}{1 - P_{f,i}} \right] + \log \frac{P(\mathcal{H}_1)}{P(\mathcal{H}_0)}. \quad (16.20)$$

The final decision rule is: if $\Lambda_0 \geq 0$, then the primary user is present; otherwise, there is no primary user.

16.2.3.2 Logical-OR (LO) Rule

The LO rule is a simple decision rule described as follows: if one of the decisions says that there is a primary user, then the final decision declares that there is a primary user. Assuming that all decisions are independent, then the probability of detection and probability of false alarm of the final decision are $P_d = 1 - \prod_{i=1}^{M}(1 - P_{d,i})$ and $P_{fa} = 1 - \prod_{i=1}^{M}(1 - P_{f,i})$, respectively.

16.2.3.3 Logical-AND (LA) Rule

The LA rule works as follows: if all decisions say that there is a primary user, then the final decision declares that there is a primary user. The probability of detection and probability of false alarm of the final decision are $P_d = \prod_{i=1}^{M} P_{d,i}$ and $P_{fa} = \prod_{i=1}^{M} P_{f,i}$, respectively.

16.2.3.4 Majority Rule

Another decision rule is based on the majority of the individual decisions. If half of the decisions or more say that there is a primary user, that the final decision declares that there is a primary user. Mathematically, define $\Lambda = \sum_{i=1}^{M} I_i$; if $\Lambda \geq \lceil \frac{M}{2} \rceil$, where $\lceil x \rceil$ denotes the smallest integer not less than x, then the primary user is present; otherwise, there is no primary user. Assuming that all decisions are independent, then the probability of detection and probability of false alarm of the final decision are, respectively,

$$P_d = \sum_{i=0}^{M-\lceil\frac{M}{2}\rceil} \binom{M}{\lceil\frac{M}{2}\rceil + i} (1 - P_{d,i})^{M-\lceil\frac{M}{2}\rceil-i}(1 - P_{d,i})^{\lceil\frac{M}{2}\rceil+i}, \quad (16.21)$$

$$P_f = \sum_{i=0}^{M-\lceil\frac{M}{2}\rceil} \binom{M}{\lceil\frac{M}{2}\rceil + i} (1 - P_{f,i})^{M-\lceil\frac{M}{2}\rceil-i}(1 - P_{f,i})^{\lceil\frac{M}{2}\rceil+i}. \quad (16.22)$$

Although cooperation can achieve better performance, there are some problems that must be addressed. First, reliable information exchange among the cooperating users must be guaranteed. In a distributed ad hoc network, this is by no means a trivial problem. Second, data fusion algorithms need to be investigated. The fusion algorithm should make a more reliable decision than all single decisions. Due to the high cost of communication, noise, multipath effects, and fading, the received information may be limited and inaccurate. The fusion algorithm then should have the ability to make decisions based on the limited information and be tolerant of some data errors. Finally, there may be untrusted/malicious users in the network who may deliberately send wrong information. A strong user authentication scheme is needed to deal with this scenario.

16.3 Dynamic Spectrum Management

In conventional systems, the main task of spectrum management is *resource allocation*. For cognitive radio networks, due to the need to protect primary users under the ever-changing spectrum utilization conditions, two other essential tasks must also be carried out, i.e., *spectrum monitoring* and *spectrum classification*.

16.3.1 *Spectrum Monitoring*

16.3.1.1 *In-Band and Out-of-Band Sensing*

A cognitive radio network needs to periodically monitor the spectrum bands it is operating on in order to detect the occurrence of primary users. This monitoring activity is termed *in-band sensing*. At the same time, the cognitive radio network needs to monitor the bands that it can potentially switch to, in case its current operating bands are reclaimed by primary services. The activity of sensing nonoperating bands is termed *out-of-band sensing*.

To achieve reliable sensing, all cognitive devices must stop transmission on a particular channel before in-band sensing for that channel can be carried out. As illustrated in Figure 16.4, quiet periods for in-band sensing

Figure 16.4 General operation sequence of a cognitive radio system, with quiet periods for sensing being inserted in between normal data transmission intervals.

are inserted in between normal data transmission intervals. We note that the insertion of quiet periods can have a serious effect on the quality of service of active connections. For example, for the case of a WRAN, it may take tens of milliseconds to sense a TV channel. This may violate the latency constraints of real-time traffic. A possible solution to this problem is to switch all real-time users to a different spectrum band while in-band sensing is carried out in the current band. Another solution is to schedule the quiet sensing in an opportunistic manner.

Once a BS determines that it is interfering with an incumbent user, it can either order a channel change, or if there is no unused channel available, it can either reduce its transmission power or to shut down the WRAN service entirely.

16.3.1.2 Opportunistic Sensing

In order to minimize disruption to data transmissions, sensing periods can be *scheduled dynamically*. Sensing periods can be scheduled into a frame if it does not cause violation to QoS of the ongoing service flows, as illustrated in Figure 16.5. A sensing period should be scheduled to occur at the same time in all devices in the WRAN, i.e., a common quiet period, so that

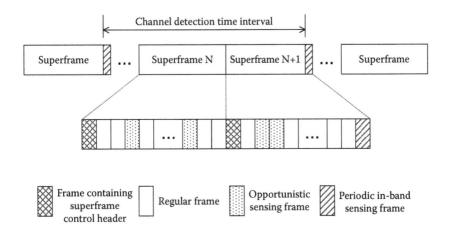

Figure 16.5 Opportunistic sensing. Note that sensing activity need not take up the whole frame in both opportunistic sensing and periodic in-band sensing frames.

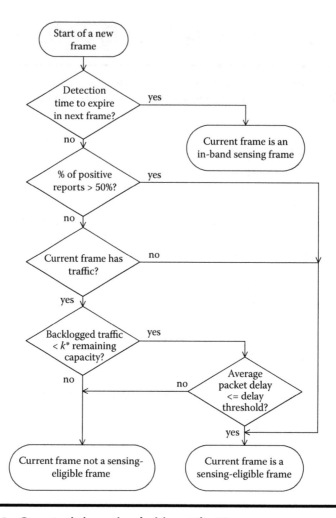

Figure 16.6 Opportunistic sensing decision at the BS.

in-band sensing can be performed. Sensing periods can also be scheduled individually for each device to carry out *out-of-band* sensing. If there is no *opportunity* available for scheduling a sensing period throughout a long duration, (e.g., because the data traffic is heavy) a sensing period can be inserted so that the *channel detection time* requirement can still be met.

Figure 16.6 shows a method the BS can use to schedule sensing periods. The BS first checks whether the previous sensing (in-band or out-of-band) was done too long ago. If so, it schedules a sensing period. If not, the BS proceeds to check whether there are more than x% positive detection reports (in the figure $x = 50$). If there are, then it is considered urgent to do more sensing and thus the BS proceeds to schedule a sensing period. If

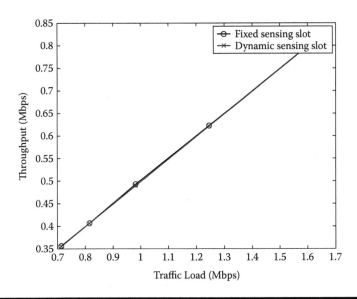

Figure 16.7   Data throughput is not affected by opportunistic in-band sensing.

there is no traffic in the next frame, then the frame can be used for sensing. If there is some traffic scheduled for the next frame, the BS checks whether it can be "backlogged"—can be scheduled onto the frame after next. If the BS finds that the data packets would not violate their QoS (such as throughput and packet end-to-end delay), the next frame can be used for sensing.

Note that under worst-case situations, opportunistic sensing essentially falls back to periodic sensing.

Since frames are opportunistically used for sensing, it has negligible effects on traffic performance. The effect on data throughput in one experiment is shown in Figure 16.7.

Since more sensing opportunities are made available with opportunistic sensing, primary users can be detected and avoided faster (see Figure 16.8). It also improves the probability of detection because more samples can be collected before the detection decision is made.

16.3.1.3   Other Considerations

Within each cognitive radio network, whenever quiet periods are scheduled for in-band sensing, care must be taken to ensure that the channel is actually clear from the signal of all cognitive radio transmissions. Due to propagation delay, signal from far away cognitive nodes can reach the receiver of another cognitive node much later, which can affect the in-band sensing reliability. As an example, consider a WRAN system with a coverage

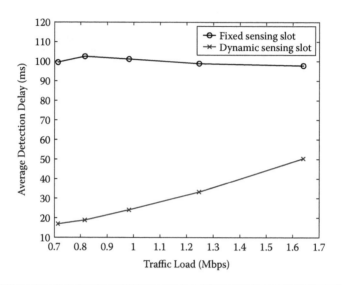

Figure 16.8   Detection delay is smaller than periodic in-band sensing.

radius of 30 km. Then the signal transmitted from a CPE at the edge of the cell can reach another CPE at the opposite side of the cell after as long as 200 microseconds. On the other hand, the signal transmitted by the BS will pass all CPEs after 100 microseconds. Another source of interference comes from other cognitive radio networks operating in nearby regions. The signal transmitted by these neighboring networks can reach a particular cognitive node during a quiet in-band sensing period.

16.3.2   *Spectrum Classification*

Spectrum classification is an important step in protecting primary users while maximizing performance of cognitive radio networks. How channels are classified depends on the requirements in protecting primary users. In the most stringent case, channels can be classified using a binary model with channels either being *available* or *unavailable*. In a more flexible situation, when primary users can tolerate a certain degree of interference, for example in terms of maximum interference temperature, then a channel can be classified as *available, available with constraints*, and *unavailable*.

Take the IEEE 802.22 functional requirements[5] as an example. Given that a TV station is currently operating on channel n, then a WRAN network will not be allowed to operate on three TV channels $n-1$, n, and $n+1$ within the protection contour of the TV station. Furthermore, if a WRAN system operates on TV channels $n-2$ and $n+2$, then some minimum distance must be kept between the WRAN BS and the TV station (operating on channel n)

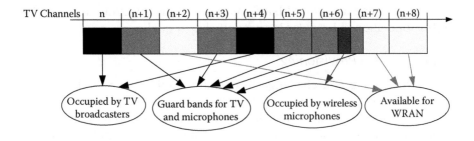

Figure 16.9 Classification of the availability of spectrum for 802.22 WRAN usage based on TV and wireless microphone operations. Note the irregularity in the available spectrum for WRAN usage.

to avoid possible interference. In general, the transmit power of CPEs will have to be limited depending on the channel separation between a WRAN and TV station.[5] For Part 74 wireless microphones, a smaller guard band is enough to protect their operation. This leads to the fact that a fraction of a TV channel can be available for WRAN usage. The spectrum availability is illustrated in Figure 16.9.

The variation in the spectrum availability profile, with respect to location and spectrum band, is termed *spectral irregularity*. This irregularity is illustrated in Figure 16.9. One of the most important requirements of the bandwidth management function is to take this spectrum irregularity into account to achieve good system performance while still protecting primary users. This will be discussed in the next section.

16.3.3   Spectrum Allocation

16.3.3.1   Scalable Bandwidth Design

Cognitive radio must be designed to operate with variations in spectrum availability. These variations are due to the fact that primary users will come on and off in the spectrum of interest over time. Furthermore, different classes of primary users will require different protection levels. Take the example of the 802.22 WRAN. At the most basic level, TV channels can have different bandwidths of 6, 7, and 8 MHz, depending on different countries' specifications. At a more detailed level, given a particular location, the bandwidth available for WRAN usage directly depends on the existence of TV and wireless microphone operations. As illustrated in Figure 16.9, for each channel occupied by TV broadcasting, the two adjacent channels must be used as guard bands. Also, for each fraction of TV channel used by wireless microphones, some adjacent fraction of the TV channel must also be set aside as a guard band. As a result, the spectrum available for WRAN operation is irregular in size and also in quality.

16.3.3.2 QoS Provisioning

QoS provisioning is a challenge in cognitive radio networks. First, a resource allocation algorithm must take into account the irregular available spectrum to support all connections while protecting primary users. The problem becomes more complicated if primary users do not collaborate, and can interfere with the operation of cognitive devices. Finally, as has been discussed, the need to schedule quiet in-band sensing periods will also have a negative effect on the delay and latency experienced by traffic of the cognitive network.

16.4 Adaptive Communications

The third component of cognitive radio is adaptive communication that allows each cognitive device to carry out data transmission based on the allocated bandwidth without interfering with the primary users. Techniques that maximize the spectrum utilization and protect the primary users are discussed.

16.4.1 Two-Layer Orthogonal Frequency Division Multiple Access

As has been mentioned, the spectrum available for a cognitive radio network varies in time and space. The communications at each cognitive device must be able to adapt to this. As a result, orthogonal frequency division multiple access (OFDMA) technology is the best modulation and multiple access technology. The major advantage of OFDMA is the ability to achieve high spectral efficiency under hostile multipath, frequency selective fading conditions. By dividing the spectrum into multiple narrow subcarriers and allowing adaptive rate and power control, OFDMA systems can exploit time, frequency, and multiuser diversities to improve performance. For an 802.22 WRAN, OFDMA technology has an added advantage in adapting to the irregularity in available spectrum. As illustrated in Figure 16.9, due to the need to protect wireless microphone operations, only a fraction of a TV channel is usable by the WRAN. In that case, an OFDMA-based system can adapt either by turning off some subcarriers that are not usable, or by using smaller FFT sizes for fractional TV bandwidths.[11]

When there are multiple contiguous or discontiguous TV channels, as shown in Figure 16.9, a two-layer OFDMA technique can be used to aggregate the available channels. To do that, this scheme will first assign a group of CPEs to a single TV channel, and within each group, OFDMA is employed for multiple access.[11] Load balancing based on traffic conditions can be considered for the first layer assignment.

16.4.2 Adaptive Time Division Duplex (TDD)

For cognitive radio networks that follow a cellular-like architecture, spectrum irregularity also affects the choice of duplexing and multiple access techniques. For example, as it may be difficult to identify paired spectrum for two-way communications, frequency division duplex (FDD) may not be a good option. In fact, based on the development within the 802.22 Working Group, time division duplex (TDD) will most likely be the duplexing technique for 802.22 WRAN. Each TDD frame usually consists of a downlink (DL) subframe and an uplink (UL) subframe that are used to transmit downlink and uplink data respectively. Compared to FDD, TDD also offers the advantage of flexible division between DL and UL throughputs, as the duration of the DL and UL subframes can vary. This is particularly useful for broadband point-to-multipoint applications with an asymmetric pattern between DL and UL traffic.

For a TDD-OFDMA system, there should be a guard interval between DL and UL subframes for synchronizing the transmissions of the farthest nodes and the nearby ones. Conventionally, the Tx/Rx transition Gap (TTG) is set equal to the sum of round-trip propagation delay to the edge of the cell, plus the modem switching time. For cellular systems, the TTG is not significant due to the small cell radius. However, for the case of 802.22 WRAN, the cell radius can go as far as 100 km, which leads to a significant loss in system efficiency due to TTG insertion. A possible solution to this problem is to employ an adaptive scheme which allows nearby nodes to start UL transmission earlier than far away nodes.[11] This scheme is referred to as adaptive TDD. As long as the symbol boundaries of all transmissions are kept aligned, reliable communication is still achieved. For a cell radius of 100 km and OFDMA duration of about 280 μs, up to three additional OFDMA symbols can be gained. Even when a single OFDMA symbol is gained, when this symbol is allocated to nearby users achieving the highest data rate, the throughput increment as compared to the average uplink throughput can be as high as 50%.

16.4.3 Adaptive Beamforming

The coverage areas of licensed incumbents are geographically localized. Using downlink beamforming with multiple BS antennas, the transmissions that may be potential sources of interference may be nulled in the

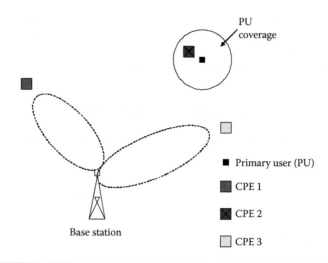

Figure 16.10   Adaptive beamforming.

direction of the coverage areas of the active primary users. As illustrated in Figure 16.10 for an 802.22 WRAN system, the beamformers direct the transmissions toward the desired locations, where CPE 1 and CPE 3 are situated, but suppress the transmitted signal in the direction of the primary user. CPE 2, which is in the coverage zone of the primary user, may employ a noninterfering channel for communication. This isolates and protects the primary user from interference and enables the reuse of the occupied frequencies in other locations. The use of beamforming for interference avoidance requires positional knowledge of the primary users. Thus, localization of the primary users is required.

During a communication process, a WRAN user might need to switch from one TV band to another TV band in case of the sudden appearance of a primary user in the band where the WRAN user is operating. This inherent nature of WRAN systems poses a great challenge to designing adaptive beamforming schemes. For WRAN systems, beamforming weights derived in one TV band might not be directly applicable to another TV band, because the frequency difference between two different TV bands could be sufficiently large to create a certain direction of arrival (DOA) shift between two beam patterns. This, in turn, may greatly degrade the performance and capacity of WRAN systems. Fortunately, the problem is similar to the beamforming problem in the frequency division duplex (FDD) scenario, so there are beam synthesis methods that can be applied.[15] However, in WRAN systems, the difference between the frequencies from one channel to another channel could be very large, which may require further joint design of antenna arrays and beam synthesis techniques.

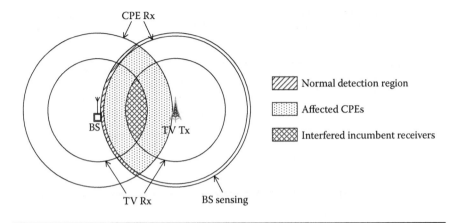

Figure 16.11 Hidden incumbent situation.

16.4.4 Hidden Incumbent Situation

The hidden incumbent situation happens when a CPE loses synchronization with its BS as a result of interference from incumbent transmissions. The BS is situated away from the incumbent transmission in such a configuration that the BS is unable to detect the incumbent transmission. Some CPEs are able to detect the incumbent transmission, but are unable to report it to the BS because they have lost synchronization with the BS. The BS continues its transmission oblivious to the presence of the incumbent user. In other words, the incumbent user is *hidden*. As a result, some incumbent receivers will be interfered with by the BS transmission, and this is a situation that should be avoided.

Figure 16.11 shows an example where TV is the hidden incumbent. The contours depict the boundaries within which the received power is above the reception threshold. For example, the contour labeled "TV Rx" around the BS demarcates the region within which the received power from the BS at a TV receiver is above its reception threshold, and therefore the TV receiver reception is interfered with. The *affected CPEs* are able to detect the TV transmission but unable to report it to the BS because they have lost synchronization with the BS due to interference from the TV transmission. Those CPEs located within the "normal detection region" are able to report the presence of the incumbent transmission, but in this example this region is relatively small and thus the probability of having a CPE located in this region is small.

In order for the affected CPEs to send the incumbent detection reports to the BS, they must first be synchronized with the BS. They must then be able to decode the MAC control messages to acquire MAC layer scheduling information so that they know when to send the reports so as not to collide

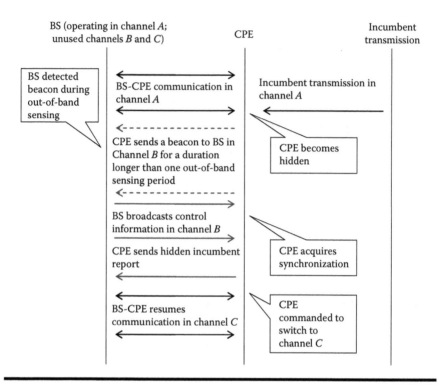

Figure 16.12 A method for an affected CPE to send a detection report to its BS under the hidden incumbent situation.

with ongoing transmissions within the WRAN. One way to achieve this is to allow the BS to broadcast control information in other unused channels, in addition to the operating channel. The affected CPEs can search for these nonoperating channels once they lose synchronization with the BS. Synchronization can then be reestablished and incumbent detection reports can be sent on one of these channels. However, this scheme requires the availability of unused channels. It also requires making periodic transmissions in the unused channels.

To circumvent these problems, the synchronization and reporting can be made *on demand*. For this to work, there must be a way for the affected CPEs to signal the BS even when they are not synchronized with the BS. One way to achieve this is for the CPEs to transmit a special tone which the BS recognizes as the signal to switch on the additional channel broadcasts. The affected CPEs can then get synchronized and send the incumbent detection report to the BS (see Figure 16.12). If there are unused channels available, the special tone can be sent on one of the unused channels so as not to interfere with the ongoing transmissions in the WRAN. If there is no unused channel, the special tone can be sent *in-band*. In this case, the

BS will have to make a decision based on just the reception of the special tone, without receiving the incumbent detection report. In both cases, the BS detects the special tone during its periodic out-of-band and in-band sensing, respectively.

16.5 Conclusions

Cognitive radio is a technology that promises to mitigate the problem of spectrum scarcity in wireless communications. As we have discussed, there are three main components that make up cognitive radio, i.e., spectrum sensing, dynamic spectrum management, and adaptive communications. In order to realize cognitive radio technology, several technical challenges need to be overcome, and it is encouraging to observe that these challenges have received great interest from academia and the industry. The development of the IEEE 802.22 WRAN, the very first standard ever based on cognitive radio, will encourage more efforts from the regulatory, research, and industry players to promote cognitive radio technology for other applications.

References

[1] Federal Communications Commission, "Spectrum policy task force report, FCC 02-155," Nov. 2002.
[2] Federal Communications Commission, "Facilitating opportunities for flexible, efficient, and reliable spectrum use employing cognitive radio technologies, notice of proposed rule making and order, FCC 03-322," Dec. 2003.
[3] J. Mitola and G. Q. Maguire, "Cognitive radio: making software radios more personal," *IEEE Personal Communications*, vol. 6, no. 4, pp. 13–18, Aug. 1999.
[4] S. Haykin, "Cognitive radio: Brain-empowered wireless communications," *IEEE J. Selected Areas in Communications*, vol. 23, no. 2, pp. 201–220, Feb. 2005.
[5] IEEE 802.22 Wireless RAN, "Functional requirements for the 802.22 WRAN standard, IEEE 802.22- 05/0007r46," Oct. 2005.
[6] A. Sahai and D. Cabric, "Spectrum sensing: fundamental limits and practical challenges," *Proc. IEEE International Symposium on New Frontiers in Dynamic Spectrum Access Networks (DySPAN) 2005*, Baltimore, MD, Nov. 2005.
[7] S. M. Mishra, A. Sahai, and R. W. Brodensen, "Cooperative sensing among cognitive radios," *Proc. IEEE International Conference on Communications (ICC) 2006*, Istanbul, Turkey, Jun. 2006.
[8] W. A. Gardner, "Exploitation of spectral redundancy in cyclostationary signals," *IEEE Signal Processing Magazine*, vol. 8, pp. 14–36, Apr. 1991.

[9] W. A. Gardner, "Spectral correlation of modulated signals, I: Analog modulation," *IEEE Trans. Communications*, vol. 35, no. 6, pp. 584–595, 1987.

[10] W. A. Gardner, W. A. Brown, III, and C.-K. Chen, "Spectral correlation of modulated signals, II: Digital modulation," *IEEE Trans. Communications*, vol. 35, no. 6, pp. 595–601, 1987.

[11] Y.-C. Liang, et al., "System description and operation principles for IEEE 802.22 WRANs," Available at IEEE 802.22 WG Web site http://www.ieee802.org/22/, Nov. 2005.

[12] G. Ganesan, Y. Li, "Cooperative spectrum sensing in cognitive radio networks," *Dyspan'2005*, pp. 137–143, Nov. 2005.

[13] A. Ghasemi and E.S. Sousa, "Collaborative spectrum sensing for opportunistic access in fading environments," *Dyspan'2005*, pp. 131–136, Nov. 2005.

[14] Z. Chair and P. K. Varshney, "Optimal data fusion in multiple sensor detection systems," *IEEE Trans. on Aerospace and Elect. Syst.*, vol. 22, pp. 98–101, January 1986.

[15] Y.-C. Liang and F. Chin, "Two suboptimal algorithms for downlink beamforming in FDD DS-CDMA mobile radio," *IEEE J. Select. Areas Commun.*, vol. 19, no. 7, pp. 1264–1275, July 2001.

[16] D. Cabric, A. Tkachenko, and R. W. Brodersen, "Spectrum sensing measurements of pilot, energy, and collaborative detection," in *Military Comm. Conf. (MILCOM)*, pp. 1–7, Oct. 2006.

[17] H.-S. Chen, W. Gao, and D. G. Daut, "Signature based spectrum sensing algorithms for IEEE 802.22 WRAN," in *IEEE Intern. Conf. Comm. (ICC)*, June 2007.

[18] H. Urkowitz, "Energy detection of unkown deterministic signals," *Proceedings of the IEEE*, vol. 55, no. 4, pp. 523–531, 1967.

[19] A. Sonnenschein and P. M. Fishman, "Radiometric detection of spread-spectrum signals in noise of uncertainty power," *IEEE Trans. on Aerospace and Electronic Systems*, vol. 28, no. 3, pp. 654–660, 1992.

[20] S. M. Kay, *Fundamentals of Statistical Signal Processing: Detection Theory*, vol. 2. Prentice Hall, 1998.

[21] R. Tandra and A. Sahai, "Fundamental limits on detection in low SNR under noise uncertainty," in *WirelessCom 2005*, (Maui, HI), June 2005.

[22] S. Shellhammer and R. Tandra, *Performance of the Power Detector with Noise Uncertainty*, in *IEEE 802.22-06/0134r0*, July 2006.

[23] A. M. Tulino and S. Verdú, *Random Matrix Theory and Wireless Communications*. Now Publishers Inc., 2004.

[24] I. M. Johnstone, "On the distribution of the largest eigenvalue in principal components analysis," *The Annals of Statistics*, vol. 29, no. 2, pp. 295–327, 2001.

[25] K. Johansson, "Shape fluctuations and random matrices," *Comm. Math. Phys.*, vol. 209, pp. 437–476, 2000.

[26] Z. D. Bai, "Methodologies in spectral analysis of large dimensional random matrices, a review," *Statistica Sinica*, vol. 9, pp. 611–677, 1999.

[27] C. A. Tracy and H. Widom, "On orthogonal and symplectic matrix ensembles," *Comm. Math. Phys.*, vol. 177, pp. 727–754, 1996.

[28] C. A. Tracy and H. Widom, "The distribution of the largest eigenvalue in the gaussian ensembles," in *Calogero-Moser-Sutherland Models* (J. van Diejen and L. Vinet, eds.), pp. 461–472, Springer, 2000.

[29] M. Dieng, *RMLab Version 0.02.* http://math.arizona.edu/ṁomar/, 2006.

[30] C. Clanton, M. Kenkel, and Y. Tang, "Wireless microphone signal simulation method," in *IEEE 802.22-07/0124r0*, March 2007.

[31] Y. H. Zeng and Y. C. Liang, "Maximum-minimum eigenvalue detection for cognitive radio," in *IEEE PIMRC*, Athens, Greece, Sept. 2007.

[32] Y. H. Zeng and Y. C. Liang, "Text on eigenvalue based sensing," in *doc.: IEEE 802.22-07/0297r3*, July 2007.

[33] Y. H. Zeng and Y. C. Liang, "Simulations for wireless microphone detection by eigenvalue and covariance based methods," in *IEEE 802.22-07/0325r0*, July 2007.

[34] Y. H. Zeng and Y. C. Liang, "Covariance based signal detections for cognitive radio," in *IEEE DySPAN*, Dublin, Ireland, April 2007.

[35] Y. H. Zeng and Y. C. Liang, "Text on covariance based sensing for wireless microphon," in *doc.: IEEE 802.22-07/0295r3*, July 2007.

[36] Y. H. Zeng and Y. C. Liang, "Text on covariance based sensing for ATSC DTV," in *doc.: IEEE 802.22-07/0294r2*, July 2007.

Chapter 17

Mission-Oriented Communications Properties for Software-Defined Radio Configuration

Brett Barker, Arvin Agah, and Alexander M. Wyglinski

Contents

17.1   Introduction

Within the wireless communications research and industry community over the last decade, the concept of software defined radios (SDRs) and cognitive radios have come to be considered the leading solutions to the problems of spectrum communication for the next century. SDRs incorporate the latest in hardware affordability, computational power, and signal analysis techniques into a device that provides nearly unlimited flexibility and portability. In addition, SDRs provide a unique solution to the growing spectrum usage challenge facing the wireless communications industry. Unfortunately, not many devices have been developed and made available for the mainstream that can yet fully meet all the demands and expectations the technology is reported to fulfill (Barker, 2007).

There are two major, broad issues that research and industry have yet to resolve which contribute to the slow production of these devices. One main area of investigation is concerned with spectrum analysis and user signal detection and interference avoidance. This problem will be discussed in a little more detail later. The other issue is more software architecture-oriented, dealing with the SDR configuration management problem. This problem deals with how a dynamic, reusable platform effectively initializes, utilizes, and analyzes its system resources to perform its operational abilities at the optimal performance level. Unfortunately, there has not been much work within this area, and it is within this domain that this chapter is centered.

We have examined the SDR configuration management problem at the initialization level and have found that in order to start thinking about configuration management, a common methodology of comparison and analysis needs to be implemented before SDRs could conceivably begin to manage their resources effectively. Thus, we have developed a set of implementation-independent properties, which we term mission-oriented communications (MOC) properties. These properties examine several common wireless signal communications properties and, both formally and mathematically, define characteristics that provide a means of discussing, and hence analyzing, SDR behavior and performance (Barker, 2007).

For the remainder of this chapter, we will discuss the major investigation areas currently being pursued by SDR resesrchers, and will examine the set of MOC properties that were developed.

17.2 Background

17.2.1 Software-Defined Radio

Most radios follow the superheterodyne receiver circuit design (Lehr et al., 2002). This configuration consists of an antenna, a component to convert the signal to or from the intermediate frequency (IF) range, a component to convert the signal to digital (ADC) or analog (DAC), and some specialized hardware to perform signal processing functionality, such as a digital signal processor (DSP) or a field programmable gate array (FPGA). An SDR is usually composed of the same configuration, except that general-purpose hardware and specialized software take the place of many of these components.

Thus, a software-defined radio can be defined as "a radio that is substantially defined in software and whose physical layer behavior can be significantly altered through changes to its software" (Reed, 2002). We adopt this definition because it emphasizes the importance of software on the behavioral output of the radio and not just as an interface. The software within an SDR is integral to the behavior of the device as a whole, as the goal is to replace dedicated circuits and hardware with high-performance software, which is more flexible, portable, and easily updatable to the latest in digital signal processing techniques.

Since the majority of an SDR is software-based, this makes upgrading the device very simple; all that is required is a download of the new firmware, which can be configured to use the general-purpose hardware with the latest in signal processing analysis. With traditional radios, upgrading is not easily achieved, i.e., the old device must be replaced with an entirely new one. Another means of flexibility is the SDR's ability to have multiple modes of operation, meaning that multiple wireless standards can be loaded into the software, allowing the device to be highly portable. Currently, this multiple-mode capability is possible with the IEEE 802.11g standard (IEEE, 2006). However, the 802.11g is only capable of handling the 802.11a, b, and g standards as they are defined. For example, adjusting the subcarriers is not possible. The SDR would allow complete customization of the standard. An additional drawback to the fixed-modes-of-operation approach is the inability to adapt to future standards that are yet to be developed. With software radios, ultimate flexibility is achieved in a device that is nearly future-proof, as firmware updates could extend the life of the device beyond its fixed-hardware-implemented counterpart.

From the developer's point of view, SDR technology provides significant cost savings and better means of user support. Developers can reduce overall costs by producing longer-lasting devices with easier forms of support. Devices would not necessarily require the latest in hardware technology, hence reducing production costs by using less expensive hardware with more sophisticated software. Additionally, the increased flexibility provided by the radios allows for easier generation of ad hoc networks. Ad hoc networks increase connectivity between devices, and extend the communication capabilities of all connected radios through the sharing of information and resources. These benefits then get passed on to the user; up-front initial costs and replacement hardware upgrades go down, customization is provided at an even deeper level, and previous geographical communications restrictions disappear.

17.2.2   Related Work

The prospects of SDR are not all theory; there has been steady work on developing SDR-inspired systems since the mid-1990s. The SpeakEASY I and II military project (Lackey and Upmat, 1995) was an Air Force research project designed to produce a standard technique of communication between different field units, branches of service, and allied armies of different nations. They generated a very early SDR that was designed to emulate 15 existing radios used by the United States military. SpeakEASY was a successful proof of concept for SDR. Continuing its success, the Department of Defense performed research for the military's warfighter's radio systems in 1999 called the Joint Tactical Radio System (JTRS) Program (Davis, 1999; Eyermann, 1999). JTRS introduced an SDR attempting to utilize an open architecture composed of commercial, off-the-shelf components combined with software to make the radio compatible with both legacy and future waveform standards.

Around the same time, researchers at MIT began exploring the idea of building SDR platforms composed of inexpensive computer hardware (Bose et al., 1999). During this period SDR projects began to get underway internationally, especially in Japan. Several Japanese university research groups and companies began to investigate the potential of SDR in regard to specific components of the architecture pipeline: signal conversion, antenna/receivers, base band processing, and prototype development (Nakajima et al., 2001).

17.2.3 Spectrum Issues

Why are SDRs being seen as the next emerging technology and why is the demand for this technology so great? One major reason is the realization that the wireless communication medium, the signal spectrum shared by all over-the-air communicating devices, is a finite resource (Kennard, 1999; DARPA, 2007). Recently, the scientists, investigators, and the Federal Communications Commission (FCC) have realized that the current means of allocating spectrum ranges to new devices is not very efficient and, if continued, will result in a severe shortage of free spectrum, and that spectrum "cramming" will start to occur (THz Science and Technology Network, 2007).

The traditional means of allocating signal spectrum space lies with the FCC, which has to date conveyed chunks of contiguous signal space to corporations for their devices to operate within. This worked well in the beginning, when communications devices were few. There is certainly a large range for signals to operate within and not interfere with one another. The problem with this is that sometimes the allocation is disproportionate to the needs of the devices. For instance, first response groups, like fire, police, and rescue, share the same or overlapping ranges of spectrum. During times of crisis, these bandwidths can experience severe strain and overloading, resulting in poor performance for all. The antithesis of this example is that sometimes the spectrum remains mostly unused the bulk of the time, and only during short periods is it actually used to its full extent. Added on top of all this are the legacy devices: devices that have a registered range of operation within the spectrum but are no longer in operation at all. These legacy devices now represent holes of unused spectrum which no one can legally use (Federal Communications Commission Spectrum Policy Task Force, 2002).

The response to these problems has come in the form of trying to turn this finite resource into more of a "free market" system, where rules on spectrum usage are loosened and made fewer, but smarter (Technology CEO Council, 2006). This free market spectrum begins with the idea of licensed and unlicensed users. Licensed users are devices that have been allocated spectrum in the standard manner by the FCC and, when operating, have the right to the use of the spectrum in the range they have been given. Unlicensed users are devices that operate within a licensed user's band and do not have the right to use the spectrum when a licensed user is operating within that spectrum range. The idea is to try and effectively use all of the bandwidth. Considering the earlier example with first responders and low-use devices, under the licensed/unlicensed spectrum community, during times of crisis, the overstrained first response spectrum can actually move into other ranges of bandwidth that they are not licensed for and use them. However, should a licensed device begin transmitting within that band, the unlicensed users (first responders) must "move out of the way" of the licensed users. It can be seen that under this approach, no amount of spectrum goes to waste and all user needs are met. The problem, however, lies in finding new operating users and avoiding them. Significant research on this topic is being conducted in both academia and industry.

17.2.4 SDR Content Management

Mitola (2000) outlined the concept for a "smart" communication device that could manage its hardware and software resources effectively enough to operate in any environment and network conditions, and still provide intelligent operations for its user. While current technology is still far from this ultimate goal, strides are beginning to be made in that direction. An interesting and complex idea that was proposed was the concept of content management.

One of the most appealing abilities of an SDR is its ability to dynamically reconfigure itself; that is, to dynamically and in real time, reconfigure its software protocols and operational algorithms to provide new or better operational performance in response to environmental, network, or user demands without interrupting current performance. This concept implies a couple of major issues which must be resolved in order to actually realize these types of systems.

One major component is real-time system analysis. Effectively, this means the device must be intelligent enough to perform analysis based on its own performance and take the necessary steps to keep transmission performance optimal. In order to provide this level of sophistication, the content management system invariably must contain an artificial intelligence (AI) component to perform resource management. There are several groups currently exploring this aspect (e.g., Virginia Tech, University of Kansas). These techniques are currently being investigated and issues involving better-performing AI techniques are being examined (Barker, 2007).

A more immediate problem lies with dynamic reconfiguration, which is a property of the system to download protocols, algorithms, or software not previously installed on the system and incorporate them into the current operation of the device. Currently, downloadable firmware is possible, and from a technological standpoint this is achievable. The challenges are knowing what content to download and when to do so. It is this problem that the proposed mission-oriented communications properties defined in this chapter seek to address.

17.2.5   SDR Configuration

One of the major concerns with software-defined radio is configuration management, which seeks to optimize a radio's configuration in response to environmental, network, and user demands. The term configuration in this sense refers to the modules, algorithms, and software components that constitute the core definition of a software-defined radio's behavior. The main issue that configuration management is concerned with is how to choose algorithms and components that will optimally meet all demands upon the radio. Traditionally, experts have utilized their specialized knowledge to build radios for specific situations. Now however, the goal is to take this expert knowledge and build a dynamic system that make these decisions based on environmental, network, and user demands.

Environmental demands are those that radios face due to the natural world, such as distance, weather, occlusion, and terrain. Depending on the operating environments, certain considerations need to be accounted for. For example, if a radio is operating in a mountainous region, more signal power may be needed in order to provide ample transmission signal. Likewise, bad weather may result in signal degradation and error-controlling algorithms may be necessary to ensure robust transmission quality.

Other types of demands are those imposed by the network. As discussed, spectrum issues can severely influence transmission capabilities of communication devices. Radios must be capable of working in crowded spectrum environments, while avoiding causing and receiving interference to and from neighboring transmitters. In addition, for sensitive transmissions, security may be of concern and a certain degree of reliability needs to be ensured. In these cases, the radio should be capable of disguising its signal or encrypting it to provide sufficient security to avoid detection and interception.

Last and probably most important, consideration must be given to the user demands upon the radio. The radio must support the end goals of the user to the best of its ability, regardless of the network and environmental demands; otherwise, the radio will fail to meet its purpose as a communication device. If users decide that they need to transmit streaming video, then the device needs to compensate to do so, regardless of the environmental and network demands. If the radio cannot meet these demands with its current configuration, then it should be able to reconfigure itself so that it can perform the necessary tasks.

17.2.6   Missions

From a software development standpoint, it is often useful to analyze behavior of software based on specific use cases. That is, behavior can be defined based on the described outcomes provided from specific cases of the device being used. We use a similar idea by denoting missions within which a communication device can be used. A mission is described as a scenario wherein certain properties of the environment, network, and user demands can be assumed or are known, and the resulting configuration that best meets this mission is the optimal configuration. Obviously, anticipating environmental conditions *a priori* is not possible, but by providing missions and analyzing configurations based on these missions, one can gain insight into how properties and behaviors are important for configuration design and decision making.

17.3 Mission-Oriented Communications Properties

The set of MOC properties, which define a mechanism for describing wireless communications behaviors, are described and formally defined in this section. These properties form only a base set and serve as an initial foundation to begin discussion of SDR comparability and performance analysis. In addition, these properties can be used as a set of metrics for AI techniques, because they provide a set of metrics that these algorithms can utilize for evaluation (Barker, 2007).

The MOC properties are as follows:

(1) Low probability of detection/interference (LPD/LPI)
(2) Avoidance/rejection of nonintentional interference
(3) Multipath mitigation
(4) Information assurance/robustness
(5) Jamming resistance
(6) Communication range
(7) Communication capacity
(8) Bandwidth efficiency

The term *interference* is used within several MOC properties and is typically defined as the amount of unwanted or spurious electromagnetic waves generated from other sources, both from outside and within the radio hardware itself. However, for our purposes, we define interference to be the amount of energy introduced only by other electronic devices operating in the common spectrum of radio frequency.

These eight properties comprise the initial base set and represent the various means of describing the end-functionality of an SDR. Table 17.1 shows which of the main radio components the MOC properties affect, where these component categories are spreading, modulation, error control, and compression.

Although Table 17.1 demonstrates the basic interrelationships between the radio components and the MOC properties, the problem of being able to compare implementations of specific components in relation to a specific

Table 17.1   Mission-Oriented Communications Properties and How They Relate to the Main Radio Components of a Software-Defined Radio

MOC	Spreading	Modulation	Error Control	Compression
1. LPD/LPI	X			
2. Avoid/reject interference	X		X	
3. Multipath mitigation			X	
4. Information assurance			X	
5. Jam resistance	X		X	
6. Communication range		X		
7. Communication capacity			X	X
8. Bandwidth efficiency		X		X

MOC remains. For example, how does one formally contrast and compare BPSK and 64-QAM with respect to bandwidth efficiency? In order to discuss and analyze tradeoffs such as these, some formal means of comparison must be established. Thus, in addition to defining the MOC properties, we will give each property what is termed a *functional relationship* (FR). These FRs will lay the groundwork for allowing analysis of different component implementation techniques in regard to an MOC property. Throughout this chapter, we use a shorthand notation to denote a particular FR for a given property by FR-X_i, where X denotes the MOC property number from Table 17.1 and i is the FR definition number.

17.3.1   Low Probability of Detection/Low Probability of Interference (LPD/LPI)

This property describes the ability of the radio to avoid detection, as well as interference, as a result of a transmission within the frequency range of the spectrum bandwidths by any third party other than the intended recipients. It also implies an ability to withstand a certain amount of interference as a result of detection.

LPD/LPI is a property that is a quantitative measure of how private the transmissions need to be for the radio. This is important for any communications device that is either operating in a sensitive area where it does not wish to be found by other persons, or transmitting sensitive material and aiming for a certain degree of extra security by trying to remain inconspicuous.

Spread spectrum modulation is one technique that can avoid detection and mitigate interference, whereby the signal is spread over a much larger frequency than would be needed to send the transmission under normal conditions. There are two types of spread spectrum: direct sequence spread spectrum (DSSS) and frequency hopped spread spectrum (FHSS) (Dixon,

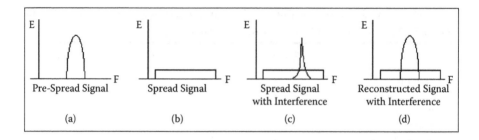

Figure 17.1 These charts demonstrate the effects of the spread signal process across a transmission cycle with interference. Graph (a) shows the shape of the signal before spreading, and graph (b) shows the effect once spreading has occurred. When a signal experiences interference, like the spike shown in graph (c), the effect on the signal is minimal when the signal is reconstructed at the receiving end, graph (d).

1994; Maxim, 2003; and Simon et al., 2002). DSSS works by transforming the transmission such that it appears as noise in the frequency domain, by widening the transmission bandwidth and lowering the power level. In the most basic form, the signal is transformed by multiplying the binary data by a pseudorandom binary sequence that makes the data sequence more random. During transmission, the signal possesses a very low transmit power level, usually below the noise floor, which makes detection harder for listeners, as they must distinguish the signal from the noise. However, even if the signal should be detected and interference applied, DSSS has the ability to withstand a certain degree of interference. When the receiver intercepts the transmission and reconstructs the original signal, most of the interference that has occurred during transmission is lost due to the inversion process in the signal recovery.

This process is demonstrated in Figure 17.1, showing a spread signal that encounters interference. As the signal is reconstructed, the interference is spread across frequency, which minimizes its impact on the desired signal.

FHSS is another spreading technique that works by sending the transmission in continuously changing frequency assignments, effectively disguising the transmission by moving the transmission around different frequencies so that it does not stay at the same frequency for long. The intent is to change the location of the transmission seemingly randomly within the spectrum so as to avoid detection. While FHSS can be an effective method of detection avoidance, it does have a weakness to interference. For example, should a transmission at a particular frequency be directly over or very near another user in the spectrum, the interference between the two signals could result in that particular transmission being lost to the receiver. Depending on the type of error robustness needed, this could be a potentially serious issue. Due to the high performance and implementation costs of FHSS, this spreading technique is not considered in this project.

A problem remains, and that is how one may distinguish between the performance of one DSSS spreading technique and that of another, or even compare with the case of no spreading. Therefore, the need for a formal means of distinction is necessary. For LPD/LPI, the first-order FR is the general relationship that describes a means of quantifying the LPD/LPI property.

The common thread with this property is the energy of the signal in the frequency spectrum. Starting with this fact, one can take advantage of the power spectral density (PSD) to be the first means of describing this property. The power spectral density is the power per unit frequency of an electromagnetic wave (Wolfram, 2006).

17.3.1.1 First-Order Functional Relationship

The amount of transmit power over any 1 MHz range is defined as

$$P_{tx} = \frac{1}{T} \int_{f_0}^{f_1} |X(F)|^2 dF \qquad (17.1)$$

where $|f_1 - f_0|$ is 1 MHz and corresponds to any 1 MHz band within the frequency spectrum and $X(F)$ is the power spectral density defined by Proakis (1995).We define the threshold $\mathbf{T_P}$ as a scenario-defined limit on the maximum power spectral density within the transmission bandwidth.

This first FR provides a clear mathematical method for calculating the power of a signal within a specified range. However, this relationship alone is problematic and is clearly not rigorous enough. One problem is that while there is a restriction in place with regard to the total power, there is no constraint on the shape of the PSD of the signal within that range.

Figure 17.2 demonstrates this issue. Here the total amount of power of the signal is below $\mathbf{T_P}$ and thus fulfills the *FR-II$_1$*, yet the shape of the signal clearly violates the intent of the property. The peak of the signal is easily identifiable within the spectrum, and this is something that must be compensated for.

17.3.1.2 Second-Order Functional Relationship

The second-order FR is defined as the maximum power level over any 1 MHz band within the spectrum bandwidth which does not exceed a threshold $\mathbf{T_P}$ where the power level is defined by (17.1).

FR-II$_2$ solves the problem demonstrated by Figure 17.2. If the limit is applied to the maximum power level within the frequency range window, then there are compliances of both the power and shape of the signal. With this second-order relationship, one can get the means of establishing the signal conformity within a 1 MHz band. 1 MHz is selected as it is a convenient step size for these calculations and provides a fine-granularity

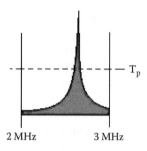

Figure 17.2 An example of the problem with the *FR-II*₁ definition. The total power between the given spectrum range is less than the stated threshold T$_p$. However, notice that the shape of the signal would still cause interference by acting as an interference spike.

level of control. It should be noted that this can be applied to any 1 MHz band within the transmission bandwidth. This then can enforce the entire signal to be spread below the established **T$_p$**.

17.3.2 Avoidance/Rejection of Nonintentional Interference

This property describes the radio's ability to avoid, handle, or reject non-intentional interference from other users within the spectrum bandwidth. Nonintentional interference is any interference that occurs from natural overlapping of transmission signals from multiple users within the spectrum bandwidth.

This MOC provides the means of describing how the radio will handle multiple users within the spectrum. This MOC does not describe how the radio detects other users in the space, but describes how it reduces interference from these other users. Thus, this property deals with the signal after spreading has already occurred, and is referred to as a post-spreading property. This distinction is made because the MOC properties can be considered at different points in the transmission process, and hence this property deals with the final spread signal. When dealing with multiple users, the radio relies upon a combination of the spreading technique and error correction controls to reduce interference. As previously discussed, spreading is an effective means of avoiding other users and interference. Error correction methods correct for the signal by adding redundant information in a specific way in order to beat interference in the hope that what is lost due to interference is overcome by the amount of information transmitted.

The problem here, as with all of the MOC properties, is that the best result is often a combination of the two radio components, but there does not exist a clearly defined means of determining beforehand what combination that is. It is possible to get a measurement of how much interference a

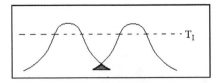

Figure 17.3 This demonstrates how *FR-II₁* can have a tolerance threshold T_I and sustain a low amount of interference while still maintaining a reasonable transmission quality. As one can see, the overlap is small and does not constitute a major source of signal corruption.

signal is experiencing through the signal-to-noise ratio (SNR). The term SNR in this context includes any undesired signals as noise. This is a measurement of the amount of unwanted interference that is mixed with the signal. Using SNR, a first-order functional relationship with the MOC property can be established.

17.3.2.1 First-Order Functional Relationship

This FR is defined as the amount of interference that the agile radio can tolerate while still providing reasonable transmission quality; this interference is limited to being below a threshold T_I.

Utilizing SNR as a reference scale, one can designate the amount of interference the radio can be expected to work within by T_I. Thus, in situations like that shown in Figure 17.3, the radio can still provide reasonable transmission quality with some nonintentional interference from nearby spectrum users. This also works for the opposite situation when there is too much interference (Figure 17.4) and the radio will have to make some adjustments in its operation.

However, two signals can overlap in a manner like that shown in Figure 17.5. In this situation, T_{INT} is the amount of interference and it is less than T_I. Here, using the provided definition, this would be acceptable, but clearly the transmission quality would suffer greatly. Thus, the

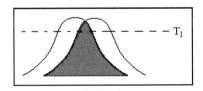

Figure 17.4 Here two signals overlap a great deal more, causing a much larger amount of interference clearly over the threshold specified by *FR-II₁*.

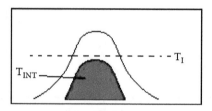

Figure 17.5 A problem that the *FR-II₁* definition fails to catch. In this situation, two signals sharing the exact same location in the spectrum but with different powers are overlapping directly. T_{INT} is the interfering signal and clearly will cause significant interference. Yet it also is below the threshold T_I. This is a problem that is addressed in *FR-II₂*.

second-order FR must incorporate not only nearby interference, but also co-location interference situations.

17.3.2.2 Second-Order Functional Relationship

T_L is defined as the tolerance level of co-location interference that the radio can operate with. T_{INT} is defined as the amount of nonintentional interference caused by third party users of the spectrum that the radio must be able to operate within and still maintain a reasonable quality of transmission. The total amount of acceptable nonintentional interference is then expressed as

$$T_L * C \geq T_{INT}, \tag{17.2}$$

where C is an operational constant. For our purposes, we defined C as 0.01.

17.3.3 Multipath Mitigation

This property describes how well the radio handles transmitting and receiving signals that have been lost, corrupted, or repeated due to the radio propagation environment in relation to time and distance between transmitter and receiver.

Multipath mitigation is the process of recovering distorted signals due to the reverberation of transmitted signals caused by interactions with objects between the transmitter and the receiver. Multipath mitigation is a significant issue that needs to be dealt with. Currently, the most effective means of manipulating the mitigation ability of a radio is by

1. Changing the data rate, i.e., modulation
2. Applying error correcting codes
3. Employing a combination of the above

The reason these techniques work is that they all directly impact the delay spread, or the amount of time between concurrently received data of the same signal due to mitigation. One measure of delay spread is root mean square (RMS) delay spread, which is defined by the following equations from Rappaport (1996):

$$\text{Root Mean Square} = \sqrt{\overline{\tau^2} - (\overline{\tau})^2} \qquad (17.3)$$

$$\overline{\tau} = \frac{\sum_k P(\tau_k)\tau_k}{\sum_k P(\tau_k)} \qquad (17.4)$$

$$\overline{\tau^2} = \frac{\sum_k P(\tau_k)\tau_k^2}{\sum_k P(\tau_k)} \qquad (17.5)$$

where $\overline{\tau}$ is the mean excess delay of a wideband multipath channel around the first moment, and $\overline{\tau^2}$ is the second central moment of the power delay profile. The square root difference of these two moments yields the RMS delay spread (Equation 17.3).

17.3.3.1 First-Order Functional Relationship

This FR is defined as the amount of RMS delay spread that is below a specified threshold, $\boldsymbol{T}_{ds}$. The value of $\boldsymbol{T}_{ds}$ is defined in terms of the symbol period T and represents the amount of delay spread tolerated by the system in order to provide an acceptable level of signal quality. Thus, the larger the value of $\boldsymbol{T}_{ds}$, the greater the delay spread that negatively influences the system. The value of $\boldsymbol{T}_{ds}$ can be chosen based on empirical measurements from channel delay profiles (Durgin, 2003).

17.3.4 Information Assurance/Robustness

This property describes the radio's ability to reliably transmit and receive data without loss or corruption.

For this MOC property, information assurance/robustness is defined to be the measured bit error rate (BER) at the end of the information sink pipeline of the radio, as shown in Figure 17.6. The information pipeline begins by the transformation of the data into symbols for transmission. After other transmission manipulation techniques have been performed, such as error control coding and compression, the signal information is transmitted across the wireless frequency channels. Once the signal is intercepted by

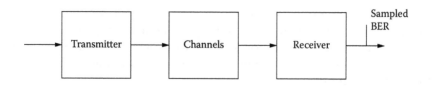

Figure 17.6   A basic schematic showing the fundamental building blocks of a digital transceiver. Each block transforms and manipulates the signal, which introduces a certain amount of error. Measuring the BER at the output of the receive block yields a metric that represents the overall error robustness of the transceiver.

the receiver, the signal is then reconstructed back to an approximation of the original data. It is at this point that the BER is determined.

Ideally, one would like to have an exact reconstruction of the original signal, but this is not possible due to the data corruption that naturally occurs during many portions of the transmission processes: interference in the channels, mitigation effects at the receiver, and errors during demodulation and decoding. Due to these effects, there are no suitable techniques for preestablishing the performance levels of given configurations without already having the combinations established. However, depending on the transmission type, certain amounts of error can be tolerated and still result in a close approximation of the original signal. The corresponding ranges that we use are given in Table 17.2, which provides the ideal robustness ranges one would like to achieve. However, one still needs to formally define robustness at the receiver portion of the transmission cycle.

17.3.4.1   First-Order Functional Relationship

Robustness is the signal-to-noise ratio as defined by the received transmission power over the received noise power.

With the first-order FR, the means of measuring the quality of transmission of the received signal through SNR is established. However, one needs to also account for the distance of the transmission. Attenuation of the signal due to distance greatly affects the received quality.

Table 17.2   Application Robustness in Sampled BER at the End of the Information Pipeline

Transmission Type	Low Robustness	High Robustness
Voice	10^{-1}	10^{-2}
Data	10^{-3}	10^{-6}
Multimedia	10^{-8}	10^{-9}

17.3.4.2 Second-Order Functional Relationship

Robustness is defined by FR-IV$_1$ times the attenuation of the signal due to the distance of the transmission, as shown by

$$R_T = A\left(\frac{P_{tx}}{N}\right) \quad \text{with} \quad A = \frac{1}{d}, \tag{17.6}$$

where P_{tx} is the power of the transmission, N is the noise, and d is the distance of the transmission.

17.3.5 Jamming Resistance

This property describes the ability of the radio to handle intentional and directed interference within the spectrum frequency range from other users without loss or with minimal disruption to operation.

This MOC property represents how well the radio deals with jamming. Jamming refers to any narrowband, high-powered signal intentionally aimed at the transmission, as demonstrated in Figure 17.7.

This MOC is designed to be a measure of how much targeted jamming the radio can handle. As a result, this FR, similar to avoidance/rejection of nonintentional interference, is a post-spreading property. The reason for this is due to the fact that the focus is on the directed, intentional interference by a third party. The effect of spreading a signal with interference after spreading was previously demonstrated. Then, the focus was on non-intentional interference, but the same effects apply even if the interference is intentional, as in jamming. Thus, this MOC needs to define how much specific jamming interference the radio can operate with.

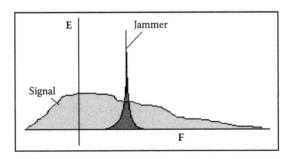

Figure 17.7 An example of a jamming spike across a signal's frequency range. The jamming spike appears as a short, high-energy burst that attempts to disrupt and overpower a transmitting signal.

17.3.5.1 First-Order Functional Relationship

This is defined as the amount of total power of an interfering signal in a small, narrow bandwidth that does not exceed a specified threshold, $\mathbf{T_J}$. The threshold is defined as a multiple of the expected noise floor level.

The threshold $\mathbf{T_J}$ is defined in order to make another distinction between interference and jamming. Although a high-powered spike in the frequency domain may occur, it does not necessarily imply that a third party is attacking the transmission. The spike could be caused by some natural phenomenon, such as a lightning strike. Additionally, the term high-powered is not specific enough—one needs to define how high an energy spike should be for it to be considered intentional.

This threshold can be defined based on the distribution of white noise present in the transmission channels. The distribution of this noise within the channel can be assumed to be continuous and normally distributed, which means the standard deviation can be calculated. Using a deviation that is very far from normal, any energy signals that are beyond this deviation can safely be assumed to be directed, intentional interference. The level of seven standard deviations was chosen to sufficiently include most distributions of white noise.

17.3.6 Communication Range

This property describes the radio's ability to transmit a signal at a reasonable degree of quality, across an approximate minimum and maximum range of distance.

The communication range and quality of a radio communications device is something that is heavily affected by the transmission data type, the data rate of transmission, and the transmission power level. By varying these independent variables, one can greatly affect the overall distance and quality of a transmission. However, the exact distance a radio can transmit cannot be empirically determined, as other local factors, such as noise and interference levels, affect the exact distance and quality. This property helps approximate the range that a communicating device could achieve under assumed ideal and worst-case operating environments; assuming there are initial ideas as to what those three independent variables are.

Since the goal of this property is to yield an approximate range of operation, the distances are derived from the following equations:

$$SNR_F = \frac{P_{rx}}{N} \tag{17.7}$$

$$P_{rx} \propto \frac{P_{nom}^{Tx}}{d^2} \tag{17.8}$$

Equation 17.7 is the standard equation for SNR with P_{rx} being the amount of received power and N as the amount of noise. Equation 17.8 is the proportional relationship between the received power, P_{rx}, and the nominal transmitted power, P_{nom}^{Tx}, over the squared distance, d. From these two equations, knowing both the nominal transmit power, SNR, and the sensitivity of the receiver, an approximate distance of operation can be determined. However, one does not necessarily know the value of the SNR at this point. Known from the application type are the acceptable ranges of BER, as listed in Table 17.2. This information can be used, along with an SNR curve, to define an acceptable value of operation to assume for the best- and worst-case modes of operation. For the best it is assumed that the SNR values for the application data type are very high, while for the worst they are low. This value is represented by SNR_F, and can be established as a fuzzy logic function of SNR and BER, based on the application type.

Using this value, the received power level of the transmission can be determined. It is assumed that the noise level for the best case is the Boltzmann's constant, $1.38 * 10^{-23}$ Joules per Kelvin degree, across the bandwidth (Clancy et al., 2006). Once the power level of the received signal is determined, the approximate value of the distance can be determined, as the nominal power level of transmission is known. As for the worst-case scenario, the noise level is not considered, as it is the worst possible situation for transmitting signals; the radio would not even attempt to transmit under the circumstances.

17.3.6.1 First-Order Functional Relationship

The maximum distance of operation of the communication device is approximately determined by Equation (17.7) and Equation (17.8), based on the application type, the nominal transmission power, and ideal noise conditions. The minimum distance of operation of the communication device is approximately determined by the application type, nominal transmission power, and the worst noise conditions.

While this first-order FR provides a good means of determining an approximate range of operation, two issues that need to be considered are (1) the amount of interference that could be in the operating environment during operation, and (2) the effect of error coding techniques on the transmission data. These factors greatly impact received quality and operating distance.

For interference, there is unfortunately no way of knowing *a priori* the amount of interference across the transmission bandwidth. However, if the interference is included as part of the noise, then SNR_F becomes the $SINR_F$. Additionally, since the best operational conditions are assumed, this only slightly changes the fuzzy SNR result at the onset of the derivation. We use Clancy's definition of interference temperature defined in Clancy et al.,

(2006) as our approximation of interference and add this to the scenario-dependent noise value.

The other issue with the first-order FR is that it assumes uncoded modulation blocks. When error-correcting codes are used on the data, these algorithms add redundancy to the data, reducing the actual data throughput but making it easier for the recipient because more redundant information is included. As a result, the use of coding can greatly improve the error performance of the transmission.

17.3.6.2 Second-Order Functional Relationship

Utilizing the first-order FR, the definition is extended by incorporating the use and effect of error-correcting codes on the data, and incorporating the interference temperature by using SINR.

17.3.7 Communication Capacity

This property defines how many cooperative transmissions are able to operate within a bandwidth, while maintaining a specified level of quality for all.

The intent of this property is to give a formal means of determining the most efficient use of bandwidth between multiple concurrent users. Within a given bandwidth, depending on the needs of the user radios, the space can be divided such that all users can operate with minimal interference simultaneously. An example of this is shown in Figure 17.8.

It is shown how a frequency bandwidth can support multiple users of varying amounts of personal bandwidth without interfering with other users. Non-used portions of bandwidth are shown as the large, single spikes. Due to the dynamic nature of spectrum usage, determining the most efficient way to discretize the space based on user presence and user

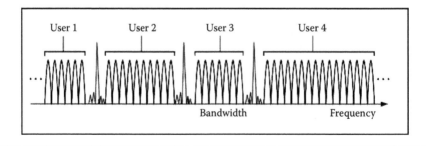

Figure 17.8 An example of multiuser capacity across a bandwidth (Wyglinski, 2006). The bandwidth is able to support several transmissions (users) without these users interfering with each other.

need is difficult. Due to this dynamic nature, the actual discretization of the bandwidth will have to be done in real time. Therefore, with this property one seeks to gain some approximation of the potential usage of the space during the mission. With information regarding the potential number of users, the type of data being transmitted, and the use of spreading, constraints regarding the bandwidth requirements can be garnered.

17.3.7.1 First-Order Functional Relationship

The capacity is the maximum number of users and the bandwidth required over a supported frequency range without spreading.

The problem with this first-order FR is that spreading is unaccounted for. This is not applicable to many situations. One thus needs to account for this with the second-order FR.

17.3.7.2 Second-Order Functional Relationship

Using the first-order definition to determine the maximum number of users, the minimum error rate produced by the least redundant error-correcting code under the effects of spreading is taken into account.

With this second-order FR, a better idea of the potential usage of the space can be achieved; the bandwidth required for the mission can be better approximated. It should be noted that while this property is concerned with bandwidth efficiency of users, there is no assurance regarding the efficient usage of the space by the users themselves. The usage of the subbandwidth space by the user could be inefficient.

17.3.8 Bandwidth Efficiency

This property describes the radio's maximum amount of bits that can be transmitted in a specified frequency band per unit time.

While communication capacity deals with user space efficiency, this property aims to determine the maximum efficiency of a single transmission. Efficiency can be easily determined based on the current usage of bits being transmitted within the total space available. Capacity can be determined using Equation (17.9) (Shannon, 1949):

$$C = B * \log_2\left(1 + \frac{S}{N}\right) \tag{17.9}$$

where C is the capacity, B is the bandwidth in hz, and S/N is the SNR of the channel. The SNR is already being utilized for other properties which thus makes this available for approximation. With this, one can define the first-order FR.

17.3.8.1 First-Order Functional Relationship

The bandwidth efficiency of a transmission is determined by the total number of bits being transmitted over the capacity of the bandwidth, provided by Equation (17.9).

17.4 Conclusion

In this chapter, we have discussed the major investigative areas concerning software-defined radio (SDR) development, and specifically looked at a set of mission-oriented communication (MOC) properties that can be used for SDR configuration management, AI analysis, and SDR evaluation and comparison.

These results have been used in the development of a software application that uses artificial intelligence techniques to define an initial configuration for a specific SDR. An expert system was designed to reason over signal properties and spectrum characteristics independently of implementation technique, demonstrating the configurability across multiple input parameters and radio components (Barker, 2007).

The properties discussed in this chapter form a preliminary set upon which to grow. Many of the properties are rather simplistic from a wireless communications perspective and need further refinement and investigation. In addition, far from all communication properties and behaviors have been included; such omissions include security and power management. We present these characteristics as a proposed set of properties upon which the designers and investigators can build in order to address the challenges in configuration management and initialization of software-defined radios.

References

[1] Bose, V., M. Ismert, M. Welborn, and J. Guttag. (1999). Virtual Radios. *IEEE Journal on Selected Areas in Communications*, Special Issue on Software Radios **17**(4): 591–602.

[2] Barker, B.A. (2007). *Applications of Expert Systems to Configuring Software Defined Radios*. M.S. Thesis, Electrical Engineering and Computer Science, University of Kansas.

[3] Clancy, T.C. and W.A. Arbaugh. (2006). *Measuring Interference Temperature*. Virginia Tech Wireless Personal Communications Symposium.

[4] DARPA. (2007). *DARPA: The Next Generation (XG) Program*. http://www.darpa.mil/ato/programs/xg/index.htm.

[5] Davis, K.V. (1999). JTRS-An Open, Distributed-Object Computing Software Radio Architecture. In *Proceedings of the 18th Digital Avionics Systems Conference*, St Louis, MO. **2**, 9.A.6-1 - 9.A.6-8.

[6] Dixon, R.C. (1994). *Spread Spectrum Systems with Commercial Applications*, John Wiley & Sons, Inc., New York, New York.

[7] Durgin, G.D. (2003). *Space-Time Wireless Channels.* Prentice Hall, Upper Saddle River, New Jersey.

[8] Eyermann, P.A. (1999). Joint Tactical Radio Systems-A Solution to Avionics Modernization. In *Proceedings of 18th Digital Avionics Systems Conference*, St. Louis, MO. **2**, 9.A.5-1 - 9.A.5-8.

[9] Federal Communications Commission Spectrum Policy Task Force. (2002). *Report of the Spectrum Rights and Responsibilities Working Group*, November 15.

[10] IEEE Computer Society LAN MAN Standards Committee. (1997). *Wireless LAN Medium Access Control (MAC) and Physical Layer (PHY) Specifications IEEE Standard 802.11.* The Institute of Electrical and Electronics, New York, New York.

[11] IEEE Standards Association. (2006). *IEEE 802.11 LAN/MAN Wireless LANS.* http://standards.ieee.org/getieee802/802.11.html, The Institute of Electrical and Electronics, New York, New York.

[12] Joint Tactical Radio System. (JTRS) Joint Program Office. (2004). *Software communications architecture specification, jtrs-5000sca v3.0.*

[13] Kennard, W.E. (1999). *Connecting the Globe: A Regulator's Guide to Building a Global Information Community.* Federal Communications Commission, Washington, D.C.

[14] Lackey, R.I. and D. W. Upmat. (1995). Speakeasy: The Military Software Radio. *IEEE Communications Magazine* **33**(5): 56–61.

[15] Lehr, W., F. Merino, and S.E. Gillett. (2002). *Software Radio: Implications for Wireless Services, Industry Structure, and Public Policy.* Massachusetts Institute of Technology Program on Internet and Telecoms Convergence.

[16] Maxim. (2003). *An Introduction to Direct-Sequence Spread-Spectrum Communications.* http://pdfserv.maximic.com/en/an/AN1890.pdf, Feb 18, 2003.

[17] Mitola, J. (2000). *Cognitive Radio: An Integrated Agent Architecture for Software Defined Radio.* Ph.D. Dissertation, Royal Institute of Technology (KTH).

[18] Nakajima, N., R. Kohno, and S. Kubota (2001). Research and Developments of Software-Defined Radio Technologies in Japan. *IEEE Communications Magazine* **39**(8): 146–55.

[19] Proakis, J.G. (1995). *Digital Signal Processing: Principles, Algorithms, and Applications.* Prentice Hall, Upper Saddle River, New Jersey.

[20] Rappaport, T.S. (1996). *Wireless Communications.* Prentice Hall, Upper Saddle River, New Jersey.

[21] Reed, J.H. (2002). *Software Radio: A Modern Approach to Radio Engineering.* Upper Saddle River, New Jersey, Prentice Hall.

[22] Shannon, C.E. (1949). Communication in the presence of noise. *Proc. Institute of Radio Engineers* **37**(17.1): 10–21.

[23] Simon, M.K., J.K. Omura, R.A. Scholtz, and B.K. Levitt. (2002). *Spread Spectrum Communications Handbook.* McGraw-Hill TELECOM Engineering, New York, New York.

[24] Technology CEO Council. (2006). *Freeing Our Unused Spectrum*. Washington, DC, February.

[25] THz Science and Technology Network. (2007). http://www.thznetwork.org.

[26] Wolfram Research. (2006). *Time Series Documentation, Power Spectral Density Function*. http://documents.wolfram.com/applications/timeseries/ UsersGuideToTimeSeries/SpectralAnalysis/1.8.1.html.

[27] Wyglinski, A.M. (2006). Effects of Bit Allocation on Non-contiguous Multi-carrierbased Cognitive Radio Transceivers. In *Proceedings of the 64th IEEE Vehicular Technology Conference*, Fall, Montreal, QC, Canada, 1–5.

Index

D

Data
  band, cognitive radio adaptive medium
    access control design, 221
  fusion algorithms, WRAN, 420
  sensitive routing, 269
  throughput, opportunistic in-band
    sensing and, 423
  transfer period (DTP), 143, 194
DCA, *see* Distributed channel assignment
    protocol
DCA, *see* Dynamic channel allocation
DCA-PC protocol, power control
    functionality, 137
DCC, *see* Default control channel
DC-MAC, *see* Decentralized cognitive MAC
DCSS-MAC, *see* Distributed coordinated
    spectrum sharing MAC
Decentralized cognitive MAC (DC-MAC),
    187, 188
Decoupled design, routing, 264
Default control channel (DCC), 229
DFH, *see* Dynamic frequency hopping
DFHC, *see* DFH community
DFH community (DFHC), 152
DHT-based routing, *see* Distributed hash
    table-based routing
Differential phase-shift keying (DPSK), 373
Diffie-Hellman-based mechanism, 336
Digital signal processor (DSP), 437
DIMSUMNet, *see* Dynamic intelligent
    management of spectrum for
    ubiquitous mobile-access network
Direction of arrival (DOA) shift, 428
Direct sequence spread spectrum (DSSS),
    443
Direct-sequence UWB (DS-UWB),
    376, 377
Distributed channel assignment protocol
    (DCA), 219
Distributed coordinated spectrum sharing
    MAC (DCSS-MAC), 133
Distributed hash table (DHT)-based
    routing, 334
DL subframe, *see* Downlink subframe
DOA shift, *see* Direction of arrival shift
DOF, *see* Duration of the flow
DOSS protocol, *see* Dynamic open
    spectrum sharing protocol
Downlink (DL) subframe, 427
DPA, *see* Dynamic packet assignment
DPSK, *see* Differential phase-shift keying
DQDCR scheme, *see* Dynamic channel
    reservation scheme

DSA, *see* Dynamic spectrum access
DSAP, *see* Dynamic spectrum access
    protocol
DSP, *see* Digital signal processor
DSSS, *see* Direct sequence spread
    spectrum
DS-UWB, *see* Direct-sequence UWB
DTP, *see* Data transfer period
Duration of the flow (DOF), 197
Dynamic channel allocation (DCA), 111
Dynamic channel reservation (DQDCR)
    scheme, 223
Dynamic cognitive radio systems, 181
Dynamic frequency hopping (DFH), 151,
    209
Dynamic intelligent management of
    spectrum for ubiquitous
    mobile-access network
    (DIMSUMNet), 152
Dynamic open spectrum sharing (DOSS)
    protocol, 195, 217–218
  data transmission, 218
  detecting presence of primary
    users, 217
  setting up three operational frequency
    bands, 217–218
  spectrum mapping, 218
  spectrum negotiation, 218
Dynamic packet assignment (DPA), 112
Dynamic spectrum access (DSA), 88, 110,
    181, 280, *see also* Medium access
    control protocols for dynamic
    spectrum access
  algorithm, POMPD-based, 158
  driven MAC, 131, 164
  model(s)
    exclusive usage model, 181
    on–off model, 183
    private commons model, 181
    public commons model, 181
  protocol (DSAP), 200
    relay, 200
    server, 200
    system model of spectrum leasing
      of, 201
  upper layer protocol design, 280
Dynamic spectrum access networks
    (DySPANs), 89, 94, 110, *see also*
    Ultra-wideband cognitive radio for
    dynamic spectrum accessing
    networks
Dynamic spectrum allocation (DSA)
    algorithms, 154
DySPANs, *see* Dynamic spectrum access
    networks

9 780367 386047